Lecture Notes in Computer Science 4415

Commenced Publication in 1973
Founding and Former Series Editors:
Gerhard Goos, Juris Hartmanis, and Jan van Leeuwen

W0091231

Paul Lukowicz   Lothar Thiele
Gerhard Tröster (Eds.)

Architecture of Computing Systems - ARCS 2007

20th International Conference
Zurich, Switzerland, March 12-15, 2007
Proceedings

 Springer

Volume Editors

Paul Lukowicz
University of Passau
IT-Center/International House
Innstraße 43, 94032 Passau, Germany
E-mail: paul.lukowicz@uni-passau.de

Lothar Thiele
Swiss Federal Institute of Technology Zurich
Computer Engineering and Networks Laboratory
Gloriastrasse 35, 8092 Zurich, Switzerland
E-mail: thiele@tik.ee.ethz.ch

Gerhard Tröster
Swiss Federal Institute of Technology Zurich
Electronics Laboratory
Gloriastrasse 35, 8092 Zürich. Switzerland
E-mail: troester@ife.ee.ethz.ch

Library of Congress Control Number: 2007922097

CR Subject Classification (1998): C.2, C.5.3, D.4, D.2.11, H.3.5, H.4, H.5.2

LNCS Sublibrary: SL 1 – Theoretical Computer Science and General Issues

ISSN   0302-9743
ISBN-10   3-540-71267-4 Springer Berlin Heidelberg New York
ISBN-13   978-3-540-71267-1 Springer Berlin Heidelberg New York

Springer is a part of Springer Science+Business Media

springer.com

© Springer-Verlag Berlin Heidelberg 2007
Printed in Germany

Typesetting: Camera-ready by author, data conversion by Scientific Publishing Services, Chennai, India
Printed on acid-free paper   SPIN: 12031580   06/3142   5 4 3 2 1 0

Preface

The ARCS series of conferences has over 30 years of tradition reporting high-quality results in computer architecture and operating systems research. While the conference is proud of its long tradition, it is also proud to represent a dynamic, evolving community that closely follows new research trends and topics. Thus, over the last few years, ARCS has evolved towards a strong focus on system aspects of pervasive computing and self-organization techniques (organic and autonomic computing). At the same time it has expanded from its roots as a German Informatics Society (GI/ITG) conference to an international event. This is reflected by the composition of the TPC which included over 30 renown scientist from 10 different countries. The conference attracted 83 submission from 16 countries across 4 continents. Of those, 20 have been accepted, which amounts to an acceptance rate below 25%.

The 20th ARCS event was a special anniversary conference. It is only fitting that it was held at a special place: the ETH Zurich. It combines one of the leading information technology schools in Europe with a beautiful location.

I would like to express my gratitude to all those who made this year's conference possible. This includes the General Chairs Lothar Thiele and Gerhard Tröster from ETH, the Tutorials and Workshops Chair Marco Platzner from the University of Paderborn, the members of the "Fachausschus ARCS" of the GI/ITG (the Steering Committee), the members of the Technical Program Committee, the Reviewers, and most of all to all the authors that submitted their work to ARCS 2007. I would also like to thank IFIP, ITG/Electrosuisse, VDE and the ARTIST2 Project for their support of the conference.

January 2007   Paul Lukowicz

Organization

Organizing Committee

Conference Chairs :   Lothar Thiele (ETH Zurich, Switzerland)
                            Gerhard Tröster (ETH Zurich, Switzerland)
Program Chair: Paul Lukowicz (University of Passau, Germany)
Workshops and Tutorials: Marco Platzner (University of Paderborn, Germany)

Program Committee

Nader Bagherzadeh, University of California, Irvine, USA
Michael Beigl, University of Braunschweig, Germany
Michael Berger, Siemens AG, Munich, Germany
Don Chiarulli, University of Pittsburgh, USA
Giovanni Demicheli, EPFL Lausanne, Switzerland
Koen De Bosschere, Ghent University, Belgium
Alois Ferscha, University of Linz, Austria
Mike Hazaas, Lancaster University, UK
Ernst Heinz, UMIT Hall i. Tirol, Austria
Paolo Ienne, EPFL Lausanne, Switzerland
Wolfgang Karl, University of Karlsruhe, Germany
Spyros Lalis, University of Thessaly, Greece
Koen Langendoen, Delft University of Technology, The Netherlands
Tom Martin, Virginia Tech, USA
Hermann de Meer, University of Passau, Germany
Erik Maehle, University of Luebeck, Germany
Peter Marwedel, University of Dortmund, Germany
Christian Mller-Schloer, University of Hanover, Germany
Stephane Vialle, Supelec, France
Joe Paradiso, MIT Media Lab, USA
Daniel Roggen, ETH Zurich, Switzerland
Pascal Sainrat, Université Paul Sabatier, Toulouse, France
Heiko Schuldt, University of Basel, Switzerland
Hartmut Schmeck, University of Karlsruhe, Germany
Karsten Schwan, Georgia Tech, Atlanta, USA
Bernhard Sick, University of Passau, Germany
Juergen Teich, University of Erlangen, Germany
Pedro Trancoso, University of Cyprus, Cyprus
Theo Ungerer, University of Augsburg, Germany
Stamatis Vassiliadis, Delft University of Technology, The Netherlands
Lucian Vintan, Lucian Blaga University of Sibiu, Romania
Klaus Waldschmidt, University of Frankfurt, Germany

Additional Reviewers

Henoc Agbota
Mohammed Al-Loulah
Muneeb Ali
Ioannis Avramopoulos
Gonzalo Bailador
David Bannach
Juergen Becker
Andrey Belenky
Mladen Berekovic
Uwe Brinkschulte
Rainer Buchty
Georg Carle
Supriyo Chatterjea
Marcelo Cintra
Philippe Clauss
Joshua Edmison
Werner Erhard
Philippe Faes
Diego Federici
Dietmar Fey
Mamoun Filali Amine
Stefan Fischer
Pierfrancesco Foglia
Thomas Fuhrmann
Martin Gaedke
Marco Goenne
Werner Grass
Jan Haase
Erik Hagersten
Jörg Hähner
Gertjan Halkes

Holger Harms
Sabine Hauert
Wim Heirman
Jörg Henkel
Michael Hinchey
Alexander Hofmann
Ulrich Hofmann
Amir Kamalizad
Dimitrios Katsaros
Bernd Klauer
Manfred Kunde
Kai Kunze
Christoph Langguth
Marc Langheinrich
Baochun Li
Lei Liu
Paul Lokuciejewski
Clemens Lombriser
Thanasis Loukopoulos
Jonas Maebe
Rene Mayrhofer
Lotfi Mhamdi
Jörg Mische
Florian Moesch
Thorsten Möller
Katell Morin-Allory
Sanaz Mostaghim
Leyla Nazhandali
Afshin Niktash
Pasquale Pagano
Thomas Papakostas

Hooman Parizi
Tom Parker
Neal Patwari
Andy Pimentel
Thilo Pionteck
Laura Pozzi
Robert Pyka
Markus Ramsauer
Thomas Schwarzfischer
Andr Seznec
Enrique Soriano
Ioannis Sourdis
Michael Springmann
Mathias Stäger
Yannis Stamatiou
Kyriakos Stavrou
Walter Stiehl
Mototaka Suzuki
Joseph Sventek
Jie Tao
Karl-Heinz Temme
Sascha Uhrig
Miljan Vuletic
Jamie Ward
Ralph Welge
Lars Wolf
Bernd Wolfinger
Markus Wulff
Olivier Zendra
Peter Zipf

Table of Contents

ARCS 2007

A Reconfigurable Processor for Forward Error Correction

Afshin Niktash, Hooman T. Parizi, and Nader Bagherzadeh

536 Engineering Tower, Henry Samueli School of Engineering,
University of California, Irvine, CA 92697-2625, USA
{aniktash,hparizi,nader}@ece.uci.edu

Abstract. In this paper, we introduced a reconfigurable processor optimized for implementation of Forward Error Correction (FEC) algorithms and provided the implementation results of the Viterbi and Turbo decoding algorithms. In this architecture, an array of processing elements is employed to perform the required operations in parallel. Each processing element encapsulates multiple functional units which are highly optimized for FEC algorithms. A data buffer coupled with high bandwidth interconnection network facilitates pumping the data to the array and collecting the results. A processing element controller orchestrates the operation and the data movement. Different FEC algorithms like Viterbi, Turbo, Reed-Solomon and LDPC are widely used in digital communication and could be implemented on this architecture. Unlike traditional approach to programmable FEC architectures, this architecture is instruction-level programmable which results the ultimate flexibility and programmability.

Keywords: Reconfigurable Processor, Processing Element, Forward Error Correction, Viterbi, Turbo.

1 Introduction

Reconfigurable architectures customize the same piece of silicon for multiple applications. While general purpose processors could not meet the processing requirements of many new applications, traditional custom ASIC dominates the design space. In wireless communication, a DSP processor is usually responsible for low data rate signal processing and is coupled with customized silicon to perform the medium and high data rate processing. The main drawback of a custom design is its long and costly design cycle which requires high initial investment and results in long time-to-market. Furthermore, lack of flexibility and programmability of traditional solutions causes frequent design changes and tape-outs for emerging and developing standards. Reconfigurable architectures on the other hand are very flexibly and programmable and could significantly shorten the design cycle of new products while even extending the life cycle of existing products. Tracking new standards is simplified to software upgrades which could be performed on-the-fly.

One of the challenging applications of a reconfigurable architecture is channel coding. Almost any digital communication system benefits from at least one form of

P. Lukowicz, L. Thiele, and G. Tröster (Eds.): ARCS 2007, LNCS 4415, pp. 1–13, 2007.

error correction coding [1]. There are four main algorithms widely used in wireless and wired communications: Viterbi, Turbo, Reed-Solomon and LDPC. However, multiple variations of each of these algorithms are employed in standards. More specifically, every standard uses a different configuration of an algorithm which makes that unique to that standard. For example, , the Turbo code used in W-CDMA standard has a different polynomial, block size, rate and termination scheme from that used in WiMAX. Viterbi coding employed in W-LAN, W-CDMA and WiMAX are not the same. This translates to having a plurality of coding accelerators for different coding algorithms and configurations which is very common in industry. In conventional approach, a separate coprocessor is employed for every FEC algorithm. Nevertheless, even one coprocessor is not programmable enough to cover all existing configurations of an FEC algorithm for multiple standards.

In this paper, we introduce RECFEC, a REConfigurable processor optimized for Forward Error Correction algorithms. RECFEC combines the programmability of a DSP processor with performance of a dedicated hardware and is architected to enable effective software implementation of FEC algorithms. The organization of this paper is as follows. Section 2 reviews the related works. Section 3 describes the RECFEC architecture and programming model. Section 4 presents two examples of algorithm implementation, Viterbi and Turbo coding and Section 5 concludes the paper.

2   Related Works

There are considerable research efforts to develop prototypes of reconfigurable architectures for channel coding. In this section, we present the features of those architectures.

A reconfigurable signal processor is introduced in [2] using an FPGA based reconfigurable processing board to implement a programmable Turbo decoder.

Viturbo[3] is among the first contributions trying to integrate Viterbi and Turbo decoders into a single architecture. Viturbo is a runtime reconfigurable architecture designed and implemented on an FPGA. The architecture can be reconfigured to decode a range of convolutionally coded data and can also be reconfigured to decode Turbo coded data. SOVA is the algorithm implemented for Turbo decoding. The target application is W-LAN, 3GPP and GSM.

A dual mode Viterbi/Turbo decoder is introduced in [4]. The component decoder in this architecture has two modes and some of the modules are shared. In Viterbi mode, and the Log Likelihood Ratio (LLR) processors are turned off. Input symbols are sent from Branch Metrics Unit (BMU) processor to Add Compare Select Unit (ACSU) processor and decoded bits are sent out after tracing back. When in Turbo mode, the decoder works as a Maximum A posteriori Probability (MAP) decoder and only the Trace Back Unit (TBU) is turned off.

A Turbo decoder on a dynamically reconfigurable architecture is introduced in [5]. The decoder is optimized for FPGA. The key power-saving technique in this design is the use of decoder run-time dynamic reconfiguration in response to variations in the channel conditions. If less favorable channel conditions are detected, a more powerful, less power-efficient decoder is swapped into the FPGA hardware to maintain a fixed bit error rate.

A custom architecture for FEC coding of 3G is introduced in [6]. It is based on a unified Viterbi/Turbo processing architecture and exploits the common trellis processing operations of two decoding algorithms providing 3GPP compliant decoding to the base station.

Several custom implementation of single mode Viterbi or Turbo decoders have been published in the past (e.g. [7],[8]).

Following key features are not effectively supported by above mentioned and similar related contributions and are in fact the design criteria of RECFEC:

- *Programmability:* Conventional designs are programmable only to the extent of few parameters needed to support a very small set of wireless standards. An example is using a limited number of polynomials for constituent encoders or supporting a few different sizes for data blocks within a Viterbi decoder.
- *Flexibility:* Only one or two FEC coding algorithms (e.g. Viterbi/Turbo or LDPC/Turbo) are integrated in related works.
- *Scalability:* The conventional implementations typically require major hardware modification to accommodate higher data rates or improved performance.
- *Choice of algorithm:* For every coding scheme, only one algorithm is selected and implemented. For example, SOVA algorithm in [3] or MAP algorithm in [4] is implemented for Turbo decoding.
- *Precision:* Data paths are fixed and tailored to the required precision. Modifying the data paths to improve the precision results in significant design changes. For example, 4-bit soft values are used for decoder inputs of Viterbi decoder.
- *Performance:* Modifying an algorithm to improve the performance or using a set of different algorithms depending on the channel quality, data rate, power consumption, etc is not allowed.
- *Multiple standards and Emerging algorithms:* The conventional approach limits us to a few standards. For a multi-standard platform, none of those designs could cover multiple wireless air interfaces or emerging algorithms without major hardware modifications.
- *Upgradeability:* Being able to upgrade a coding engine via software and even on-the-fly is an appealing feature that can not be achieved in related works.

Having in mind the aforementioned features of an ideal architecture, RECFEC is built on a reconfigurable processor design methodology to make it a multi-standard instruction-level programmable FEC engine.

3   RECFEC Architecture

RECFEC architecture is comprised of a parallel array of Processing Elements (PE) controlled by a controller. Fig. 1 demonstrates the architecture of RECFEC. The PE Pool is a two dimensional array of Processing Elements performing parallel

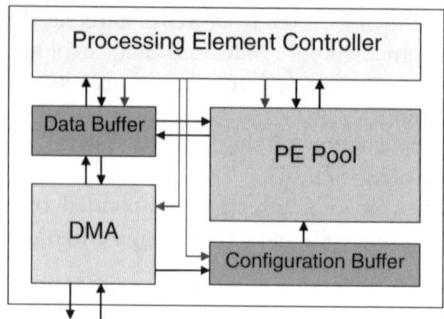

Fig. 1. RECFEC architecture

execution of instructions, which are called *Configurations*. Data and Configuration information are stored in the *Data Buffer (DB)* and the *Configuration Buffer (CB)*.

During the execution, *Configuration Words* which accommodate the configuration information of PEs are broadcast from the CB to the PE Pool and are stored in *Configuration Registers*. DB is embedded data memory that interfaces with external memory pumps data to the PE Pool. The high throughput data network facilitates the supply of data to and collection of the results from the PE Pool. Reconfiguration of the network connecting the DB to the PE Pool facilitates the support of different data movement patterns. All data transfers between the DB or the CB and the external memory are handled by the *DMA Controller*. The *PE Controller* is a general purpose 32-bit RISC processor which controls the sequence of operations. In following sections, components of this architecture are elaborated upon.

3.1 Processing Elements Pool

The PE Pool is an array of reconfigurable processing elements. Considering the implementation of FEC algorithms, PE Pool is organized as an 8x8 array of PEs. The PE Pool follows the *SIMD* model of computation. All PEs in the same row/column share the same Configuration Word. However, each PE operates on different data. Sharing the configuration across a row/column is useful for data-parallel FEC algorithms. The PE Pool has a mesh interconnection network, designed to enable fast data exchange between PEs. Each PE encapsulates four 8-bit functional units, as shown in Fig. 2:

- *ALU:* The Arithmetic Logic Unit (ALU) is designed to implement basic logic and arithmetic functions. The functionality of the ALU is configured through control bits. These control signals are generated by the PE decoder based on the information in the Configuration Register
- *Add, Compare and Select (ACS) unit:* This unit performs ACS, Max and Max* operations required in Viterbi and Turbo coding. In the Viterbi algorithm, branch metrics are updated using following equation:

$$PM_{i',t+1} = Min(PM_{i,t} + BM_{ii'}, PM_{j,t} + BM_{ji'}) \qquad (1)$$

where PM and BM represent the path metrics and branch metrics of a trellis butterfly as shown in Fig. 3. This operation is performed in a single cycle using the dedicated ACS unit shown in Fig. 4. The unit is also capable of calculating Max and Max* operations which are frequently used in Turbo coding:

$$\text{Max}^*(x,y)= \text{Max}(x,y)+\log(1+e^{-|y-x|})=\text{Max}(x,y)+f_c(|y-x|) \tag{2}$$

where $f_c(.)$ is the correction factor that could be implemented using the integrated lookup table in ACS unit.

- *GF Accelerator:* The Galois Field accelerator is a dedicated engine that performs modulo and modulo multiply operations used in Reed-Solomon decoding [9]. Modulo add instruction is implemented using the XOR integrated in ALU.
- *Lookup Table:* It is a 16-byte local memory in PEs used for lookup table and temporary storage.

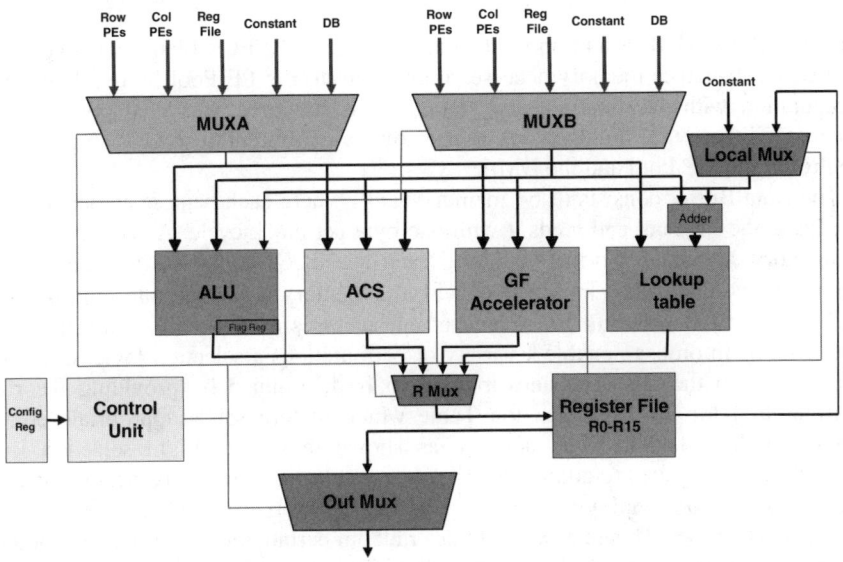

Fig. 2. Processing element architecture

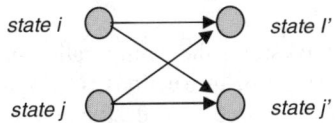

Fig. 3. Trellis butterfly

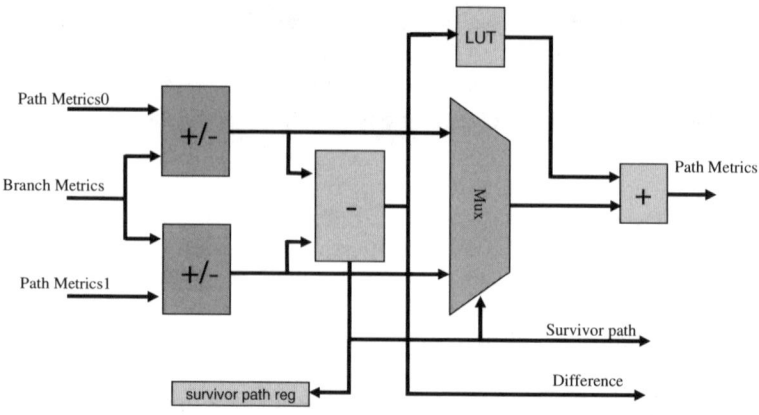

Fig. 4. ACS Unit

3.2 Data Buffer

The Data Buffer (DB) is an important component of RECFEC which is analogous to a data cache. It makes memory accesses transparent to the PE Pool by overlapping of computation with data loading and storage. It is designed as a dual port memory which facilitates the simultaneous access by two different resources, e.g. the PE Controller, the PE Pool and the DMA.

The Data Buffer consists of 64 memory banks where each bank is associated with one PE in the PE Pool and feeds it with one byte per clock cycle. A DB bank is a set of memory segments. Normally, during each access 64 banks are activated which provide 64×8 bits of data for PE Pool with the memory organized into multiple 512-bit lines. The DB supports 16, 32 and 64-bit accesses as well. Fig. 5 illustrates the Data Buffer. In order to enable a variety of permutations a selection logic network is integrated into the DB. The network pattern is determined by providing the right configuration for the Configuration Table which in turn sends appropriate control signals to the selection logic network as shown in Fig. 6. Eight selection logic networks are integrated to address 8 rows. Every selection logic network provides the data for 8 PEs. By employing two levels of multiplexers every PE is able to access any bank in the DB. This feature facilitates random permutation, interleaving and data shuffling. Consequently, the design of PE Pool interconnection network is relaxed and complicated data movement is handled by selection logic in the DB.

3.3 Configuration Buffer

The Configuration Buffer (CB) stores the configuration program of the PE Pool. The Configuration Program is the list of instructions (Configuration Words) of every PE including the operation and its sources and destination. Configuration Words are 32 bits each. In every cycle, Configuration Words of PEs are broadcast to the PE Pool and stored in the Configuration Registers of the PEs. Broadcast patterns can be horizontal or vertical. In horizontal/vertical mode, all PEs in the same row/column

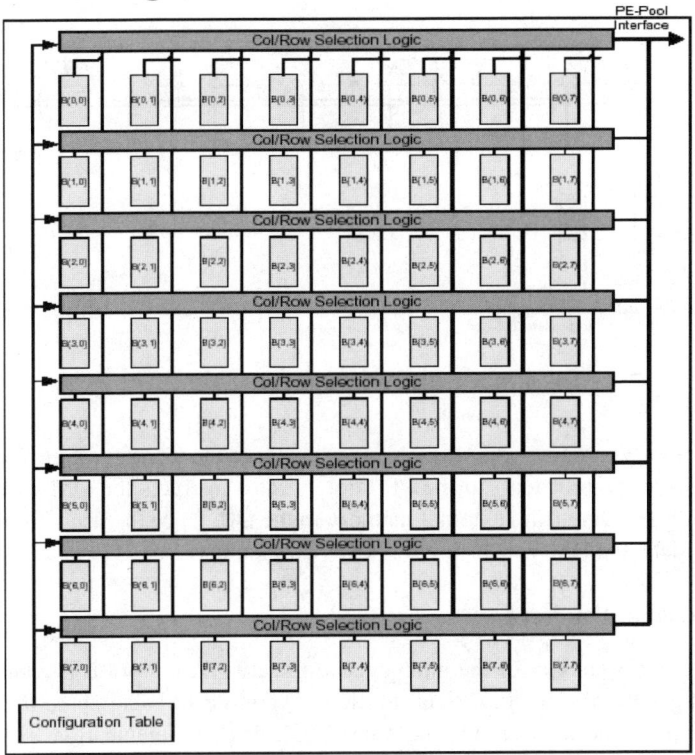

Fig. 5. Data Buffer organization

shares the same Configuration Word and perform the same operations on different input data (SIMD). Furthermore, Configuration Words can initially be stored in the Configuration Registers of the PEs. Later, the PEs can execute the stored Configuration Words repeatedly (MIMD).

3.4 PE Controller

The PE Controller is a 32-bit MIPS-like RISC processor with a four stage pipeline which orchestrates the operation of the PE Pool. It includes sixteen 32-bit registers and three functional units, a 32-bit ALU, a 32-bit shift unit and a memory unit as well as a 4-state pipeline. The instruction set of this RISC processor is augmented with special instructions to control the PE Pool. The RISC processor works at 250MHz.

3.5 DMA Controller

The DMA controller is programmed to transfer data and Configuration Words to the DB and the CB. The PE Controller commands the DMA controller to bring the

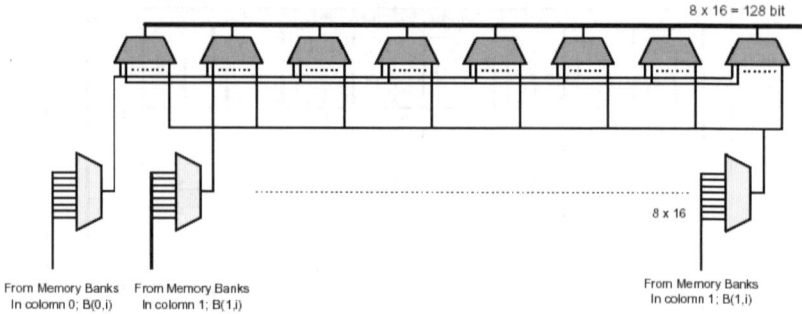

Fig. 6. Row/column selection logic in Data Buffer

Configuration Words to Configuration Memory. Once the Configuration Words are in place, they can be broadcast to the PE Pool. It also instructs the DMA to bring the raw data for decoding from external memory to the DB. Once the decoding is done on a block of data, the DMA sends it out to external memory.

3.6 Interconnection Network

A mesh design is chosen for the interconnection network. Every PE is connected to 4 neighboring PEs: top, bottom, right and left. Therefore PEs can share their internal registers with adjacent ones. On the other hand, data communication between non-adjacent PEs is handled via the DB. In many algorithms, high speed data movement happens between adjacent PEs. For example, ACS inputs in Viterbi or Turbo will be originated from neighboring cells which will be performed in single cycle. In this case mesh network could effectively support this type of traffic. For more complicated patterns, data is written to the DB and reordered while reading back to the PE Pool. This will require an extra cycle.

3.7 Programming Model of RECFEC

In RECFEC, FEC algorithms can be implemented in software. An algorithm is partitioned into two segments: *control segment* and *processing segment*. In FEC algorithms, the processing segment is usually very parallel and is mapped to the PE Pool. There are some control and scheduling parts in algorithms that are handled by the PE Controller. A RECFEC program consists of two sections: *PE Controller Code (PEC Code)* is the main control and scheduling program that orchestrates the operation of the PE Pool, the Data Buffer, the DMA and the Configuration Buffer. *PE Pool Code (PEP Code)* is the parallel Configuration Words for PEs. Configuration Words are broadcast to PE Pool using special instructions in PEC Code pointing to corresponding Configuration Words in PEP Code. The programming model of RECFEC is illustrated in Fig. 7.

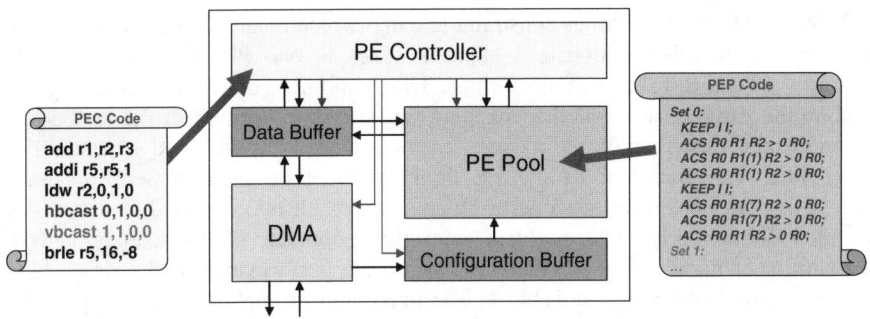

Fig. 7. Programming model

4 Implementation of FEC Algorithms

In this section we present an example of implementation of the Viterbi and Turbo algorithms on RECFEC.

4.1 The Viterbi Algorithm

The Viterbi decoder [10],[11] is employed in many different wired and wireless applications ranging from disk drives to W-LAN, W-CDMA, WiMAX and DVB-H.

In order to utilize the maximum parallelism of the Viterbi algorithm, an efficient mapping approach is of great importance. A systolic implementation on a reconfigurable SIMD architecture is shown in [12]. However due to the interconnection network of the PE Pool and between the PE Pool and the DB, a parallel implementation has a better performance on RECFEC. The Viterbi decoder used in W-LAN has the constraint length $K=7$ which corresponds with 64 trellis states. By assigning one state to every PE, 64 parallel trellis states are mapped to 64 PEs. Following steps summarize the algorithm:

- *Branch metrics calculation:* Branch metrics are calculated in parallel in all 64 PEs. The 4-bit soft value symbols are stored in the DB. 128 symbols are broadcast to the PE Pool. Every PE reads two symbols and calculates appropriate branch metrics. Four branch metrics are required per trellis stage. Two of them are calculated. By changing their signs, other two branch metrics are generated.
- *Path metrics calculation:* ACS is the main instruction in this step. 64 PEs perform 64 parallel ACS operations in every cycle. Path metrics are updated and then stored back to the DB.
- *Survivor paths storage:* Survivor paths are represented by single bit flag of ACS instruction and stored in the DB to be used for Trace Back.
- *Trace back:* Due to the sequential nature of trace back, it is performed on the PE Controller. The PE Pool is disabled in this step.

Fig. 8 shows the processing schedule of the Viterbi decoder and the mapping of trellis diagram to the PE Pool as performed on RECFEC. This mapping is applicable for the Viterbi decoder used in W-LAN which has 64 states.

In W-CDMA, the coder has constraint length 9 which results 256 trellis states. One efficient mapping is to allocate 4 adjacent states to one PE. Therefore every PE performs 4 back-to-back ACS operations. This mapping requires 4 internal registers to store the path metrics. Another mapping scenario is to use half of the PE Pool for the trellis and consequently ACS operations of 8 states are performed sequentially in every PE. The other half of the PE Pool can be used then to perform similar operations, but on a different Viterbi block. As a result two encoded blocks will be decoded concurrently. The problem with this approach is the latency and the availability of parallel blocks. We summarized the performance of the single block implementation of Viterbi in Table 1. The maximum throughput in this approach is good enough for W-LAN application. For W-CDMA, the maximum throughput is much more than mandated date rate. In W-CDMA standard, Viterbi algorithm is only used for low data rate voice and not for high speed data (HSDPA).

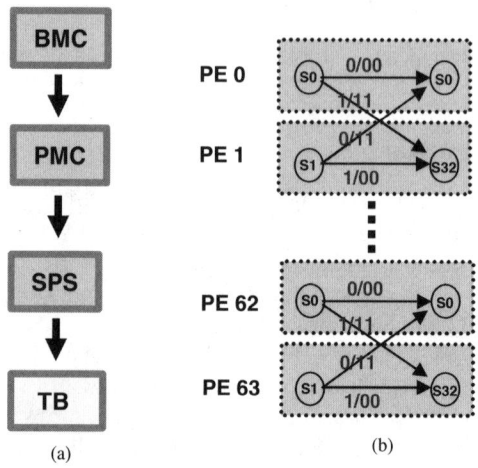

(a)   (b)

Fig. 8. a. Processing schedule of the Viterbi algorithm, b. Trellis butterflies mapped to PEs

Table 1. Performance of the Viterbi decoder implementation

K	R	Brach Metrics Calculation (cc/bit)	Path Metrics Calculation (cc/bit)	Survivor Path Storage (cc/bit)	Trace Back (cc/bit)	Total (cc/bit)	Mbps
7	1/2	0.03	3	1.06	0.42	4.51	55.43
7	1/3	0.05	3	1.06	0.42	4.53	55.24
9	1/2	0.03	12	4.25	0.63	16.91	14.79
9	1/3	0.05	12	4.25	0.63	16.92	14.77

4.2 Turbo Algorithm

By using iterative techniques, Turbo decoders [13] provide large coding gains and are widely used in digital communication standards like W-CDMA, HSDPA, WiMAX as well as satellite and deep-space communications.

An iteration of Turbo decoder consists of two Soft Input Soft Output (SISO) decoders corresponding to the encoder network which provide a measure of reliability of the decoded bits.   There are two types of soft decision decoding algorithms which are typically used, the first being a modified Viterbi algorithm which produces soft outputs and hence is called a soft output Viterbi algorithm (SOVA) [14]. A second algorithm is the maximum *a posteriori* (MAP) algorithm [15, 16] which results a better performance especially in low SNR conditions with the penalty of higher computational complexity. Log-MAP which simplifies the MAP algorithm computation by taking that to log domain and its approximation, MAX-Log-MAP are typically used in hardware implementation [17, 18].

RECFEC supports the implementation of either of these algorithms in software. As shown in Fig. 3, a powerful ACS unit is embedded in PEs which can perform MAX or MAX* operations. It can also generate the difference of path metrics to be used in soft Viterbi implementation.

The Turbo decoder in W-CDMA has 8 states. Two rows of the PE Pool are effectively utilized to implement a Turbo decoder. A MAX-LOG-MAP implementation requires the mapping of four major functions of a SISO decoder:

- *Gamma calculation*: It is similar to branch metrics calculation in the Viterbi algorithm. 4 γ values are required for every trellis stage: γ_{00}, γ_{01}, γ_{10}, γ_{11}. Two values are calculated and the other two are derived by changing the sign of the first two values. 8 PEs in the first row and the second row calculate γ_{00} and γ_{01} of 8 trellis stages respectively in 6 cycles. The state assignment to the PEs is shown in Fig. 9.

- *Alpha calculation:* Using the ACS accelerator, MAX* operation which is the main operation of α and β calculations is performed effectively on PEs. 8 PEs on the first row is allocated to handle α calculation of 8 states of a trellis stage in 4 cycles.

- *Beta calculation:* It is similar to alpha calculation and is mapped to the second row of PEs and is performed concurrently with α calculation.

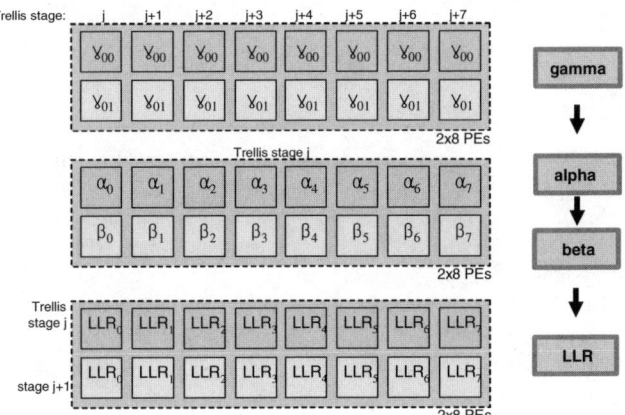

Fig. 9. Mapping of a Turbo decoder on PEs

- *Log Likelihood Ratio (LLR) calculation:* This step requires reading of α, β and γ of all states and computation of log likelihood which is add, subtract and comparison. Once LLR values of all states are compared, the maximum is selected. Two rows perform the LLR computation of two separate stages in 9 cycles.

Fig. 9 shows the mapping of α, β and γ calculations. Interleaving function is performed by programming the customized pattern to the Configuration Table of the DB. Hard decision bits are generated after the final iteration on two rows.

Multiple levels of parallelism in a Turbo decoder can be address by this mapping using the PE Pool: *SISO level* (multiple blocks in two concurrent SISO decoders), *Decoder level* (multiple SISO decoders) and System *level* (multiple decoders to process parallel FEC blocks).

Table 2 captures the performance of a Turbo decoder used for W-CDMA on a 3856- bit block using MAX-LOG-MAP algorithm.

Table 2. Performance of the Turbo decoder implementation

K	R	\multicolumn{5}{c}{SISO Decoder (cc/dec/bit)}	Hard Decision (cc/bit)	Total (cc/bit/iter)	Mbps (per iter)
		γ	α	β	LLR	Int/Deint			
4	1/3	0.75	2	2	4.50	0.50	0.38	19.58	51.09

5 Conclusion

In this paper, a novel reconfigurable processor is introduced that is optimized for forward error correction. The architecture is parallel and flexible and compared to conventional implementation of FEC decoders, it is instruction-level programmable and can accommodate multiple implementations of FEC algorithms. The architecture can efficiently host evolving algorithms or multiple algorithms for the same coding scheme and enables adaptive choice of algorithms in different situations. A mapping scenario of Viterbi and Turbo coding is presented as a case study and performance metrics are presented.

References

1. A.J Viterbi, "Wireless digital communication: a view based on three lessons learned," IEEE Com. magazine, Vol. 29, pp 33-36, Sep. 1991.
2. S. Halter, M.Oberg, P.M. Chau, P.H.Siegel,"Reconfigurable signal processor for channel coding and decoding in low SNR wireless communications," IEEE workshop on signal processing systems, pp. 260-274, 1998.
3. J.R. Cavallaro, M. Vaya, "VITURBO: a reconfigurable architecture for Viterbi and turbo decoding," in proceedings of ICASSP '03, Vol. 2, pp. 497-500, 2003
4. K. Huang, F.M. Li, P.L. Shen, A.Y. Wu, "VLSI design of dual mode Viterbi/Turbo decoder for 3GPP," in proceedings of ICAS '04, Vol. 2, pp. 773-776, 2004.
5. J. Liang, R. Tessier, D. Goeckel, "A dynamically-reconfigurable, power-efficient Turbo decoder," in proceedings of 12th Annual IEEE Symposium on Field-Programmable Custom Computing Machines, pp. 91-100, 2004.

6. C. Thomas, M.A. Bickerstaff, L.M. Davis, T. Prokop, B. Widdup, G. Zhou, D. Garrett, C. Nicol, "Integrated circuits for channel coding in 3G cellular mobile wireless systems," IEEE Com. Magazine, Vol. 41, pp. 150-159, Aug. 2003.
7. J. Ertel, J. Vogt, A. Finger, "A high throughput Turbo Decoder for an OFDM-based WLAN demonstrator," in proceedings of 5th International ITG Conference on Source and Channel Coding (SCC), Jan. 2004.
8. A.J. Viterbi, "An intuitive justification and a simplified implementation of the map decoder for convolutional codes," IEEE Journal on Selected Areas in Communications, Vol. 16, pp. 260-264, Feb. 1998.
9. J. H. Lee, J. Lee, M. H. Sunwoo, "Design of application-specific instructions and hardware accelerator for Reed-Solomon codecs," EURASIP Journal on Applied Signal Processing , Vol 2003 , pp 1346-1354, 2003.
10. A. Viterbi, "Error bounds for convolutional coding and an asymptotically optimum decoding algorithm," IEEE Trans. Info. Theory, Vol. IT-13, pp. 260-269, Apr. 1967.
11. G.D. Forney, "The Viterbi algorithm," in proceedings of the IEEE, Vol. 61, pp.268-278, Mar. 1973.
12. A. Niktash, H. Parizi, N. Bagherzadeh, "A Multi-standard Viterbi Decoder for mobile applications using a reconfigurable architecture," in proceedings of VTC, Fall 2006.
13. C. Berrou, A. Glavieux, P. Thitimajshima, "Near Shannon limit error-correcting coding and decoding: Turbo codes," in proceedings of ICC '93, pp. 1064-1070, 1993.
14. J. Hagenauer, P. Hoeher, "A Viterbi algorithm with soft-decision outputs and its applications," in proceedings of GLOBECOM '89, pp. 1680-1686, 1989.
15. L. R. Bahl, J. Cocke, F. Jelinek, J. Raviv, "Optimal decoding of linear codes for minimizing symbol error rate," IEEE Trans.Info. Theory, Vol. 20, pp. 284-287, Mar. 1974.
16. S. Pietrobon, S. A. Barbulescu, "A simplification of the modified Bahl decoding Algorithm for systematic convolutional codes," Int. Symp. Info. Theory and its Applications, pp. 1073–1077, Nov. 1994.
17. J. A. Erfanian, S. Pasupathy, G. Gulak, "Reduced complexity symbol detectors with parallel structures for ISI channels," IEEE Trans. on Com., Vol. 42, pp. 1661-1671, 1994.
18. W. Koch, A. Baier, "Optimum and sub-optimum detection of coded data disturbed by time-varying inter-symbol interference," in proceedings of GLOBECOM '90, pp. 1679-1684, Dec. 1990.

FPGA-Accelerated Deletion-Tolerant Coding for Reliable Distributed Storage

Peter Sobe and Volker Hampel

University of Luebeck
Institute of Computer Engineering
{sobe, hampel}@iti.uni-luebeck.de

Abstract. Distributed storage systems often have to guarantee data availability despite of failures or temporal downtimes of storage nodes. For this purpose, a deletion-tolerant code is applied that allows to reconstruct missing parts in a codeword, i.e. to tolerate a distinct number of failures. The Reed/Solomon (R/S) code is the most general deletion-tolerant code and can be adapted to a required number of tolerable failures. In terms of its least information overhead, R/S is optimal, but it consumes significantly more computation power than parity-based codes. Reconfigurable hardware can be employed for particular operations in finite fields for R/S coding by specialized arithmetics, so that the higher computation effort is compensated by faster and parallel operations. We present architectures for an application–specific acceleration by FPGAs. In this paper, strategies for an efficient communication with the accelerating FPGA and a performance comparison between a pure software-based solution and the accelerated system are provided.

1 Introduction

Distributed and parallel computations require globally accessible, fast and reliable storage. To avoid bottlenecks and to overcome the relatively low performance of single magnetic disks, storage architectures serving these requirements are distributed systems itself. The basis technique is to parallelize storage activities on a group of storage units in combination with a striping layout of the data. Storage systems that make use of this principle by software–based layers are, for example, PVFS2[5], Lustre[6] and NetRAID[8]. Fault-tolerant codes take benefit from the striping layout by providing check information across many storage resources that allows to tolerate the loss of storage units. Performance and the degree of fault-tolerance provided by a code usually underlie a tradeoff, i.e. improving reliability increases computation effort for coding which slows down access rates. This tradeoff can get mitigated by parallelizing coding on a proper hardware architecture. FPGAs can be such a platform for parallel coding. This is motivated by the recent advances in this technology that have resulted in powerful devices with reasonable high clock rate and a high number of logic blocks. Recently, FPGAs have been integrated in pioneer High Performance Computer (HPC) systems, such as the Cray XD1. These systems distinguish from others

P. Lukowicz, L. Thiele, and G. Tröster (Eds.): ARCS 2007, LNCS 4415, pp. 14–27, 2007.

by a closer connection of CPUs and FPGAs, which allows low latency and high bandwidth communication. When FPGAs are configured as customized coprocessors, particular applications may gain an execution acceleration of 20 to 100, yet, achieving these values proofs to be non-trivial. The presented work is based on the NetRAID [8,9] storage system that is designed for cluster systems with distributed memory and local disks. If available, FPGAs can be interfaced to speed up R/S coding. The experiments documented in this paper were run on a Cray-XD1 machine, equipped with Xilinx Virtex 4 FPGAs.

The paper is structured as follows. In Section 2, a distributed storage system is introduced with a focus on redundancy coding together with a preliminary estimation of the acceleration gain. Accelerator designs, beginning with a FPGA-based Galois Field (GF) multiplier array and yielding to an architecture of a R/S coprocessor are explained in Section 3, followed by an experimental performance evaluation in Section 4. A section positioning the presented work within related research concludes the paper.

2   Distributed Reliable Storage

2.1   Overview

A software-based distributed storage system, called NetRAID is used to compare the pure software approach with an FPGA-accelerated one. In NetRAID, a group of storage servers establish parallel access to data. Fig. 1 illustrates for example a group of 6 servers, accessed by 2 clients. Data is striped across N servers, additionally M servers can be configured to store redundancy. For the depicted example, parameters (N,M) would be (5,1) for a parity code, and (4,2) for a R/S code, the latter with the ability to tolerate two failed storage devices. The computations for encoding and decoding are completely done by the clients that use a library for file system access. Alternatively, an integration into a filesystem is reached using a FUSE-based daemon. Together with several variants of parity-based codes, a Reed/Solomon code according to [2,3] is integrated. By data striping and parallel storage activity, the access bandwidth to data is improved significantly. Ideally, the bandwidth can be scaled with the number of storage servers until the network capacity is saturated. Practically, 8 to 16 storage devices work efficiently in parallel. Access rates up to 280 MByte/s were measured for a single client process and up to 640 MByte/s for collective access by many clients to a set of 16 storage servers, both without applying any redundancy codes. Redundancy codes slow down access (a) when data is written and (b) when data is read from a storage system containing a single or a few failed storage devices. Still a good performance is achieved by parity-based codes. Operations for en- and decoding are bit-wise XORs of 32/64 Bit operands that are properly supported in all common microprocessors. The situation changes when Reed/Solomon (R/S) codes have to be calculated on common microprocessors. The arithmetics necessary for R/S are not supported by instructions directly and thus require complex and time-consuming calculations.

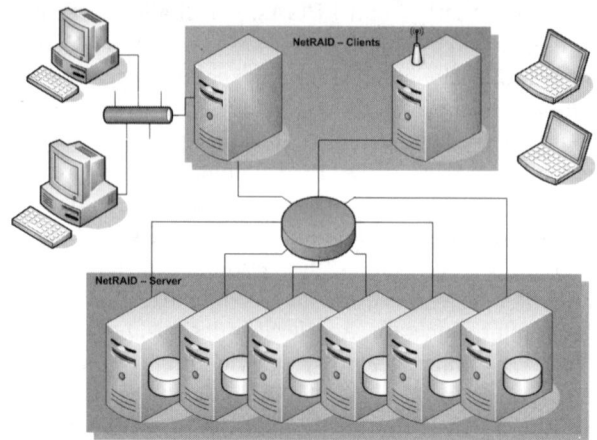

Fig. 1. The NetRAID storage system

2.2   Reed/Solomon Coding

Conceptually, the creation of $N + M$ data words and check words (the redundancy) to store on $N + M$ locations is done by multiplying a constant matrix A with the vector D that consists of N data words:

$$AD = E, \qquad A = \begin{bmatrix} I \\ G_{sub} \end{bmatrix}, E = \begin{bmatrix} D \\ B \end{bmatrix} \tag{1}$$

where I represents the $(N \times N)$-identity-matrix, G_{sub} is a sub-matrix of the code generator matrix. The structure of the generator matrix guarantees the invertibility when any M rows get discarded from A. The result vector E contains the check words in B, with M check words. For encoding, solely redundancy has to be calculated, dropping the part of the equation system related to the identity matrix. The check words are calculated word-wise as a linear combination of N data words and factors, as expressed by equation (2), where b_i $(i = 1, \ldots, M)$ are the words on the i-th storage unit among the M redundant ones, d_j $(j = 1, \ldots, N)$ denote original data words on a data storage unit j. The factors g_{ij} are elements of a $(M \times N)$ matrix G_{sub}.

$$b_i = \sum_{j=1}^{N} d_j g_{ij}. \tag{2}$$

R/S is able to recover up to M failures at all. If an error occurs, the missing elements of vector D can be recalculated by using data at the non faulty storage devices. E' represents the vector with at least N remaining words and A' is the subset of N rows of A belonging to the non faulty elements in E'. The data vector can be recovered with help of the matrix-vector multiplication:

$$D = A'^{-1} E' \tag{3}$$

A' and E' can be taken directly from A and E by omitting the rows/elements that correspond to failed storage devices. The inverse A'^{-1} is calculated once after fault detection. Later on, all missing data words can be reproduced by computing the missed elements in D according to formula (3).

This approach, based on a linear equation system, requires Galois Field (GF) arithmetics for an exact reproduction of missed data. Integer and floating point operations are not capable of an exact recovery, due to the inability to operate within a finite value range and due to rounding errors, respectively. The word size of GF elements depends on the field used, n=8 for $GF(2^8)$ fits well to practical system sizes and leads to elements that can be addressed in a Byte–aligned way.

Practically, redundancy is calculated by multiplying a word from each storage device with a constant taken from G_{sub}. After that, the products have to be added, which fortunately maps to the XOR operation in the Galois Field. This leads to N multiplications and $N-1$ additions in GF arithmetics. The number of operations is dedicated to a single word on a single redundancy storage. Thus, the N GF-multiplications and $N-1$ XORs must be executed M times each, as $M \geq 2$. The effort as number of operations is summarized in Table 1 for parity, R/S and for a computationally optimal MDS code, which is available only for a few special cases (e.g. Evenodd [1] with $M = 2$). Maximum distance separable codes[4] (MDS) allow to tolerate any M failures with M additional storage devices, such as R/S.

Table 1. Computation effort in number of operations

	XOR	GF-Mult
Parity (M=1)	$N - 1$	0
R/S	$M \times (N - 1)$	$M \times N$
optimal	$M \times (N - 1)$	0

As expected, Reed/Solomon codes are expensive in terms of computation cost. This coding effort appears for the write operation, because data is to encode always, even on a group of fault free storage units. The read operation is still fast for a fault free system, but the access rate drops to a few ten MByte/s in case of a storage group with failed units. Particularly, the slow write access is an undesired effect for applications that have to rely on fault-tolerant data storage. A detailed performance analysis can be found in [9], where for example a write rate slowdown of about 50 % for $M = 2$ is reported. R/S coding executed on a hardware-accelerated platform promises to keep up with fast storage access and still to achieve the fault-tolerance properties of a MDS code.

2.3   Preliminary Analysis

An analysis shall show, under which system parameters an acceleration by a FPGA will be beneficial. A limiting factor must be seen in the lower clock frequency of the FPGA, thus a frequency factor T is introduced.

$$T = \frac{f_{FPGA}}{f_{CPU}} \tag{4}$$

A microprocessor is able to complete a XOR operation in each clock cycle. An accelerating coprocessor can deliver results of a GF multiplication and further logic operations each clock cycle as well, a pipelined architecture taken into account (see 3.2).

When a software-based execution of a GF multiplication is A times slower than a XOR operation, the coprocessor would do a comparable computation A times faster than software, assumed processor and FPGA would run with the same clock frequency. The effect of the lower FPGA clock rate is taken into account by factor T. This approach allows to measure A by comparing the XOR bandwidth (B_{XOR}) with the GF multiplication bandwidth (B_{GFM}) in the CPU.

$$A = \frac{B_{XOR}}{B_{GFM}} \tag{5}$$

To be comparable, the accelerated R/S encoding includes GF products on a number of operand pairs, filling the word size of the processor (32/64 Bit). Additionally, subsequent XOR operations to add the products can get easily included. Thus, more functions are provided by the accelerator compared to a single instruction on the CPU. The factor A' extends A by expressing these additional computations. N multiplications and N-1 XOR operations provided by the coprocessor have to be set into relation with N-1 XOR operations that would be calculated by software for a computationally optimal MDS code.

$$A' = \frac{A \cdot N + (N - 1)}{N - 1} \tag{6}$$

With the following criterion,

$$S = T \times A' \tag{7}$$

the speedup is quantified. Acceleration by a FPGA is worth when S is significantly larger than one. Experiments on a Cray XD1 node let us measure the calculation rates and determine $T = 200\text{MHz}/2200\text{MHz} = 0.\overline{09}$. The results shown in Table 2 demonstrate that an acceleration related to software-based R/S can be achieved.

Table 2. Calculation bandwidth (MByte/s) for XOR and GF-multiplications

block size	B_{XOR}	B_{GFM}	A	A' for $N{=}8$	S
512 Byte	7721	180	42.9	50.0	4.55
2048 Byte	8533	182	46.9	54.6	4.96
8192 Byte	8684	169	51.4	59.7	5.42

3   FPGA-Accelerated Reed/Solomon Coding

3.1   Cray XD1 Environment

As described in [10], the Cray XD1 is a parallel computer with multiprocessor nodes that can be optionally equipped with several types of Xilinx FPGAs. All communication between CPUs and FPGAs is handled by the Cray RapidArray interconnect that directly interfaces the CPUs via HyperTransport. AMD Opteron CPUs provide a direct access to the HyperTransport, and thus to the RapidArray. For the FPGA logic, appropriate blocks have to be integrated to control the communication via RapidArray. For this purpose Cray provides an intellectual property core to access the RapidArray via two independent communication interfaces for system and FPGA initiated requests, respectively. As the FPGA can be clocked with up to 200 MHz and each of the communication channels being 64 Bits wide, a maximum data rate of 1.6 GB/s per channel can be achieved. The programmed logic in the FPGA is utilized by processes running on the CPUs via an application programming interface provided by Cray. To transfer data from the application into the FPGA, a mapped memory section is used. The FPGA can issue requests to the application as well. To respond to these requests the application must register a memory section to which the FPGA requests are mapped.

3.2   Galois Field Multiplier

As a building block, a finite field or Galois Field multiplier (GFM) based upon the Mastrovito Multiplier architecture [11] is implemented in the FPGA logic. The product of two polynomials $A(y)$ and $B(y)$ in $GF(2^n)$ is calculated using a product matrix Z. This product matrix depends on the polynomial coefficients of the multiplicand and a matrix representation of the field polynomial $Q(y)$, yielding $Z = f(A(y), Q(y))$. The multiplication is carried out as a matrix-vector multiplication in $GF(2^1)$ of the product matrix and the multiplier polynomial and thus can be implemented with AND and XOR gates. The whole calculation can be expressed by the following equations:

$$A(y) \cdot B(y) = C(y) \, mod \, Q(y)$$

$$C(y) = \begin{pmatrix} c_0 \\ c_1 \\ \vdots \\ c_{n-1} \end{pmatrix} = Z \cdot B(y) = \begin{pmatrix} z_{0,0} & \cdots & z_{0,n-1} \\ \vdots & \ddots & \vdots \\ z_{n-1,0} & \cdots & z_{0,n-1} \end{pmatrix} \cdot \begin{pmatrix} b_0 \\ b_1 \\ \vdots \\ b_{n-1} \end{pmatrix}$$

This scheme allows to build a highly parallel and pipelined architecture. In the first stage all the coefficients of the product matrix are calculated according to the current field polynomial and the multiplicand. In a second stage, all product matrix rows are multiplied with the multiplier polynomial. Adding the element's products forms the third stage, and writing the result the fourth. This leads to a latency of four clock cycles for the whole multiplication.

Our implementation is fixed to a width of 8 Bit but variable regarding the finite field to work on. To configure a particular Galois Field, the appropriate matrix representation of the field polynomial can be written to the multiplier.

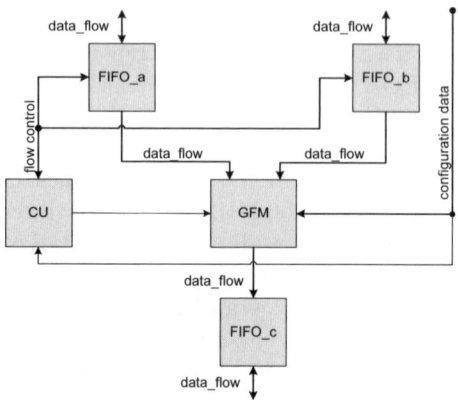

Fig. 2. Block diagram of the buffered Galois Field multiplier (BGFM)

As depicted in Fig. 2, factors are buffered in two separate FIFOs before getting multiplied. Another buffer keeps the results. A control unit synchronizes interactions between all elements, i.e. creates configuration signals and the buffer control signals. The control unit activates the calculation as soon as both input buffers are non-empty and as long as the result buffer is not full.

3.3   GF Multiplier Array

As the RapidArray IP-core can issue one quadword, i.e. 64 bits, per cycle four buffered GF multipliers can work parallel, each with 2×8 Bit input width. Doing so completely utilizes the bandwidth provided by the RapidArray. The multiplier array receives the input factors via one communication channel and sends the results via the other. To handle delays or high traffic on the RapidArray, two modules TXC and RXC have been implemented that control the incoming and the outgoing data streams. Based on the signaling of the RapidArray and the filling state of the buffers they may halt and resume communication.

3.4   GF Multiplier Array with Implicit Factors (GFMA-IF)

With binding the logic more specifically to R/S encoding and decoding, obviously one factor of each multiplication can be held constant. Thus, one factor of each multiplication must be retrieved from the data flow only, while the other is held fixed in the accelerator. Correspondingly, each clock cycle 64 bits of input data, eight data words for multiplication, can be processed, using the full capacity of the FPGA's input data bus. These considerations apply to the FPGA's output data bus as well. Each clock cycle eight products are released and have to be returned. The architecture shown in Fig. 3 utilizes eight GF multipliers in parallel.

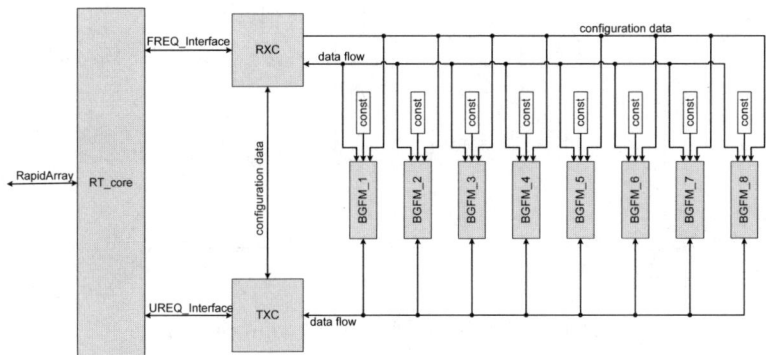

Fig. 3. Multiplier array with implicit input factors (GFMA-IF)

3.5   R/S Coprocessor

In 3.3 and 3.4, two variants of a finite field multiplication acceleration were presented. So far, the algorithms for encoding and decoding still must be executed in software. When moving the R/S coding completely into the FPGA, by combining parallel GF multipliers and XOR logic, the CPU can get completely released from most of the coding operations. The encoding and decoding algorithm, both represented by a matrix-vector multiplication, can be directly translated into a hardware structure. Figure 4 depicts a R/S coprocessor layout of this structure. Data words from N storage units are multiplied with constant factors by N GF-multipliers and summed by a bitwise N-input parity logic. This data path generates a single check word for encoding. Several multiplier-parity paths have to be instantiated for multiple check words that are commonly required. Similarly to the GF multiplier array with implicit factors, the computation bandwidth gets fully effective as encoding rate. The code generator matrix G_{sub} is written to the coprocessor in advance, and hence, the factors do not have to be transfered during coding anymore. Redundancy information is transferred with a lower rate, $\frac{M}{N}$ of the data input rate. Decoding, which is only necessary when data of failed storage devices has to be reconstructed, is supported by the presented hardware structure as well. For decoding, the matrix memory has to be filled with the rows of A'^{-1} that correspond to the indexes i of the storages d_i to reconstruct. The reconstruction requires to fetch N data and redundancy words, to multiply it with the matrix memory values and to sum the products. Up to M data words can be reconstructed by the M parallel data paths.

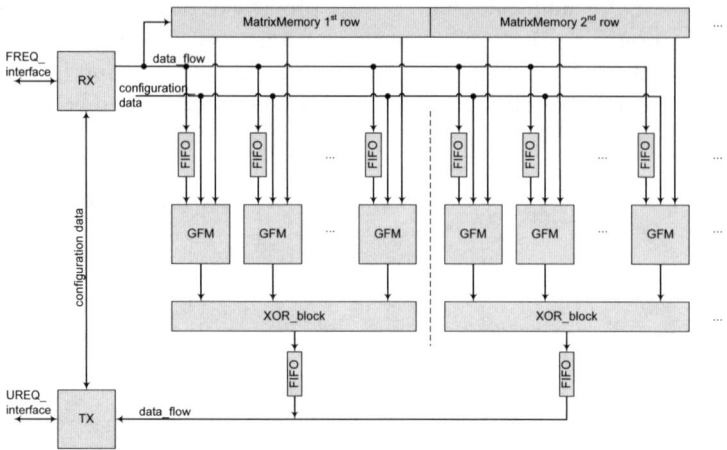

Fig. 4. Structure of the R/S coprocessor

3.6   Comparison

All the three architectures were synthesized on a XC4VLX160 FPGA with 67584 slices in total. As can be seen from the synthesis results in Table 3, the number of logic cells is not a limiting factor. The multiplier array designs limit the clock speed to 160 MHz and to 115 MHz for the more complex design of the R/S coprocessor. The theoretically reachable transfer bandwidth is determined by the product of clock frequency and the interface width. For the multiplier array, the computed value is only effective by a half, due to the necessity to transfer both factors. The actually reached bandwidth depends on the interaction with the CPUs memory and is slightly lower.

Table 3. Synthesis results

Design	Slices used	Clock speed	Theoretical bandwidth (data)
GF multiplier array	8 %	160 MHz	640 MB/s
GF mult. array with implicit factors	5 %	160 MHz	1280 MB/s
R/S coprocessor	24 %	115 MHz	920 MB/s

From a functional point of view, the R/S coprocessor embeds most logic functionality in the FPGA and reduces the data stream between CPU and FPGA to a minimum. Each word written to the data storage units is transferred to the FPGA once; and also each check word originated by the coding algorithm traverses the output link from the FPGA to the CPU once. No further processing on the CPU is necessary when the R/S coprocessor is utilized. A drawback of the R/S coprocessor approach is that data is accessed non-contiguously. To encode a R/S code word, original data must be collected from different positions in the input data that is commonly organized as a linear data structure. The multiplier array with implicit factors (GFMA-IF) corresponds closer to the data structures in a software-based environment. Input data presented in large blocks (e.g. 6 × 24 kByte) can be transferred sub block-wise (24kByte-wise) to the GFMA-IF and multiplied with a particular factor. The product array must be transfered back to the CPU and there XOR-ed to an array of check words. The GFMA-IF architecture (i) does not embed the entire functionality within the FPGA and (ii) produces more output data than the R/S coprocessor. Despite of these disadvantages, the GFMA-IF architecture with a contiguous data access is still eligible as an accelerator. Eventually, the choice which architecture performs better - the multiplier array with implicit factors or the R/S coprocessor - depends on the granularity of data distribution and on the performance of the memory hierarchy for non-contiguous access.

4   Performance Evaluation

In this section we present and discuss the performance of the FPGA-based accelerators. The results are set in relation with the software-based GF multiplication.

4.1   Evaluation Method

For evaluation, the computation bandwidth of the GF multiplication in software as well as for the FPGA-based accelerator has been measured. The FPGA-accelerated variant includes the transfer of data to the logic and the transfer of the results back into the address space of the software process. This process writes a block of factors to the FPGA, where computation of the data starts immediately and the results are written to the transfer region in the memory. Transfer to the FPGA, the processing in the multiplier arrays and the transfer of results back to the memory is fully pipelined. A block-based invocation of the accelerator functionality has been set up to keep allocated memory at reasonable sizes, while large block sizes lead to a sufficient overlap of transfer and communication.

To calculate multiplication rates, the time between writing the first quadword and receiving the last result is taken. In conjunction with the block size in quadwords, different data streams were formed. We chose to run four sessions in which four different amounts of data, being 4 MB, 40 MB, 160 MB, and 640 MB, were multiplied. The block size was varied in each session and the number of iterations was consequently adapted. Each of these combinations of block size and iterations was measured ten times and later analyzed statistically.

4.2   Computation Bandwidth

The multiplication rates for the software implementation are shown in Fig. 5. Two software variants have been evaluated - an ordinary product table lookup

(Product Lookup) and another product lookup (Index Products) that uses indexes of data and check storages to access a product of a distinct $g_{i,j}$ with a data word directly. Computation rates of about 130 and 180 MByte/s for each variant were measured. As expected, the indexed lookup performs better. The deviations from the mean values are very small, compared to the FPGA–accelerated GF multiplication.

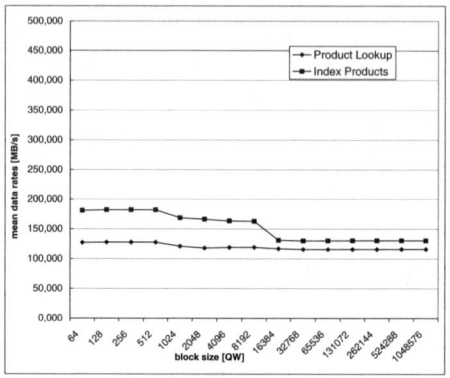

Fig. 5. Software GF(2^8) multiplication: Mean throughput (MByte/s) for a varied block size in number of quadwords

The left part in Fig. 6 shows the mean computation bandwidth, measured for the GF multiplier array with implicit factors (GFMA-IF). Slightly lower mean values, but with very similar progression have been observed for the R/S coprocessor (710 MByte/s maximum instead of 795 MByte/s). The four different plots correspond to the four amounts of data multiplied, the sessions mentioned above. The plot in the right part of Fig. 6 shows the error of the mean values against the block size. It was calculated by dividing the root mean square deviation by the square root of the number of measurements for one combination. When varying the transfer block size in a range from 64 to 32K quadwords, the data rate heavily depends on the block size. With further increasing block sizes, the rate stabilizes at about 800 MByte/s. The threshold block size for fast communication with the FPGA-coprocessor is 16384 quadwords, which equals 128 KByte. The errors of the data rates indicate that the transfer rates fluctuate heavily for small blocks. Errors get very small as soon as a block size of 32768 quadwords is reached, which equals 256 KByte. The achieved computation bandwidth depends on the utilization of both of the buses. An implementation of the architecture can be clocked at 160 MHz. On the input bus a contiguous data stream can be implemented, allowing a maximum transfer rate of 1280 MByte/s. On the output bus a wait cycle has to be implemented after each eight byte block, reducing the maximum transfer rate by 8/9 to 1138 MByte/s. Another aspect is the larger data amount to return in combination with a practically slower output link of the FPGA. We learned from debugging, that communication on the output link

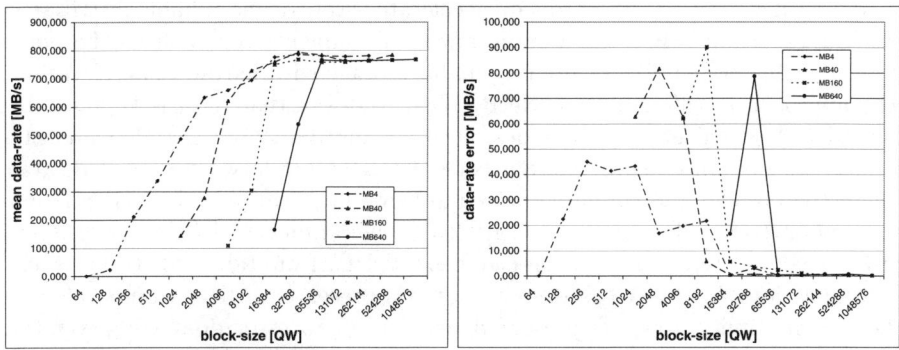

Fig. 6. FPGA-based GFMA-IF: Mean values and errors of the coding throughput

is stalled more frequently than on the input link. The R/S coprocessor is not affected by that slightly lower output rate. The number of results words is M per N input words. Practically, M is less than N and the lower output rate is fully compensated by that.

4.3 Insights

From the interpretation of the measured computation rates, three points shall summarize our insights.

(1) Coarse granular software-accelerator interaction - To achieve high data throughput the coprocessor should process blocks of at least 16384 quad-words. In distributed storage systems such block sizes are usual, e.g. 24 KByte per block for striping across eight nodes in NetRAID yields to 24576 quadwords (192 KByte) per invocation of the coding function.

(2) Estimated speedup has been reached - Comparing the hardware-accelerated GF multiplication with the software variant, at least a speedup of four is reached. We compared the fastest software function with the slowest variant of FPGA-accelerated coding on reasonable large blocks. This corresponds to the preliminary estimated speedup in Section 2.

(3) GFMA-IF and R/S coprocessor are both appropriate solutions - The R/S coprocessor and the GFMA-IF architecture both utilize the input bandwidth for GF multiplications fully. The R/S coprocessor adds the products internally and less output data has to be transferred. This compensates a lower transfer rate back to the CPU and releases the memory from traffic. Still a disadvantage is the non-contiguous data access by the R/S coprocessor. Thus, the multiplier array with implicit factors (GFMA-IF) is another proper solution.

5 Related Work

Reliable Storage is commonly supported with RAID adapters that work at the I/O bus level. Except expensive RAID controllers for storage area networks, there is still a gap between hardware accelerated storage and reliable distributed storage, especially when the storage system is implemented with COTS components. ClusterRAID[7], combines reliable storage with hardware acceleration. The approach is different from NetRAID in the way that data is kept locally on disks. A group of storage nodes, each storing data objects independently, is protected by an erasure–tolerant code. Updates are forwarded to dedicated redundancy nodes that update the redundancy information. ClusterRAID implements a R/S code as suggested in [2]. Acceleration by FPGAs and graphic processors were reported that directly base on logarithm-tables for Galois Field multiplication. A FPGA-based coprocessor is presented in [13], comparable to the GF multiplier array as presented here. The processor does not cover the entire coding data path for R/S; instead it can be configured to perform GF multiplications in a SIMD mode.

GF arithmetic is supported by a few digital signal processors (such as Texas Instruments TMS processors) by simple GF multiplication instructions. Complex GF instructions, such as multiply-add which is often needed for R/S coding, are suggested in [14] for a SIMD implementation in a processor. There are generally numerous systems that implement Galois field arithmetics for error correction in digital communication systems or digital storage with correction of arbitrary Bit corruption and burst errors in the data stream. In comparison, less development is directed to storage systems with an erasure characteristic. For distributed systems, most R/S systems still base on software coding. Despite advances in code construction such as Cauchy R/S codes that allow faster software coding than classical R/S codes, only a few systems make use of specialized hardware. Only recently, FPGAs can be coupled close enough to the CPUs that an acceleration effect can be reached. Today, the trend is visible to add more specific functionality to processors or to provide configurability. In this context our work covers the variant of CPU-FPGA cooperation for fast coding. Other variants of sophisticated coding solutions become feasible, e.g. using configurable coprocessors in parallel on a chip for R/S coding.

6   Conclusion

Sophisticated codes for reliable distributed storage, such as Reed/Solomon, are compute-intense and bound the data access bandwidth. Acceleration by customized logic is a way to reach fast encoding and decoding, keeping up with higher data rates of aggregated storages. The design space, spanning from a Galois Field multiplier array to a R/S coprocessor that covers the entire data path for code word processing has been analyzed. It could be shown that the bandwidth of the FPGA-based accelerator is noticeable higher compared to software coding. The coding bandwidth is bounded by the transfer bandwidth to and from the accelerator. Despite, for the current implementation on a Cray XD1 we could reach a factor four for coding bandwidth acceleration. With such a FGPA acceleration, coding in a distributed data storage architecture is faster than the aggregated storage bandwidth and nearly as fast as high-speed networks.

References

1. Blaum, M., Brady, J., Bruck, J., Menon, J.: EVENODD: An Efficient Scheme for Tolerating Double Disk Failures in RAID Architectures. IEEE Transactions on Computers. Vol. 44, No.2, February 1995
2. Plank, J. S.: A Tutorial on Reed-Solomon Coding for Fault-Tolerance in RAID-like Systems. SOFTWARE - PRACTICE AND EXPERIENCE. pp. 995-1012, September 1997
3. Plank, J. S., Ding Y.:, Note: Correction to the 1997 Tutorial on Reed-Solomon Coding. Technical Report, University of Tennessee, UT-CS-03-504, April 2003
4. Mac Williams, F.J., Sloane, N.J.A.: The Theory of Error–Correcting Codes. Part I, North Holland Publishing Company, Amsterdam, New York, Oakland, 1977
5. Carns, P. H., Ligon, W. B., Ross, R. B., Thakur R.: PVFS: A Parallel File System for Linux. Proc. of the 4th Annual Linux Showcase and Conference, pp. 317-327, 2000
6. Braam, P.J. et al.: The Lustre Storage Architecture., Cluster File Systems Inc., http://www.lustre.org/docs/lustre.pdf, 2004
7. Wiebalck, A., Breuer, P. T., Lindenstruth, V., and Steinbeck, T. M.: Fault-Tolerant Distributed Mass Storage for LHC Computing. CCGrid Conference, 2003
8. Sobe, P.: Data Consistent Up- and Downstreaming in a Distributed Storage System. Proceedings of Int. Workshop on Storage Network Architecture and Parallel I/Os, pp. 19-26, IEEE Computer Society, 2003
9. Sobe, P. and Peter, K.: Comparison of Redundancy Schemes for Distributed Storage Systems. 5th IEEE International Symposium on Network Computing and Applications, pp 196-203, IEEE Computer Society, 2006
10. Cray Inc.: Cray XD1 FPGA Development. Release 1.4 documentation, 2006
11. Paar, C.: A New Architecture for a Parallel Finite Field Multiplier with Low Complexity Based on Composite Fields. IEEE Transactions on Computers, Vol. 45, No. 7, pp. 856-861, 1996
12. Gilroy, M. and Irvine, J.: RAID 6 Hardware Acceleration. Proc. of the 16th Int. Conference on Field Programmable Logic and Applications 2006, pp. 67-72, 2006
13. Lim, W.M. and Benaisse, M.: Design Space Exploration of a Hardware-Software Co-designed $GF(2^m)$ Galois Field Processor for Forward Error Correction and Cryptography. In: International Symposium on System Synthesis 2003, pp. 53-58, ACM Press, 2003
14. Mamidi, S., Iancu, D., Iancu, A., Schulte, M. J. and Glossner, J.: Instruction Set Extensions for Reed-Solomon Encoding and Decoding. Proceedings of the 16th. Int. Conf. on Application-specific Systems, Architecture and Processors (ASAP'05), IEEE, 2005

LIRAC: Using Live Range Information to Optimize Memory Access

Peng Li[1], Dongsheng Wang[2], Haixia Wang[3], Meijuan Lu[4], and Weimin Zheng[5]

National Laboratory for Information Science and Technology
Research Institute of Information Technology
Tsinghua University,Beijing, China
{p-li02[1], lmj[4]}@mails.tsinghua.edu.cn,
{wds[2], hx-wang[3], zwm-dcs[5]}@tsinghua.edu.cn

Abstract. Processor-memory wall is always the focus of computer architecture research. While existing cache architecture can significantly mitigate the gap between processor and memory, they are not very effective in certain scenarios. For example, when scratch data is cached, it is not necessary to write back modified data. However, existing cache architectures do not provide enough support in distinguishing this kind of situation. Based on this observation, we propose a novel cache architecture called LIve Range Aware Cache (LIRAC). This cache scheme can significantly reduce cache write-backs with minimal hardware support.

The performance of LIRAC is evaluated using trace-driven analysis and simplescalar simulator. We used SPEC CPU 2000 benchmarks and a number of multimedia applications. Simulation results show that LIRAC can eliminate 21% cache write-backs on average and up to 85% in the best case.

The idea of LIRAC can be extended and used in write buffers and CMP with transactional memory. In this paper, we also propose LIve Range Aware BUFfer (LIRABuf). Simulation results show that the improvement of LIRABuf is also significant.

Keywords: LIRAC, Live Range, Cache, Write Buffer, Memory Hierarchy.

1   Introduction

Processor-memory wall [1] is always one of the most important problems in computer architecture research. While processor follows the well-known Moores law, doubling performance every 18-24 months, the speed of memory access only grows about 7% every year [2]. In the upcoming Chip Multi-Processor (CMP) era, the presence of more processors on a single chip significantly increases the demand of off-chip bandwidth and exacerbates this problem even more.

Modern computers are often equipped with various buffers, such as write-back cache, write buffer and commit buffer in CMP with transactional memory [3][4]. We observe that in computers with buffering technology, actual write operation

P. Lukowicz, L. Thiele, and G. Tröster (Eds.): ARCS 2007, LNCS 4415, pp. 28–42, 2007.

happens after write instruction commitment. By the time memory write operation executes, the data written might already go out of its live range which causes a useless write operation. However, existing buffer architectures cannot effectively take advantage of this observation.

In this paper, we explicitly distinguish writes in processor domain and memory domain, and further propose LIve Range Aware Cache (LIRAC) and Live Range Aware Buffer (LIRABuf). LIRAC and LIRABuf focus on reducing the number of write-back operations, and they do not affect read operations. With write operations decreased, LIRAC and LIRABuf can reduce both execution time and energy.

Trace-driven simulations show that LIRAC can reduce 21% write-backs on average and up to 85% in the best case, and LIRABuf can reduce 23% write-backs on average and up to 84% in the best case.

The rest of this paper is organized as follows. Section 1 reviews related work. Section 3 describes the architecture of LIRAC. Section 4 explains software support for LIRAC architecture. Section 5 and Section 6 present methodology and simulation results of LIRAC architecture. Section 7 shows that live range architecture can be used in other buffers such as write buffer and CMP with transactional memory. Finally Section 8 draws conclusions and proposes future work.

2   Related Work

Various efforts have been made to optimize memory access. New memory technologies like DDR (Double Data Rate) and QDR (Quad Data Rate) are developed to speedup memory access. Some researches integrate memory and processor onto a single chip to minimize access time [5][6] . Memory compression [7] compresses data to reduce memory bandwidth consumption. While previous studies focus mainly on the reduction of average memory access time, this paper focuses on reducing the number of memory access operations, particularly cache write-back operations.

The live range of register has been studied by previous research. In [8], Franklin and Sohi studied the lifetime of register instances and concluded that many registers were short-lived. Efforts have been made to reduce register commitments in superscalar processors to ease the pressure of register allocation and save energy [9][10][11][12]. This paper studies the live range of memory address and focuses on reducing memory write operations. Compared with previous register live range analysis, memory optimization is more important because memory access is far more expensive than register access.

Lepak et al. observed that many store instructions have no effect since the written value is identical to the content already existing in memory [13]. Based on this observation, silent store architecture was proposed in which each store is converted into a load, a comparison, followed by a store (if the store is non-silent). Silent store architecture reduces write operations at the expense of increasing read operations. In the proposed LIRAC and LIRABuf architectures, the number of write operations can be reduced without increasing read operations.

In some SOC systems, addition to a data cache that interfaces with slower off-chip memory, a fast on-chip SRAM called Scratch-Pad memory, is often used to guarantee fast access to critical data. However, to efficient utilize on-chip memory space, carcfully allocation must be made manually by sophisticated programmers.

3   Live Range Aware Cache Architecture

In this section, we will articulate the architecture of LIRAC in detail. We first distinguish two types of writes: writes in processor domain and writes in memory domain (Section 3.1), then we define the live range and dead range of memory address (Section 3.2). We propose the architecture of LIRAC in Section 3.3, and then in Section 3.4 we show that LIRAC architecture can also be used in multi-level cache. In section 3.5, we discuss debugging support for LIRAC.

3.1   Write in Processor and Memory Domain

We observe that there are two types of writes in a computer system, writes in processor domain (Wp) and writes in memory domain (Wm). Wp refers to write instructions committed by processor and Wm refers to write operations performed to memory (e.g. cache write-back). Wp is the output of processor while Wm is the input of memory system. In computers without buffering technology, Wp and Wm is roughly the same. The number of Wp and Wm is the same and they are in the same order. However, in computers with buffering technology, both the occurrence and sequence of Wp and Wm may be different, as shown in Fig. 1.

Store R1, ADDR0   //Wp of ADDR0

Store R2, ADDR1   //Wp of ADDR1

Store R3, ADDR0   //Wp of ADDR0

Load R4,ADDR1'   //Wm of ADDR1

//ADDR1' and ADDR1 are mapped to the same cache line

Load R5,ADDR0'   //Wm of ADDR0

//ADDR0' and ADDR0 are mapped to the same cache line

Fig. 1. Writes in Processor and Memory Domain

Fig. 1 shows the assembly code of an example program. The first instruction writes R1 to memory address ADDR0. The value is temporally saved in cache and not written to memory. At this time, cache and memory are incoherent. Cache holds the current value while memory holds the stale value. The following two instructions write R2 to ADDR1 and R3 to ADDR0 respectively. Up till this time, all values are buffered in the write-back cache and no actual memory write operation has occurred. The fourth instruction reads memory address ADDR1, which is different from ADDR1 but mapped to the same cache location. The cache line containing ADDR1 is swapped out and replaced by the new line. At this time, the dirty cache line containing ADDR1 is written back to memory. Similar operation happens when the fifth instruction is executed.

Wp and Wm of ADDR0/ADDR1 are shown in the figure. From the figure, we can tell that both the occurrence and sequence of Wp and Wm are different. For a given address, Wm always lags behind Wp.

Wp is the inherent property of software. Given a program, the occurrence and sequence of Wp is fixed, so the number of Wp cannot be reduced. On the other hand, Wm is determined by both software and hardware. A program may generate different Wm sequences with different memory hierarchy. This paper focuses on how to reduce the number of Wm.

3.2 Live Range and Dead Range

To reduce Wm, the purpose of Wm is reexamined first. A write operation is useful only if the memory address might be read again. If a write operation is definitely followed by another write operation to the same memory address, then the first write becomes useless.

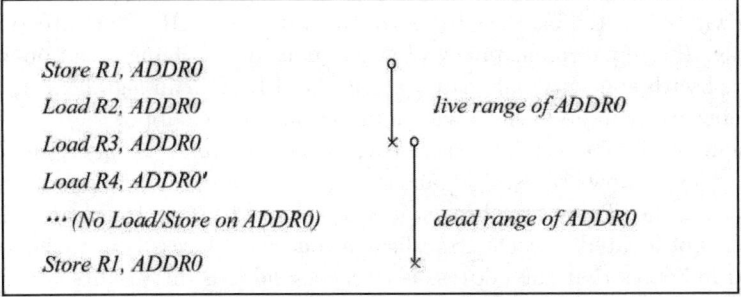

Fig. 2. Example of Live Range and Dead Range

Fig. 2 shows a sequence of instructions. The first instruction writes R1 to ADDR0. If it is executed on a computer with no buffers or write-back caches, the value must be written to memory to ensure correctness. If the program is executed on a computer with write-back cache, the value can be buffered in cache temporally.

Thereafter the second instruction reads the value from cache. In conventional write-back cache, when the fourth instruction is executed, the dirty cache line containing ADDR0 will be swapped out and written back to memory. Nonetheless, if we can know in advance that the next memory access to ADDR0 is definitely write, as shown in the figure, we can discard the dirty line without writing it back to memory. This observation can be used to reduce Wm.

To achieve this, the live range and dead range of a memory address are defined. The word Live Range is borrowed from compiler technology. A live range of a variable in compiler refers to an isolated and contiguous group of nodes in the flow graph in which the variable is defined and referenced [14]. Here we define the live range and dead range of a memory address. The live range of a memory address is from a write of the memory address to the last read of the address before another write of the same address. Similarly, the dead range of a memory address is from the last read of the memory address to the next write of the address. The definitions are also illustrated in Fig. 2.

Notice that live range of a memory address is always ended with a read instruction. A new instruction, **LastRead**, is added to support live range identification. LastRead works similarly to normal read instruction, moving data from memory to register, except that it indicates the end of live range of the memory address. In Section 4, we will explain how to generate LastRead instructions for a given program.

3.3   Architecture of LIRAC

In conventional write-back cache, a cache line will be written back to memory if the data has been modified. In our proposed LIRAC architecture, a cache line is written back to memory if the data has been modified AND the modified address is in its live range. In other words, when the data is in its dead range at eviction, nothing will be written back regardless of the dirty flag. LIRAC architecture does not change the replacement policy of cache, nor does it change read operations. Compared with conventional write-back cache, LIRAC can significantly reduce the number of write operations with minimal hardware support.

For simplicity, we first assume that each cache line contains only a single cache word. Fig. 3(a) shows the state transition graph of a cache line in conventional write-back cache. There are three states: M, S and I. State I indicates that the address is not located in cache. S indicates that the address is in cache and it is clean. M indicates that the address is in cache and it is dirty. Three operations, R, W and P, standing for Read, Write and rePlace respectively, can act on these states. If operation P acts on the state S, the cache line is just replaced without write back to memory. If P acts on the M, the cache line is written back to memory before replaced by another cache line.

Fig. 3(b) shows the state transition of a cache line in a LIRAC system. A new operation, LR, is added to indicate the LastRead instruction mentioned above. LR acts similarly to normal read operation except on state M. If LR acts on M, the next state is S because the live range of the given address ends. In other words, if a cache line in dead range is replaced by others, it is simply thrown

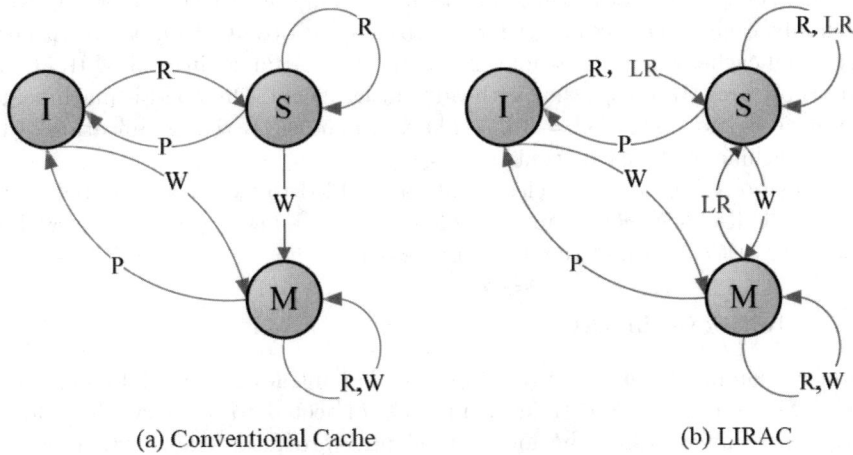

(a) Conventional Cache   (b) LIRAC

Fig. 3. Cache State Transition Graph

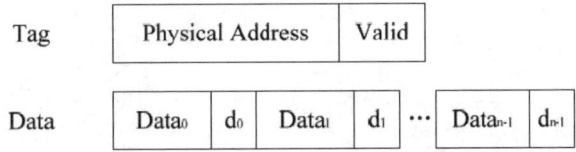

Fig. 4. Organization of a LIRAC Cache Line

away rather than written back to main memory. With the additional transition, the number of Wm can be reduced.

When a cache line contains multiple data words, each word is associated with a dirty bit to indicate whether the word has been modified or not, as shown in Fig. 4. When the cache line is swapped out, all the dirty bits in the cache line are ORed together. If the result is 1, then the cache line is written back, otherwise it is thrown away. When LastRead instruction is executed, the dirty bit of the corresponding data word is cleared to zero.

Apparently there is a tradeoff between the granularity of dirty flag and the chance of reducing write-back. When the whole cache line is tracked with one dirty flag, the scheme may not reduce Wm since the compiler may not generate many LastRead instructions due to multiple objects being mapped into the same cache line. On the other hand, if each byte is tracked with a dirty flag, then the compiler can generate more LastRead instructions, possibly reducing the number of Wm. However, there is more overhead caused by the dirty flag bookkeeping. In this paper, we assume data word granularity.

In order to implement the LIRAC, three hardware modifications are needed. First, a new instruction, LastRead, is added to the machine ISA. LastRead instruction acts just like normal Read instruction, transferring data from memory to registers, except that it will clear the dirty bit of the corresponding data word. Secondly, each cache word must have a dirty bit in LIRAC. Compared with conventional cache, where the whole cache line share a single dirty bit, LIRAC will take more area. However, the overhead is insignificant. On a 32-bit machine, the overhead is less than $1/32 = 3.1\%$. The third change is the modification of the state machine of the cache controller.

Transitions from dirty to clean state are added to reduce the probability of memory write backs. This can be implemented by a few logic gates and will not affect the critical path of the whole processor.

3.4 Multi-level LIRAC

Modern computers are often equipped with multi-level caches. In multi-level cache hierarchy, the highest level directly connected to memory often adopts write-back policy, while the lowest level near processor may be writeback or write-through.

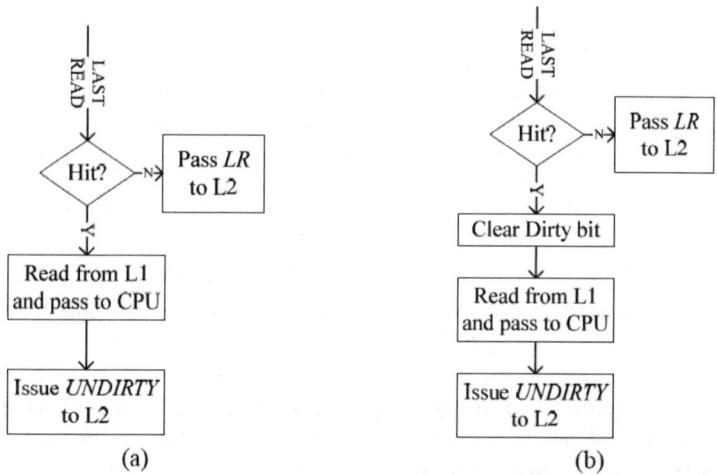

Fig. 5. Cache Reaction on LastRead Command

We first discuss LIRAC with write-through L1 cache and write-back L2 cache, as illustrated in Fig. 5(a). In this case, when processor decodes the LastRead instruction and issues LastRead command to L1 cache, if the address hits L1 cache, correct data is read from L1 cache and passed to processor. Meanwhile L1 cache issues an Undirtify command to L2 cache to clear the corresponding dirty bit. If the address misses L1 cache, the command LastRead is just passed to L2 cache. L2 cache will act as the previous single-level LIRAC architecture, i.e. returning read data and clearing the corresponding dirty bit.

Fig. 5(b) shows LIRAC with write-back L1 cache and write-back L2 cache. Compared with Fig. 5(a), the dirty bit in L1 cache is also cleared, hence both L1 and L2 caches can benefit from reduced write-backs.

3.5 Debugging Support for LIRAC Architecture

During program debugging, a programmer may want to examine the value of overdue variables that have already gone out of their live range. In this case, LIRAC can be disabled so that it will roll back to conventional write-back cache.

4 Software Support For LIRAC

As discussed earlier, live range and dead range of a memory address is determined only by program behavior and has nothing to do with memory hierarchy (i.e. whether the computer has a cache or what kind of cache structure it has). Therefore software needs to convert some normal Read instructions to LastRead instructions. There are three ways to support LIRAC architecture in software: compiler analysis, binary transformation and memory tracing.

4.1 Compiler Analysis

Apparently using compiler to add LastRead instructions is the most desirable way to support LIRAC, for it can be both accurate and transparent to programmer. In this paper, we touch the subject by only outlining the principles. In some modern compilers (e.g. gcc 3.4.1 and later versions), variables are expressed in SSA (Single Static Assignment) form [15], where every variable is assigned exactly once. If the live range of every variable in SSA form can be calculated, live range of each memory address can be determined. As a matter of fact, there is already a live range analysis pass in the register allocation, as shown in Fig. 6. Before register allocation, all the variables are expressed in pseudo-register form. Register allocation pass is used to allocate and assign registers from pseudoregister to architecture registers. Therefore it is quite feasible to support LIRAC architecture with minor modification.

There is another common case where live range can be determined. When a function call or an interrupt occurs, context needs to be saved and restored later. Registers are saved to memory on the entrance of function call or interrupt and read back from memory at exit. Normally, context restore is the last read, so LastRead instructions can be used in restoring context for function call or interrupt.

4.2 Binary Transformation

Although compiler can generate live range information accurately, it needs source code of the program. For programs without source code, binary code can be analyzed to determine the live range. Binary transformation is more difficult

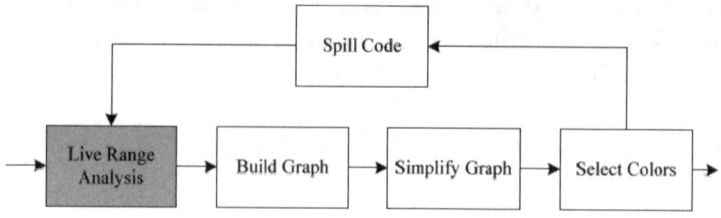

Fig. 6. Compiler Register Allocation Process

and less accurate than compiler analysis because less information can be obtained. Context restore at the end of function call can be easily converted by binary transformation. However, using binary transformation to determine the live range of local variable is much more difficult.

4.3   Memory Tracing

The third approach to generate live range information is memory tracing. This is achieved by collecting all memory traces and then analyzing the traces to find the live range of each memory address. Tracing is much simpler to implement than the other two methods aforementioned, but it may not be as practical. Trace studies depend on many factors such as program input and thus is unreliable. Moreover, trace analysis is an oracle heuristics approach (i.e. perfect branch prediction, perfect memory disambiguation, etc.) and thus the result can only be used as an upper bound. However, the result of trace analysis can be helpful to compiler analysis and binary transformation. Since trace analysis is simple, we use this method in this paper to evaluate the performance of LIRAC architecture.

4.4   Two Types of LastReads

A read instruction can be executed many times and leave multiple instances in memory trace. For some read instructions, all instances mark the end of live range of the corresponding address. We define thsese reads as Type 1 LastRead, as shown in Fig. 7(a). For some other read instructions, only some instances are LastReads while other instances are normal reads. For example, in Fig. 7(b), the shaded instruction is executed 10 times; only the last instance is LastRead. We define this type of partial LastReads as Type 2 LastRead.

Type 1 LastRead can be exploited by both compiler and binary transformation with simple substitution. Type 2 LastReads cannot be simply converted to LastRead instructions, but sophisticated compiler can change program to expose LastReads, as shown in Fig. 7(c).

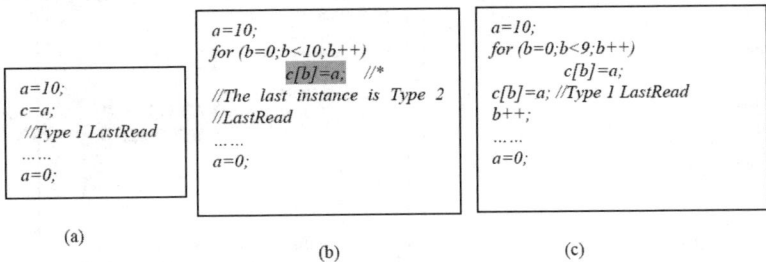

Fig. 7. Two Types of LastReads

5 Methodology

Memory trace simulation contains three steps: trace generation, trace analysis and trace execution. 3 multimedia applications and 10 SPEC 2000 applications are selected to evaluate the performance of LIRAC architecture. All benchmarks are compiled with -O2 option.

Simplescalar simulator [16] is used to generate traces. Trace generation is to get memory access sequence in processor domain, so sim-fast mode is selected because no detailed hardware implementation is necessary. Each trace item contains three fields: memory access type, memory access address and instruction address. The size of the generated trace files are huge. For example, running gcc in test mode will generate a trace file larger than 5GB. To save disk space and simulation time, scaled inputs instead of standard inputs are used in some benchmarks including gcc, gzip, and bzip2. Sampling technique is not used because accurate live range information is not available in sampled traces.

Trace analysis is used to find live range of every memory address. The algorithm is sketched in Fig. 8(a). Live structure is a hash map with address as key and trace number as value. It is used to track the last read. Trace file is fed to trace analyzer sequentially. For a read trace item, Live structure is updated; for a write item, trace analyzer will look up Live structure and mark previously recoded read as LastRead.

Type 1 and Type 2 LastRead can be distinguished using algorithm shown in Fig. 8(b). The first pass picks all read instructions with normal read instances and records them in set NotType1LastRead. The second pass distinguishes the two types of LastRead instances by looking up the set NotType1LastRead.

In trace execution, both the original trace and optimized trace are run on a trace simulator to evaluate the effect of LIRAC architecture. Dinero IV simulator [17] is a fast, highly configurable trace-driven cache simulator developed by Wisconsin University. It supports multi-level cache, different replacement policy and sub-block organization. Less than 10 lines are modified to support LIRAC architecture in Dinero IV simulator.

```
map< int, int> Live;

while(!trace_end)
{
    traceno++;
    fetch_trace();
    if (tracetype == READ)
        Live[addr]=traceno;
    else //Write
        if (find(Live, addr))
        {
            LastRead.insert(index->traceno);
            Live.erase(index);
        }
}
For_each_traceitem()
    LastRead.insert(index->traceno);
```

(a)

```
set< int> NotType1LastRead;

while(!trace_end())
{
    fetch_trace();
    if (type ==NORMAL_READ)
        NotType1LastRead.insert(pc);
}
trace.restart();
while(trace_end())
{
    fetch_trace();
    if (type== LASTREAD)
            if (find(NotType1LastRead, pc))
            type = TYPE2_LASTREAD;
    else
            type = TYPE1_LASTREAD;
    Write_trace();
}
```

(b)

Fig. 8. Algorithm for Trace Analysis

6   Results

Fig. 9 shows the result of memory trace analysis. From the figure, we can see that about 4-28% of total memory accesses are LastReads.

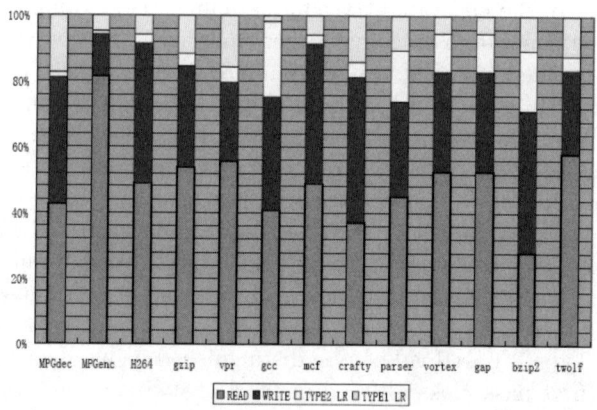

Fig. 9. Memory Access Trace Breakdowns

Fig. 10 shows the reduction of LIRAC with direct mapped cache and 4-way set-associative cache. The baseline cache structure is single-level, write-back, write-alloc with LRU replacement policy. For each benchmark, there are 6 stacked bars, respectively for cache sizes 1KB, 4KB, 16KB, 64KB, 256KB and 1MB. The total height of each stacked bar shows the maximum reduction from binary transformation in percentage(i.e. optimizing both type-1 and type-2 Las-tReads), whereas the lower portion of the stacked bar represents the limit of compiler analysis (i.e. optimizing type-1 LastReads only).

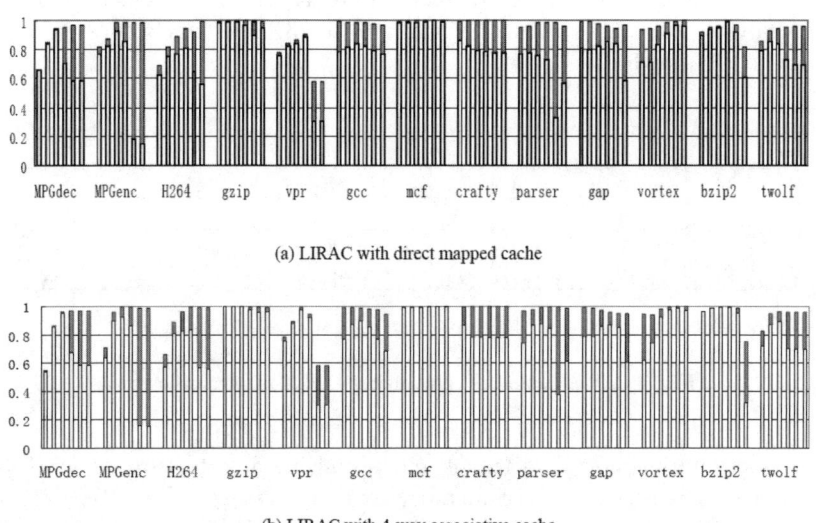

(a) LIRAC with direct mapped cache

(b) LIRAC with 4-way associative cache

Fig. 10. Memory Access Trace Breakdowns

From the experiment results, following conclusions can be made: 1. LIRAC architecture can reduce the number of memory write operations (Wm) in all cases. LIRAC architecture can be used in various cache structures.

2. The effect of LIRAC depends heavily on the method of software support. Compiler analysis can achieve much better result than binary transformation. Trace-driven simulations show that LIRAC can reduce 21% memory writes on average and up to 85% in the best case. Through binary transformation, LIRAC can only reduce 6% memory writes on average and up to 45% in the best case.

3. The effect of LIRAC varies from program to program. For some programs like mcf, very few memory writes can be reduced; for other programs like crafty, binary transformation is of little use while compiler analysis can be very helpful. Both methods are helpful for multimedia benchmarks.

7   Live Range Aware Buffer

In addition to write-back cache, live range aware architecture can be applied to all kinds of buffers such as write buffer. In Live Range Aware write Buffer(LIRABuf), if buffered data finishes its live range, commitment can be omitted. Software support for LIRABuf is exactly the same as LIRAC since live range information is determined by program only.

Figure 11 shows the benefit of live range aware write buffer using trace-driven simulation. For each benchmark, from left to right, buffer size is 1KB, 4KB, 16KB, 64KB and 256KB respectively. Interpretation of the stacked bars is the same as in Figure 10. With binary transformation (entire bar), LIRABuf can reduce 7% memory writes on average and up to 44% in the best case. With compiler analysis (lower bar), LIRABuf can reduce 23% memory writes on average and up to 85% in the best case.

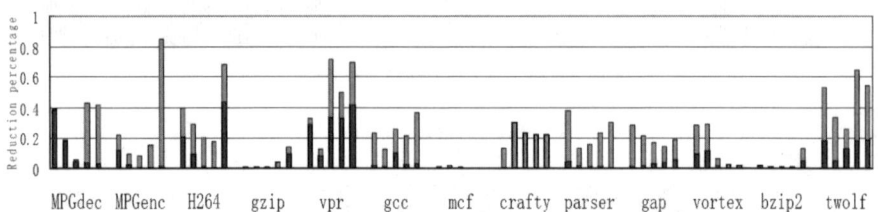

Fig. 11. Reduction of Wm using LIRABuf architecture

Transactional memory is a recent and promising architecture for CMP[3][4]. A transaction is a sequence of memory loads and stores that either commits or aborts with atomicity, isolation and consistency. When a transaction ends, a bunch of write operations are committed to memory and system interconnect to maintain coherence. Contention tends to be unavoidable when multiple transactions end at the time.

Table 1. Taxonomy of Data in CMP with Transactional Memory

		To Later Transactions in Local CPU	
		Useful	Not Useful
To Other CPUs	Useful	Global Data	Communication Data
	Not Useful	Local Data	Invalid Data

In this paper, we propose a taxonomy for data in CMP with Transactional Memory based on usefulness to other CPUs and later transactions in local CPU (shown in Table 1). If different types of data can be discriminated and treated differently, both bandwidth and power consumption can be saved. For example, local data can be written to memory without broadcasting to other CPUs and invalid data can be just discarded. How to distinguish local data from global data is beyond the scope of this paper, but with live range information, invalid data can be distinguished from local data. When a transaction commits, if written data is local and out of live range, it can be thrown away to save energy and alleviate bus contention.

8   Conclusions and Future Work

In this paper, we made the following contributions:

1. Explicitly distinguished two kinds of writes: write in processor domain (Wp) and writes in memory domain (Wm) and defined live range and dead range of a memory location.

2. Proposed the architecture of LIve Range Aware Cache (LIRAC) and Live Range Aware BUFfer (LIRABuf).

3. Proposed a taxonomy for data in CMP with Transactional Memory that could possibly save energy and alleviate bus contention.

We also discussed the principle of software support for live range identification. Simulation results show that the potential benefit of LIRAC and LIRABuf can be great. While the initial results are promising, a lot more work needs to be done. In the future, we plan to study and work on following issues:

1. Compiler implementation to support LIRAC and LIRABuf.

2. Live range aware transactional memory architecture.

Acknowledgments. We would like to thank the anonymous reviewers for their valuable feedback. This material is based on work supported by National Science Foundation of China(No.60673145), National Basic Research Program of China (2006CB303100), Basic Research Foundation of Tsinghua National Laboratory for Information Science and Technology (TNList) and Intel / University Sponsored Research.

References

1. Win. A. Wulf and Sally A. McKee. Hitting the memory wall: implications of the obvious, ACM SIGARCH Computer Architecture News, Volume 23 , Issue 1 March 1995.
2. Semiconductor Industry Association, International Technology Roadmap for Semiconductors, http://www.itrs.net/Common/2004Update/2004Update.htm. 2004.
3. Lance Hammond, Vicky Wong, Mike Chen, Ben Hertzberg, Brian D. Carlstrom, John D. Davis, Manohar K. Prabhu, Honggo Wijaya, Christos Kozyrakis, and Kunle Olukotun, Transactional Memory coherence and consistency, Proceedings of 31st Annual International Symposium on Computer Architecture (ISCA-31) 2004, 102-113.
4. Kevin E. Moore, Jayaram Bobba, Michelle J. Moravan, Mark D. Hill and David A. Wood, LogTM: Logbased Transactional Memory, Proceedings of 12th Annual International Symposium on High Performance Computer Architecture (HPCA-12), 2006.
5. Maya Gokhale, William Holmes, Ken Iobst, Processing in memory: the Terasys massively parallel PIM array. IEEE Computer,28(4):23 31, April 1995.
6. Christoforos Kozyrakis, Joseph Gebis, David Martin, Samuel Williams, Iakovos Mavroidis, Steven Pope, Darren Jones, David Patterson and Katherine Yelick, Vector IRAM: A Media-oriented Vector Processor with Embedded DRAM. 12th Hot Chips Conference, August, 2000.

7. Jenlong Wang and Russell W. Quong, The feasibility of using compression to increase memory system performance, Proc. of IEEE International Symposium on Modeling, Analysis, and Simulation of Computer and Telecommunication Systems, pages 107-113, 1994.
8. Manoj Franklin and Gurindar S. Sohi, "Register Traffic Analysis for Streamlining Inter-Operation Communication in Fine-Grain Parallel Processors", Proceedings of 25th International Symposium on Microarchitecture, 1992, 236–245.
9. Luis A. Lozano C. and Guang R. Gao, "Exploiting Short-Lived Variables in Superscalar Processors", Proceedings of 28th International Symposium on Microarchitecture, 1995, 292–302.
10. Guillermo Savransky, Ronny Ronen and Antonio Gonzalez, "A Power Aware Register Management Mechanism". International Journal of Parallel Programming, Volume 31, Issue 6, December 2003, 451–467.
11. Dmitry Ponomarev, Gurhan Kucuk, Ponomarev, Oguz Ergin and Kanad Ghose, "Isolating Short-Lived Operands for Energy Reduction", IEEE Transaction on Computers, Vol. 53, No. 6, June 2004, 697–709.
12. Milo M. Martin, Amir Roth, and Charles N. Fischer, "Exploiting Dead Value Information". Proceedings of 30th International Symposium on Microarchitecture, 1997, 125–135.
13. Kevin M. Lepak, Gordon B. Bell, and Mikko H. Lipasti, Silent Stores and Store Value Locality, IEEE Transactions on Computers, Vol. 50, No. 11, November 2001.
14. Frederick Chow and John L. Hennessy, Register Allocation for Priority based Coloring, Proceedings of the ACM SIGPLAN 84 Symposium on Compiler Constructions, 1984, 222-232.
15. Ron Cytron, Jeanne Ferrante, Barry K. Rosen, Mark N. Wegman and F. Kenneth Zadeck, Efficiently Computing Static Single Assignment Form and the Control Dependence Graph. ACM Transactions on Programming Languages and Systems, 13(4): 451-490, October 1991.
16. Todd Austin, Eric Larson and Dan Ernst, "SimpleScalar: An Infrastructure for Computer System Modeling", IEEE Computer 35(2), 2002, 59–67.
17. Jan Edler and Mark D. Hill, "Dinero IV Trace-Driven Uniprocessor Cache Simulator", http://www.cs.wisc.edu/ markhill/DineroIV, 2003.

Optimized Register Renaming Scheme for Stack-Based x86 Operations

Xuehai Qian, He Huang, Zhenzhong Duan, Junchao Zhang, Nan Yuan,
Yongbin Zhou, Hao Zhang, Huimin Cui, and Dongrui Fan

Key Laboratory of Computer System and Architecture
Institute of Computing Technology
Chinese Academy of Sciences
{qianxh,huangh, duanzhenzhong,jczhang,
yuannan,ybzhou,zhanghao,cuihm,fandr}@ict.ac.cn

Abstract. The stack-based floating point unit (FPU) in the x86 architecture limits its floating point (FP) performance. The flat register file can improve FP performance but affect x86 compatibility. This paper presents an optimized two-phase floating point register renaming scheme used in implementing an x86-compliant processor. The two-phase renaming scheme eliminates the implicit dependencies between the consecutive FP instructions and redundant operations. As two applications of the method, the techniques used in the second phase of the scheme can eliminate redundant loads and reduce the mis-speculation ratio of the load-store queue. Moreover, the performance of a binary translation system that translates instructions in x86 to MIPS-like ISA can also be boosted by adding the related architectural supports in this optimized scheme to the architecture.

1 Introduction

X86 is the most popular ISA and has become the de facto standard in microprocessor industry. However, the stack-based floating point ISA has long been considered as a weakness of the x86 in compare with the competing RISC ISAs. The addressing of FP stack register is related to a Top-Of-Stack (TOS) pointer. Every x86 FP instruction has implicit effects on the status of the floating point stack, including the TOS pointer and incurs implicit dependencies between consecutive floating point instructions. Furthermore, the stack-based architecture requires one of the operands of an FP instruction comes from the top of the stack, so some transfer or swap operations are needed before real computations.

The AMD x86 64-bit processor attacks the above problems by using a flat register file. It uses SSE2 to replace the stack-based ISA with a choice of either IEEE 32-bit or 64-bit floating point computing precision. However, even the 64-bit mode can not obtain the totally same results as the original double extended FP computing precision. Another fact is that a lot of legacy libraries are written in highly optimized floating point assembly; rewriting of these libraries takes a long time. Therefore, the replacement of the stack-based ISA is not easy. We should seek for novel techniques to bridge the FP performance gap between x86 and RISC architecture.

P. Lukowicz, L. Thiele, and G. Tröster (Eds.): ARCS 2007, LNCS 4415, pp. 43 – 56, 2007.

This paper first presents a comprehensive and solid methodology of implementing x86-compliant processor based on a generic RISC superscalar core. The techniques to handle x86-specific features such as, complex instruction decoding, Self Modified Code (SMC), non-aligned memory access are outlined. After giving a motivating example, we emphasize on the key elements of our methodology, 2-phase register renaming scheme for attacking the stack-based FP operations. The first phase of the renaming scheme eliminates the implicit dependencies imposed to the consecutive FP instructions by maintaining speculative stack-related information in the instruction decode module. The second phase eliminates almost all of the redundant operations by value "short-circuiting" in rename table, which actually achieves the effect of a flat register file. Critical issues in processor design such as branch misprediction and exception handling are carefully considered.

The scheme has two applications. First it can be used to reduce the mis-speculation ratio of loads in implementing POP instructions in x86 ISA. The results of stores may be directly "short-circuited" to the loads in the renaming module in some cases, so that the redundant loads are eliminated and will not incur mis-speculations in the load-store queue. Secondly, the generic RISC core that originally supports x86 ISA by binary translation can be augmented with the 2-phase renaming scheme, so that the performance of the translated code is boosted. The codes generated by the original binary translation system have a poor performance for FP programs due to the significant semantic gap between the RISC and x86 FP ISA.

The rest of the paper is organized as follows. The design methodology of the prototype processor architecture is presented in section 2. Section 3 provides an example which motivates our optimized scheme. Details about the proposed scheme are presented in section 4. Section 5 briefly summarizes the two applications of the scheme. Section 6 describes the simulation environment and methodology. The simulation results are presented in section 7. The related work is outlined in section 8. We conclude the paper in section 9.

2   Overview of the GodsonX Architecture

Fig. 1 shows the architecture of the proposed prototype x86 processor GodsonX which is built based on the core of Godson-2C[1] core, a typical RISC 4-issue out-of-order superscalar. The architectural supports for x86 architecture are presented as shaded blocks.

The front end of x86 processor is much more complex than the RISC processor. It is organized in 5 pipeline stages. The first stage generates instruction addresses. The second stage is composed of five modules providing the content of instructions, pre-decode information and performing branch prediction. The third stage, decode-0, consists of four sub-modules and is responsible for instruction alignment and queue management. The decode-1 module reads the aligned x86 instructions from the first instruction register (IR1 in the figure), translates them into the "RISC style" intermediate instructions and some micro-ops, and puts them in the second instruction

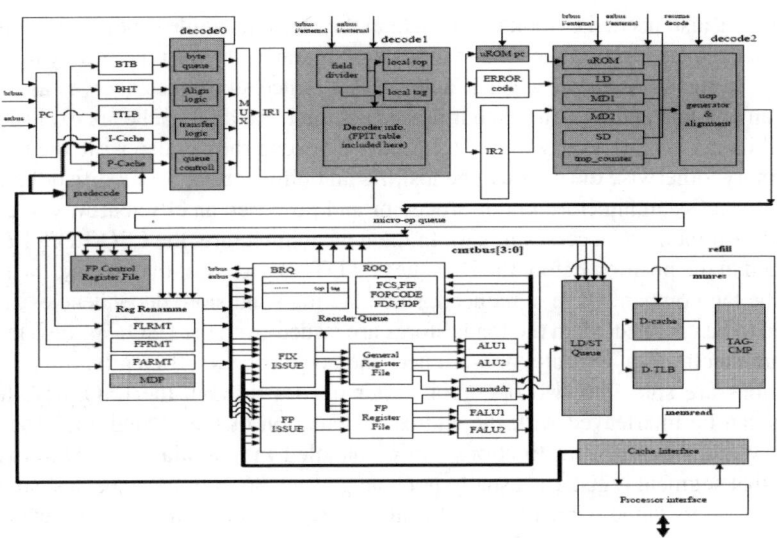

Fig. 1. GodsonX architecture

register (IR2 in the figure). The decode-2 module generates sequences of micro-ops and put the sequences in the micro-op queue (uop in the figure). The methods that decode-2 uses to generate micro-op sequences include: short micro-op sequence Decoder (SD), Long micro-op sequence Decoder (LD), Memory access Mode decoder (MD) and micro-op Rom (urom). The uop queue acts as a buffer between decode-2 and the register rename module inside the RISC core, enabling stable instruction issue from the front end.

The architectural supports to the FP stack are distributed. These modules work cooperatively to guarantee the correct stack status while ensuring efficient execution of floating point instruction. The decode-1 module maintains a local copy of TOS and Tag, and has a table storing the x86 FP instruction information. Each micro-op should carry the TOS after dispatched from the register renaming stage. The reorder queue (ROQ) handles FP exceptions specified in x86 ISA. The branch queue (BRQ) keeps the local TOS and Tag of the branch instruction for misprediction handling. The floating point data type (Tag) of the result value is computed in FALU1/2, and passed to ROQ in write back stage. When a floating point instruction commits, ROQ writes the TOS and Tag into the architectural FP status and tag word. Under exceptions, the decode-1 module recovers the TOS and Tag from these registers.

We tried to handle some burdensome x86 ISA features, such as Self Modified Code (SMC) and non-aligned memory accesses, etc. by architectural modifications or supports. The method to handle SMC is in the granularity of cache line, so it has the advantage over the method adopted by Intel, which detects and handles the event in the granularity of page. Specifically, we added a Simplified Victim Cache (SVIC) to the decode module. As a store writes back, we use the address to look up the ICache and SVIC, if an entry is found, then an SMC occurs, then the pipeline need to be flushed. Similarly, the miss queue is also looked up as a store writes back. When the ICache misses, it is also needed to look up the data in the DCache, while sending the request to miss queue, trying to find the data in memory. Similarly, it is also necessary to look up the data in the store queue when ICache misses, since store queue serves as a buffer between pipeline and DCache. In our processor, the L2 and L1 cache are exclusive. If a missed L1 access hits a dirty line in L2 cache, L2 cache should also write the line back to memory, otherwise the data will be lost. As an optimization, we modified the former memory access architecture in our processor and proposed an efficient way to execute non-aligned memory accesses in x86. In the new architecture, the LSQ(LD/ST Queue) is placed in the position before the DCache and DTLB accesses. In this way, the latency in cache tag compare stage is reduced, since the load and store dependencies are only needed to be checked when the loads/stores are issued from the LSQ. We also leverage this architecture to execute non-aligned memory access efficiently, this kind of operations are split into two operations after it goes through the LSQ, and the first access can be interleaved with the address calculation of the second one. The 80-bit memory accesses in FP instructions can be handled in a similar way. Moreover, we found that segment register is usually not changed as x86 program executes, so we can speculate on its value, which makes the address calculation simpler. The performance of memory access module can also be optimized by making the common operations run faster, which reduces the hardware cost. We found that in 2 issue memory access pipeline, one-port TLB is enough in the common case. The MDP in register renaming module is a simple Memory Distance Predictor, which can reduce the possibility that the pipeline flush due to load/store conflicts.

3   Motivating Example

A floating point computation example is presented in this section, showing the advantages of the optimized scheme. Consider the computation: atan((a+b)/(a*c)), a possible x86 instruction sequence and two possible corresponding micro-op sequences are given below.

X86 instructions	naïve micro-op sequence	optimized sequence
FADD ST(1);	fadd fr(5), fr(5), fr(6)	fadd fr(5), fr(5), fr(6)
FXCH ST(4);	fmov fr(9), fr(5)	fxch fr(5), fr(1)
FMULP ST(1),ST(0);	fmov fr(5), fr(1)	fmul fr(6), fr(6), fr(5)
FLD ST(3);	fmov fr(1), fr(9)	fmov fr(5), fr(1)
FPATAN;	fmul fr(6), fr(6),fr(5)	fpatan fr(5), fr(5), fr(6)
	fmov fr(5), fr(1)	
	fpatan fr(5), fr(5), fr(6)	

Fig. 2 shows the change of FP stack status when the instructions are executed. We assume the left-most the initial state, where TOS is 5. The FXCH instruction swaps the content of ST(4) and ST(0). The suffix "P" in the multiplication instruction implies that a "pop" operation is required. The "pop" increases the TOS and empties the former top element. The last instruction computes the arc-tangent function of (ST(0)/ST(1)), the source operands position are fixed, so there is no need to specify them in instruction. It reflects a feature of stack-based architecture. Also, the FPATAN instruction includes a "pop" operation.

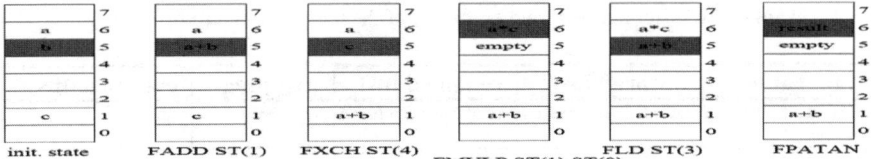

Fig. 2. Stack operations of the x86 instructions

We use four FP register number address spaces. FP registers used in FP instructions are called **relative FP stack registers** (ST(i), i ∈ [0,7]), the relative FP registers are mapped to **absolute FP stack registers** (st(i), i ∈ [0,7]) by adding ST(i) to TOS (module 8). The FP registers used in micro-op are **FP logical registers** (fr(i), i ∈ [0,15]). The FP registers after renaming in the RISC core are **FP physical registers** (pr(i), i ∈ [0,63]). st(0)-st(7) are directly mapped to fr(0)-fr(7). Temporary registers are used to hold intermediate results. The micro-op sequences can use up to 8 temporary FP registers which are fr(8)-fr(15). In the naïve sequence, we use three transfer operations ("fmov") to implement FXCH. In the optimized scheme, a dedicated swap operation ("fxch") is used.

Table 1 and table 2 show the execution and register mapping processes of the two micro-op sequences. We find that the number of micro-ops that would go into the issue queue is 7 in the naïve sequence and 3 in the optimized sequence; while the numbers of physical registers consumed are 10 and 6, respectively. Both the micro-ops and the mapped physical registers are reduced. Due to the nature of stack-based architecture, there are inherently a large amount of FXCH instructions in x86 FP program. Reducing the number of this kind of operations can directly reduce the execution time of the program. Moreover, it will alleviate the burden to the physical register file and issue queue, which makes other operations execute faster as well.

In the next section, we will present our novel 2-phase renaming scheme. The first phase of renaming releases the serial requirement in decoding x86 FP instructions via speculative local copy of certain floating point information in decode-1 module. The second phase adopts an optimized RAM-based approach, which can support the optimization.

Table 1. Register mapping and execution of the naive micro-op sequence

Map-Table	orig. inst	renamed inst	mapping	executed ops
fr1->pr1,fr5->pr5, fr6->pr6	fadd fr5,fr5,fr6	fmul pr9,pr5,pr6	fr5->pr9	(pr5+pr6)->pr9
fr1->pr1,fr5->pr9, fr6->pr6	fmov fr9,fr5, fmov fr5,fr1 fmov fr1,fr9	fmov pr10,pr9 fmov pr11,pr1 fmov pr12, pr10	fr9->pr10 fr5->pr11 fr1->pr12	pr9->pr10 pr1->pr11 pr10->pr12
fr1->pr12, fr5->pr11, fr9->pr10, fr6->pr6	fmul fr6,fr6,fr5	fmul pr13, pr6, pr11	fr6->pr13	(pr11*pr6)->pr13
fr1->pr12, fr5->pr11, fr9->pr10, fr6->pr13	fmov fr5,fr1	fmov pr14, pr12	fr5->pr14	pr12->pr14
fr1->pr12, fr5->pr14, fr9->pr10, fr6->pr13	fpatan fr5, fr5, fr6	fpatan pr15, pr14, pr13	fr5->pr15	atan(pr13/pr14) ->pr15

Table 2. Register mapping and execution of the optimized micro-op sequence

Map-Table	orig. inst	renamed inst	mapping	executed ops
fr1->pr1,fr5->pr5, fr6->pr6	fadd fr5,fr5,fr6	fmul pr9,pr5,pr6	fr5->pr9	(pr5+pr6)->pr9
fr1->pr1,fr5->pr9, fr6->pr6	fxch fr5, fr1	**eliminated**	**fr5->pr1** **fr1->pr9**	Swap the FP physical register that fr5 and fr1 mapped
fr1->pr9, fr5->pr1, fr6->pr6	fmul fr6,fr6,fr5	fmul pr10, pr6, pr1	fr6->pr10	(pr1*pr6)->pr10
fr1->pr9, fr5->pr1, fr6->pr10	fmov fr5,fr1	**eliminated**	**fr5->pr9**	Make fr5 map to the physical register that fr1 mapped to, it is pr9
fr1->pr9, fr5->pr9, fr6->pr10	fpatan fr5, fr5, fr6	fpatan pr11, pr9, pr10	fr5->pr11	atan(pr10/pr9) ->pr11

4 Optimized 2-Phase Register Renaming Scheme

4.1 Mapping from Stack Registers to Logical Registers

In the first phase mapping, the FP stack registers are mapped to FP logical registers. We adopt a speculative decoding technique in this process. The decode-1 module maintains a local copy of partial TAG and TOS. The TAG is partial since it just indicates if a FP register is empty. The decode stage determines the absolute FP register based on the local TOS and update these information after each FP instruction is decoded, according to the specifications of each FP instruction. In this way, the decoding of FP instructions can be pipelined, since the decode module does not have to wait for the committed TOP and TAG information. Here we should note that the effects of each FP instruction on the stack are totally predictable.

If no exceptions, the local TOS and Tag is synchronized with the architectural TOS and Tag in the FP status and tag word when the instruction commits. In case of exceptions, the changes to the local TOS and Tag at the decode stage have to be recoverd to the architectural state. We explain the branch misprediction case by an example.. Fig. 3(a) presents a branch misprediction scenario. The nodes 1 and 2 represent the committed instructions, the nodes 3 and 4 represent the instructions executed in the correct path but not committed, node 4 is the branch instruction, and nodes 5-7 are the instructions in the wrong path, executed and need to be cancelled. From this scenario, if the TOS and Tag are recovered from the FP status word and the Tag word, the decode stage will hold TOS and Tag information of the last instruction committed before the branch, that is node 2. But this is incorrect, since what we need is the TOS and Tag after the execution of the branch instruction, the node 4. Therefore we need to keep the TOS and Tag of each branch instruction in the BRQ shown in Fig. 1. When a branch misprediction occurs, the decode stage recover the TOS and Tag from BRQ.

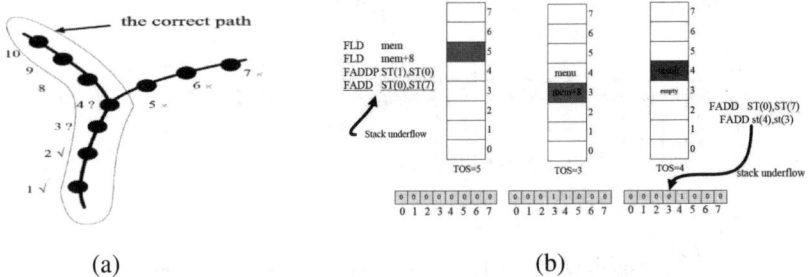

(a)   (b)

Fig. 3. (a) Branch misprediction scenario. (b) An example: detection of FP "stack underflow".

The partial TAG information can be used in the detection of stack overflow/underflow. An FP stack underflow occurs when an instruction references an empty FP stack register, a stack overflow occurs when an instruction loads data into a non-empty FP stack register. Fig. 3(b) shows an example of detecting FP stack underflow conditions. The partial TAG is the 8-bit vector at the bottom; "0" indicates the related position in FP stack is empty, "1" means the position is non-empty.

The Floating Point Instruction Table (FPIT) is used as an effective and low cost way to maintain the information of each FP instruction in decode module. After analyzing the bit codes of each instruction, we found certain group of instructions has similar bit codes and similar effects to the stack. We can represent information for these FP instructions by just one entry in FPIT. This technique makes the table smaller. In each entry, we store the effects to the stack. More details about this table are out of the scope of the paper.

4.2   Optimized Register Mapping in the RISC Core

There are two ways to implement the register renaming, the RAM-based design and CAM-based design, they use separate or merged architectural and rename register files. The former design of Godson-2C processor adopts the CAM approach, which can not support the optimization in the paper. We propose a new RAM-based register renaming design. This design allows one physical register to be mapped to more than one logical registers. Fig. 4 gives the outline of this architecture.

We use three mapping tables to maintain the relationship between FP logical registers and physical registers. The Floating point Logical Register Mapping Table (**FLRMT**) is used to rename logical registers to physical registers. It has 16 entries representing 16 floating point logical registers. The field "pname" indicates which physical register the logical register is mapped to. Each FLRMT entry contains 8 lastvalid items, corresponding to 8 BRQ entries. The lastvalid(i) keeps the mapped physical register number when the branch instruction in BRQ(i) is mapped. The Floating point Physical Register Mapping Table (**FPRMT**) merely maintains the state of each physical register. It has 64 entries corresponding to 64 physical registers. There three fields in each entry. The state records the state of the physical register, brqid is used in branch misprediction to recover the correct register mapping, and counter indicates how many logical registers are mapped to this physical register. When the register renaming stage finds an entering instruction an "fmov", it directly maps the

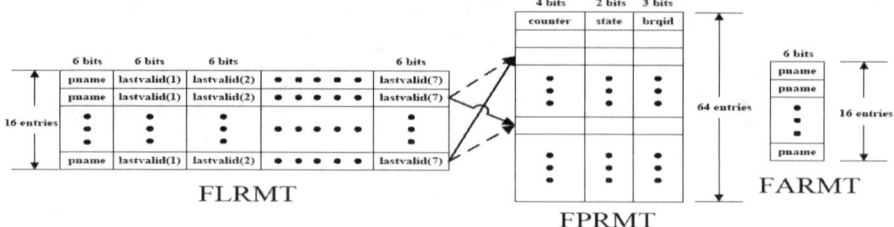

Fig. 4. Optimized register renaming inside RISC core

destination register to the physical register of the source register, and increases the counter for the physical register. When "fxch" is encountered, the source and destination registers are simply swapped. When an instruction committed, the counter for the destination physical register is decreased. We need four bits for the counter field, since at most 16 logical registers can be mapped into a physical register. Note that the change of mapping between logical and physical registers at the register renaming stage by fmov or FXCH is speculative, the instructions can be canceled later because of exception. The Floating point Architectural Mapping Table (**FARMT**) records the committed physical register which the logical register is mapped to, it is used in exception recovery. This table has 16 entries corresponding to 16 logical registers. Each entry just records the physical register that the logical register mapped to. The table is updated when an instruction is committed. When an exception is encountered, the renaming stage can recover the mapping relationship from FARMT. This optimized architecture simplifies lookup logic compared with the CAM-based implementation and is more scalable.

TAG update should also be considered in the optimization. When an instruction committed, the tag for the destination register should be updated, reflecting the latest status. In our design, the tag is computed in FALU, but the eliminated "fxch" and "fmov" will not enter FALU. This is not a problem for "fxch", since its two operands are both architectural visible registers, so the only thing to do as an instruction commits is to swap the tags for the two operands in the FP tag word. For "fmov", it is more difficult because the source operand may be a temporary register. To make the design simple, the optimization is not applied to the special case. As the statistics data shows, the special case is rare.

5   Applications of the Renaming Scheme

The proposed scheme has two applications. First, it can be used as the supports for the Godson-2C processor that implements MIPS-like ISA and runs application-level binary translator to support x86 applications. The binary translator that we conducted the experiments is Digital Bridge[6]. It works in a Godson based LINUX server, and translates the elf file of x86 ISA to Godson ISA (MIPS-like). Although the translator works well for fix point programs the performance of floating point applications suffers. It is mainly due to the remarkable ISA semantic difference between x86 and general propose RISC in floating point specification. The existing method on binary translation is not efficient enough to bridge such a gap, architectural supports are needed to narrow the gap. Without architectural supports, the Bridge translator use static FP registers in Godson processor to emulate FP stack operations. When loading data into the FP stack register, for example, ST(2), we must dynamically determine the corresponding absolute register and put the value in fr(2). The process will incur a lot of swap operations in the target Godson code. This approach still needs the help of memory. The valid values on the stack should be loaded from and stored into memory at the beginning and the end of each basic block. Finally, this approach assumes that the TOS is the same and TAGs are all valid at entry of each basic block. Only under this assumption, ST(i) can always correspond to fr(i), regardless of the preceding path from which the code arrives the entry. But a large amount of extra code must be added at the head of the translated code for each block to judge if the above speculation is held. From the experiment results, the condition is satisfied almost all the time. It is obviously a waste to execute a large segment of extra code for rare conditions. We add our architectural support for FP stack to Godson-2C processor without x86 features. With these supports, we can directly use the relative FP registers in the translated code, making the burden of maintaining status of FP stack to hardware.

As the second application of the method, it can also used to eliminate redundant loads. An impediment to Godson-X performance is the high miss rate of load speculation. After analyzing the program execution behavior, we found that the problem came from the x86 PUSH and POP instructions for parameter passing in function calls. These two instructions are mapped to store and load micro-ops. In a function call, the store and load come in pair and close to each other. Godson-X always speculates on the value of the load before the store commits. Therefore when the store commits, mis-speculation occurs. We added a 4-entry table to the register renaming module to forward the store value to the loads. The table maintains the source register numbers and memory addressing information of the 4 most recent store instructions. If a load instruction's memory address matches to one of the entries, it can be eliminated by modifying the register mapping relationship to directly get the stored value. Moreover, we are trying to extend this technique to eliminate redundancy in control flow with some hardware support. It is out of the scope of this paper.

6 Experimental Infrastructure

We have developed a cycle accurate full-system simulator for x86-compliants. Unlike the Simplescalar-based performance simulators, which decouple the execution and timing logic and can only provide an estimation of the performance, our simulator models the exact signals and timing except inside the ALU/FALU. This makes the result more accurate. Table 3 shows the detailed configuration of the simulator. For the latency of FP operations, we make following assumptions on latency: absolute, negation, comparison and branch take two cycles; addition, subtraction and conversion take three cycles; multiplication takes four cycles; division and square root take 4 to 16 cycles to complete; and transcendental functions take 60 cycles to complete. The real computation is carried out by a modified library of standard FP software implementation, the main modifications are FP exception handling.

Table 3. Configuration of GodsonX processor

decode width	at most 2 x86 instructions each cycle
functional units	2 fix point ALU, 2 floating point ALU, 1 memory
ROQ	32 entries
BRQ	8 entries
fix issue queue	16 entries
float issue queue	16 entries
branch predictor	Gshare: 9-bit ghr, 4096-entry pht, 128-entry BTB,direct mapped
L1-ICACHE	64KB 4-way set associative
L1-DCACHE	64KB 4-way set associative
memory access latency	50 cycles for the first sub block, 2 cycles for consecutive sub blocks

We use X86 emulator Bochs[9], which can boot LINUX and Window XP, as a reference in validating our design. Every time an instruction is committed, the whole architectural state is compared with Bochs. Due to this method, we debugged and validated our design. Finally, our simulator can boot the LINUX and Window XP.

SPEC CPU2000 is used as our benchmark. First, we find the representative region of each program by a SimPoint-like performance simulator for GodsonX processor, which is built from the counterpart for Godson-2C processor[5]. We fast-forward each program to its representative region and run 1 billion cycles using the cycle-accurate simulator to get precise results.

7   Simulation Results and Discussion

7.1   Performance and Characteristics of x86 Programs

Fig. 5(a) presents the performance comparison between our processor and the 2.4 GHz Intel Celeron processor. The IPC of the latter is obtained as follows. We first run each program in Bochs and record the instruction count, them we execute it in a real Celeron machine, record the execution time. We compute the IPC of each program and then compare it with the IPC of representative region in GodsonX. For some programs such as wupwise, swim, facerec and swim, performance of GodsonX is much better than that of Celeron, for programs like applu, equake, ammp and apsi, performances are similar. However, for some programs, especially sixtrack, GodsonX's performance is worse. Fig. 5(b) shows the latency distribution of micro-ops in GodsonX, the height of each bar represents the absolute number of cycles from map to commit. We can see that most of the cycles are due to register mapping or waiting for commit. This indicates that a large physical register file or reorder buffer is needed. Fig. 5(c) shows the cycle distribution with respect to the number of micro-ops committed in a cycle. For every program more than one micro-op is committed each cycle on average, especially for ammp, in most of the time four micro-ops are committed each cycle. It indicates that the efficiency of GodsonX is quite good. Fig. 5(d) presents the average number of micro-ops per x86 instruction, which indicates the quality of our micro-op mapping. On average about 2 are needed to implement an x86 instruction.

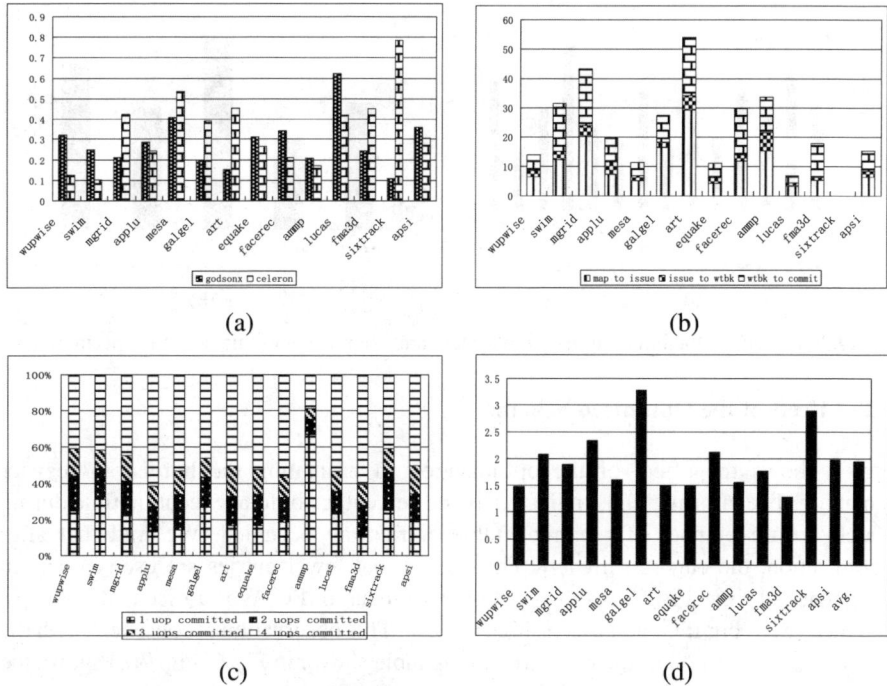

(a)   (b)

(c)   (d)

Fig. 5. Execution results of x86 program on our processor

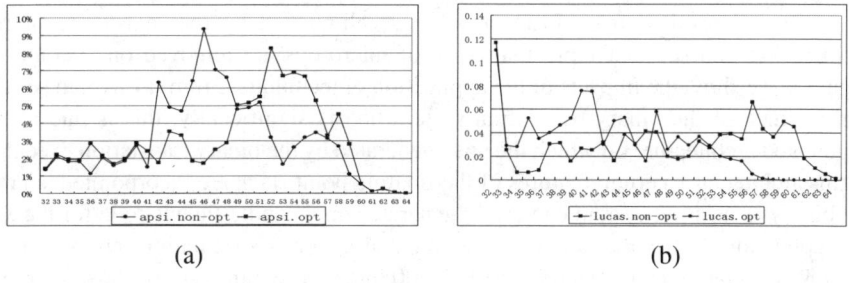

(a)   (b)

Fig. 6. Comparisons of register renaming

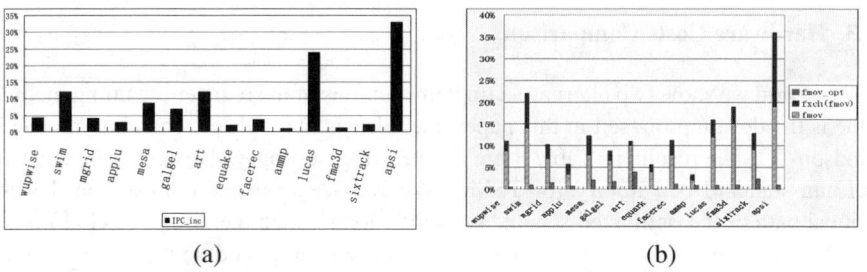

(a)   (b)

Fig. 7. IPC increse and the ratio of eliminated operations

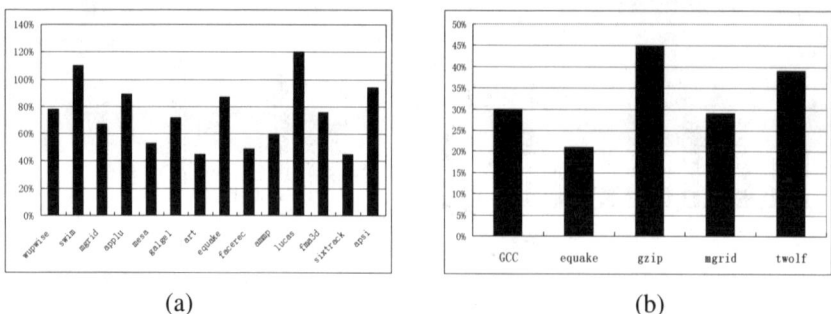

(a)   (b)

Fig. 8. Effects on the binary translation system and the ratio of eliminated mis-speculation

7.2 Effects of the Optimized Scheme

Fig. 6 shows the effect of the optimization in mitigating the burden to register renaming. The information is indicated by the percentage of total execution time during which certain number of renaming table entries are occupied. We find that after optimization, the curves shift left. This means that fewer entries are used in certain percentage of time. Fig. 7(a) shows the improvement to IPC. We can see that both apsi and lucas get a big IPC increase, as high as 30%. This observation is consistent with the optimization effect on usage of the renaming table shown in Fig.6. Fig. 7(b) shows the number of "fmov" before and after the optimization. The elimination of "fmov" comes from two sources. First, the "fmov" for FXCH are completely eliminated by simply swapping the source and destination registers mappings. Secondly, a large portion of other "fmov" resulted from FLD or FST are eliminated by the register renaming module. We can see that more than 10% of micro-ops are removed on average. In Figure 8, we show the impacts of the application of technique. From (a) we can see the performance of the binary translation system boosts significantly, this is due to the augmented architecture on which the code generated by the binary translation executes. In this architecture, certain features of the floating point stack are incorporated, so that the binary translation system can easily generate simple and efficient code for the x86 FP application. In (b), the ratio of the eliminated load mis-speculations are presented, the mis-speculation is frequently in SPEC2000 Integer programs, so we show the ratios for several ingeter programs.

7.3 Hardware Costs Comparison

This section we show two alternative implementations of the register renaming module. One is the design proposed in this paper, the other is the register renaming module in Godson-2C. The results in Table 4 are derived from Synopsys Design Compiler with 0.13um standard cell library for TSMC. We can see from the comparison that the critical path of the optimized scheme is slightly longer than the design of Godson-2C, but the area comsumed by the GodsonX register renaming module is greatly reduced. The main reason to the decrease is in the Godson-2C design, a large combinational

Table 4. Hardware cost comparison

	Lat. (ns)	Area(um^2)
GodsonX	1.25	616841.937500
Godson-2C	1.23	981161.687500

logic is used to generate a table that maps the logical registers to physical registers, this part of logic consumes a lot of area. Although the proposed scheme has a longer critical path than that of Godson-2C, the RAM-based design apporach has better scalability. When the number of physical registers increases, the proposed scheme will show more advantages over the former design.

8 Related Work

The implementation of FP stack is a critical issue in x86-compliant processor design. Some mechanisms have been patented by Intel[4] and AMD[3], but they are different from the scheme proposed in this paper. The main distinctions are that they normally adopt multiple tables to hold the stack related information, and the structures to hold the information are distributed in the processor. The synchronizations under exceptions and branch mispredictions are much more complicated. The modification to the RISC core in our scheme is trivial and the handling of exceptions or mispredictions is easy to understand and implement. More important is that we present an applicable methodology of implementing the FP stack based on a generic RISC core efficiently.

The elimination of FXCH has been used in some x86 processor, but it can only be done when FXCH comes with certain types of instructions, in those conditions, it can be combined with the surrounding instructions and eliminated. Both Intel and AMD employ a dedicated unit to execute FXCH instruction. Our scheme is more general and has lower cost. We only incorporate some simple functions in register renaming stage to detect the optimization opportunities. As in our scheme, not all FXCH in Intel or AMD processors can be eliminated. For example, under stack error (stack overflow or underflow), AMD processor will generate 5 micro-ops for the FXCH instruction. Moreover, from P4 processors this operation will have 3-cycle execution time again. Due to the elegant style of eliminating such operations in our scheme, the optimization will exist continually in our processor.

Our attempt is the first effort to implement a full x86-compliant processor based on a typical RISC core. The methodology presented in this paper can be applied to build processor in different ISAs. We also provide some x86 program characteristics and behaviors on our processor. IA-32 execution layer[7] and transmeta morphing software[8] are two efforts to translate x86 programs to other ISAs. Software based approaches are adopted in these systems, and the underlying architectures are VLIW, while our methodology is based on hardware architectural support to an existing and more general superscalar architecture. It is also the first work to investigate the impact of architectural support to binary translation.

9 Conclusion

This paper presents an optimized floating point register renaming scheme for stack based operations used in building an x86-compliant prototype processor. We compared the hardware cost of two register renaming designs; the proposed scheme has a slightly longer critical path but greatly reduced area. We find a large amount of swap and data transfer instructions in FP programs, and most of them can be eliminated by our proposed scheme. The IPC improvements due to the optimization are as high as 30% for some programs, and near 10% on average. Similar techniques in the scheme can also be extended to eliminate redundant loads and used as the architectural supports for RISC superscalar core to boost the performance of the binary translation system which run on that architecture. Our future work includes finding the optimal design trade-off in the co-designed x86 virtual. We will implement the x86 features that are critical to the performance and easy to be supported in hardware, for example the supports for the floating point stack, while implementing the complicated and unusual features in software.

Acknowledgement

This work is under the support of the National Basic Research Program, also called 973 program in China (grant number: 2005CB321600) and Innovative Program of ICT (grant number: 20056610).

References

1. Weiwu Hu, Fuxin Zhang, Zusong Li. Microarchitecture of the Godson-2 processor. Journal of Computer Science and Technology, 3 (2005) 243-249.
2. David Patterson, John Hennessy. Computer Architecture: A Quantitative Approach. Morgan Kaufmann Publishers, Inc. 1996.
3. Michael D. Goddard, Scott A. White. Floating point stack and exchange instruction. US Patent Number: 5,857,089. Jan.5, 1999
4. David W. Clift, James M. Arnold, Robert. P. Colwell, Andrew F. Glew. Floating point register alias table fxch and retirement floating point register array. US Patent Number: 5,499,352. Mar.12, 1996
5. Fuxin Zhang. Performance analysis and optimization of microprocessors. PHD Thesis, Institute of Computing Technology, Chinese Academy of Sciences. (6) 2005
6. Feng Tang. Research on dynamic binary translation and optimization. PHD Thesis, Institute of Computing Technology, Chinese Academy of Sciences. (6) 2006
7. Leonid Baraz, Tevi Devor, Orna Etzion, Shalom Goldenberg, Alex Skaletsky, Yun Wang, Yigai Zemach. IA-32 execution layer: a two-phase dynamic translator designed to support IA-32 applications on Itanium®-based systems. MICRO-2003 (11) 2003
8. Dehnert. J.C., Grant. B.K., Banning. J.P, Johnson, R.; Kistler. T., Klaiber. A., Mattson. J., The transmeta code morphing software: using speculation, recovery, and adaptive retranslation toaddress real-life challenges. CGO-2003 (3) 2003
9. Bochs: The Open Source IA-32 Emulation Project . http://bochs.sourceforge.net/

A Customized Cross-Bar for Data-Shuffling in Domain-Specific SIMD Processors

Praveen Raghavan[1,2], Satyakiran Munaga[1,2], Estela Rey Ramos[1,3],
Andy Lambrechts[1,2], Murali Jayapala[1],
Francky Catthoor[1,2], and Diederik Verkest[1,2,4]

[1] IMEC vzw, Kapeldreef 75, Heverlee, Belgium - 3001
{ragha, satyaki, reyramos, lambreca, jayapala, catthoor,
verkest}@imec.be
[2] ESAT, Kasteelpark Arenberg 10, K. U. Leuven, Heverlee, Belgium-3001
[3] Electrical Engineering, Universidade de Vigo, Spain
[4] Electrical Engineering, Vrije Universiteit Brussels, Belgium

Abstract. Shuffle operations are one of the most common operations in SIMD based embedded system architectures. In this paper we study different families of shuffle operations that frequently occur in embedded applications running on SIMD architectures. These shuffle operations are used to drive the design of a custom shuffler for domain-specific SIMD processors. The energy efficiency of various crossbar based custom shufflers is analyzed and compared with the widely used full crossbar. We show that by customizing the crossbar to implement specific shuffle operations required in the target application domain, we can reduce the energy consumption of shuffle operations by up to 80%. We also illustrate the tradeoffs between flexibility and energy efficiency of custom shufflers and show that customization offers reasonable benefits without compromising the flexibility required for the target application domain.

1 Introduction

Due to a growing computational and a strict low cost requirement in embedded systems, there has been a trend to move toward processors that can deliver a high throughput (MIPS) at a high energy efficiency (MIPS/mW). Application-domain specific processors offer a good trade-off between energy efficiency and flexibility required in embedded system implementations. One of the most effective ways to improve energy efficiency in data-dominated application domains such as multimedia and wireless, is to exploit the data-level parallelism available in these applications [1,2]. SIMD exploits data-level parallelism at operation or instruction level. Prime illustrations of processors using SIMD are [3,4,5], Altivec [6], SSE2 [7] etc.

When embedded applications like SDR (software defined radio), MPEG2 etc., are mapped on these SIMD architectures, one of the bottlenecks, both in terms of power and performance, are the shuffle operations. When an application like GSM Decoding using Viterbi is mapped on Altivec based processors, 30% of all instructions are shuffles [8]. Functional unit which can perform these shuffle operations, known as

P. Lukowicz, L. Thiele, and G. Tröster (Eds.): ARCS 2007, LNCS 4415, pp. 57–68, 2007.

shuffler or permutation unit, is usually implemented as a full crossbar, which requires a large amount of interconnect. It has been shown in [9] that interconnect will be one of the most dominant parts of the delay and energy consumption in future technologies[1]. Hence it is important to minimize the interconnect requirement of shufflers to improve the energy efficiency of future SIMD-based architectures.

Implementing a shuffler as a full crossbar offers extreme flexibility (in terms of varieties of shuffle operations that can be performed), but such a flexibility often is not needed for the applications at hand. Only a few specific sequences of shuffle operations occur in embedded systems and the knowledge of these patterns can be exploited to customize the shuffler and thus to improve its energy efficiency. To the best of our knowledge, there is no prior art that explores different shuffle operations in *embedded systems* and exploits these patterns to design energy-efficient shufflers.

In this paper, we first study different families of shuffle operations or patterns that occur most frequently in embedded application domains, such as wireless and multimedia, and later use them to customize crossbar based shuffler. Customization exploits the fact that shuffle operations of target application domains does not require all inputs be routed to all outputs, which is the case in full crossbar, and thus reduces both the logic and interconnect complexity.

This paper is organized as follows: Section 2 gives a brief overview of related work on shuffle networks in both the networking and SIMD processor domain. Section 3 describes different shuffle operations that occur in embedded systems. Section 4 shows how crossbar can be customized for required shuffle operations and to what extent such customization can help. Section 5 presents experimental results of custom shufflers for different datapath and sub-word sizes. Finally we conclude the paper in section 6.

2   Related Work

A large body of work exists for different shuffle networks in the domain of networking switches and Network-on-Chips [10]. These networks consists of different switches like Crossbar, Benes, Banyan, Omega, Cube etc. These switches usually have only a few cross-points, as the flexibility that is needed for NoC switches is quite low. When a large amount of flexibility is needed, a crossbar based switch is used. Research like [11,12,13,14,15,16] illustrates the exploration space of different switches for these networks. In case of network switches, the path of the packet from input to output is *arbitrary* as communication can exist between any processing elements. Therefore the knowledge of the application domain cannot be exploited to customize it further. In case of networks also, other metrics like bandwidth, latency, are important and hence the optimizations are different.

Other related work exists in the area of data shuffle networks for ASICs. Work like [17,18] and [19], which customize different networks for performing specific applications, like FFT butterflies, cryptographic algorithms etc. [20] customize the shuffle network for linear convolution. They are too specific to be used in a programmable processor and none of them have focused on power or energy consumption. To the best of

---

[1] In our experiments using 130nm technology, we observe that roughly 80% of the crossbar dynamic power consumption is due to inter-cell interconnect.

our knowledge, there is no work which explores the energy efficiency of shuffle networks for SIMD embedded systems. The crossbar is picked over other shuffle networks (like Benes, Banyan etc.) as it can perform all kinds of shuffle operations. Also the data routing from inputs to outputs is straightforward which eases the control word (or MUX selection) generation, design verification, and design upgrades [2].

Table 1. Different Shuffle Families for a 64-bit Datapath and 8-bit Sub-word Configuration that occur in Embedded Systems. ';' denotes the end of one shuffle operation. '|' denotes the end of one output in case of a two outputs family. '-' denotes a don't care.

Family Name	Occurs in Domain	Description	Shuffle Operations
64_m8_O1_F_FFT	Wireless	FFT Butterflies	$a_0 b_0 a_2 b_2\ a_4 b_4 a_6 b_6$; $a_1 b_1 a_3 b_3\ a_5 b_5 a_7 b_7$
			$a_0 a_1 b_0 b_1\ a_4 a_5 b_4 b_5$; $a_2 a_3 b_2 b_3\ a_6 a_7 b_6 b_7$
			$a_0 a_1 b_0 b_1\ a_2 a_3 b_2 b_3$; $a_4 a_5 b_4 b_5\ a_6 a_7 b_6 b_7$
			$a_0 a_1 a_2 a_3\ b_0 b_1 b_2 b_3$; $a_4 a_5 a_6 a_7\ b_4 b_5 b_6 b_7$
64_m8_O1_F_GSM	Wireless	GSM Decode (Viterbi)	$a_0 a_2 a_4 a_6\ b_0 b_2 b_4 b_6$; $a_1 a_3 a_5 a_7\ b_1 b_3 b_5 b_7$
			$a_0 a_1 a_0 a_1\ b_0 b_1 b_0 b_1$; $a_1 a_0 a_1 a_0\ b_1 b_0 b_1 b_0$
			$a_2 a_3 a_2 a_3\ b_2 b_3 b_2 b_3$; $a_3 a_2 a_3 a_2\ b_3 b_2 b_3 b_2$
			$a_4 a_5 a_4 a_5\ b_4 b_5 b_4 b_5$; $a_5 a_4 a_5 a_4\ b_5 b_4 b_5 b_4$
			$a_6 a_7 a_6 a_7\ b_6 b_7 b_6 b_7$; $a_7 a_6 a_7 a_6\ b_7 b_6 b_7 b_6$
64_m8_O1_F_Broadcast	Multimedia	Broadcast for masking	$a_0 a_0 a_0 a_0\ a_0 a_0 a_0 a_0$; $a_1 a_1 a_1 a_1\ a_1 a_1 a_1 a_1$
			$a_2 a_2 a_2 a_2\ a_2 a_2 a_2 a_2$; $a_3 a_3 a_3 a_3\ a_3 a_3 a_3 a_3$
			$a_4 a_4 a_4 a_4\ a_4 a_4 a_4 a_4$; $a_5 a_5 a_5 a_5\ a_5 a_5 a_5 a_5$
			$a_6 a_6 a_6 a_6\ a_6 a_6 a_6 a_6$; $a_7 a_7 a_7 a_7\ a_7 a_7 a_7 a_7$
64_m8_O1_F_DCT	Multimedia	DCT	$a_0 b_0 a_1 b_1\ a_2 b_2 a_3 b_3$; $a_4 b_4 a_5 b_5\ a_6 b_6 a_7 b_7$
			$a_0 a_1 b_0 b_1\ a_2 a_3 b_2 b_3$; $a_4 a_5 b_4 b_5\ a_6 a_7 b_6 b_7$
64_m8_O1_F_Interleave	Multimedia and Wireless	Interleaving two inputs	$a_0 b_0 a_1 b_1\ a_2 b_2 a_3 b_3$; $a_1 b_1 a_2 b_2\ a_3 b_3 a_4 b_4$
			$a_2 b_2 a_3 b_3\ a_4 b_4 a_5 b_5$; $a_3 b_3 a_4 b_4\ a_5 b_5 a_6 b_6$
			$a_4 b_4 a_5 b_5\ a_6 b_6 a_7 b_7$;
64_m8_O1_F_Filter	Multimedia and Wireless	Filtering, Correlators, Cross-correlator	$a_1 a_2 a_3 a_4\ a_5 a_6 a_7 b_0$; $a_2 a_3 a_4 a_5\ a_6 a_7 b_0 b_1$
			$a_3 a_4 a_5 a_6\ a_7 b_0 b_1 b_2$; $a_4 a_5 a_6 a_7\ b_0 b_1 b_2 b_3$
			$a_5 a_6 a_7 b_0\ b_1 b_2 b_3 b_4$; $a_6 a_7 b_0 b_1\ b_2 b_3 b_4 b_5$
			$a_7 b_0 b_1 b_2\ b_3 b_4 b_5 b_6$;
64_m8_O2_F_FFT	Wireless	Two adjacent FFT butterflies	$a_0 b_0 a_2 b_2\ a_4 b_4 a_6 b_6$ $\mid$ $a_1 b_1 a_3 b_3\ a_5 b_5 a_7 b_7$
			$a_0 a_1 b_0 b_1\ a_4 a_5 b_4 b_4$ $\mid$ $a_2 a_3 b_2 b_3\ a_6 a_7 b_6 b_7$
			$a_0 a_1 b_0 b_1\ a_2 a_3 b_2 b_3$ $\mid$ $a_4 a_5 b_4 b_5\ a_6 a_7 b_6 b_7$
			$a_0 a_1 a_2 a_3\ b_0 b_1 b_2 b_3$ $\mid$ $a_4 a_5 a_6 a_7\ b_4 b_5 b_6 b_7$

3 Shuffle Families

A shuffle operation takes two input words and produces one or two outputs with the required composition of input sub-words, which is represented by the control or selection lines. The choice of two outputs has both advantages and disadvantages on the processor architecture. The usage of two output based shuffle unit implies that lower number

---

[2] The instructions and their encoding remain the same, even when the shuffler specification (in terms of set specific shuffle operations to be implemented) changes during the design process, as long as the encoding of MUX selection lines remains unchanged in the customization.

of instructions are required for performing the shuffles required for an application, but at the cost of increased control overhead. The two output shuffle would also require that shuffler uses two ports of the register file to write back the results. In this paper we present both a single output shuffler as well as two output based shufflers. But furthur details on the implications of using one or two output based shuffler unit on the full system is beyond the scope of this paper. The required shuffle operations vary across application kernels, sub-word sizes, and datapath sizes. To illustrate the different shuffle operations, we first introduce a set of definitions:

- *Shuffle Operation*: For a given set of sub-word organized inputs, a particular output sub-word organization.
- *Family*: A set of closely related shuffle operations that are used in an application kernel for given sub-word and datapath sizes
- *Datapath/Word size*: The total number of bits the datapath operates on at a given time.
- *Sub-word Size*: The size of an atomic data element e.g 8-bit and 16-bit.

The different families use the following naming convention: *(Datapath Size)_m(Sub-word Size)_O(# of Outputs)_F_Type*. For example *128_m8_O2_F_FFT* is a collection of shuffle operations required by an "FFT" kernel operating on 8-bit size data elements and implemented on a datapath of size 128-bit.

3.1 Families of Shuffle Operations

1. *FFT*: The FFT family includes all the butterfly shuffle operations that are needed for performing an FFT.
2. *Interleave*: The Interleave family includes the shuffle operations required for interleaving the two inputs words in different ways.
3. *Filter*: The Filter family includes the shuffle operations required to perform various filter operations, correlators and cross-correlators.
4. *Broadcast*: The Broadcast family includes the shuffle operations required for broadcasting a single sub-word into all the sub-word locations.
5. *GSM*: The GSM family includes the shuffle operations required for the different operations during the Viterbi based GSM decoding.
6. *DCT*: The DCT family includes the shuffle operations required for performing a two-dimensional DCT operation.

Table 1 shows the shuffle operations required by the aforementioned application kernels operating on 8-bit sub-words and implemented on a 64-bit datapath. The table also indicates the domain in which these shuffle operations occur. It is assumed that the two inputs to the shuffler are two words $a_0 a_1 a_2 a_3 a_4 a_5 a_6 a_7$ and $b_0 b_1 b_2 b_3 b_4 b_5 b_6 b_7$ respectively, where each of these a_0 to b_7 are sub-words of size 8-bit. Similarly the operations that correspond to other datapath sizes and sub-word modes can be derived. The two-output (*O2*) shuffle operations are similar to the one-output (*O1*) shuffle operations except that they perform two consecutive permutations that are needed by the algorithm simultaneously. For example in case of the FFT, two butterflies that are needed in the same stage are done together. As the shuffle operation for two-output operation can be obtained by concatenating two adjacent shuffle operations of one output operation, only one example is shown in the table.

4 Crossbar Customization

Figure 1 shows a typical full-crossbar implementation, where all the inputs are connected to all the outputs. Used in a 32-bit datapath, it can perform all varieties of one-output shuffle operations with both 8-bit and 16-bit sub-words. The hardware required to implement this is four 8-bit 8:1 multiplexers (MUXes) and the interconnections from the different sub-word inputs to the MUXes. It is clear that this is extremely flexible, but requires a large amount of interconnect. Therefore the power consumption of this full-crossbar implementation is extremely high[3].

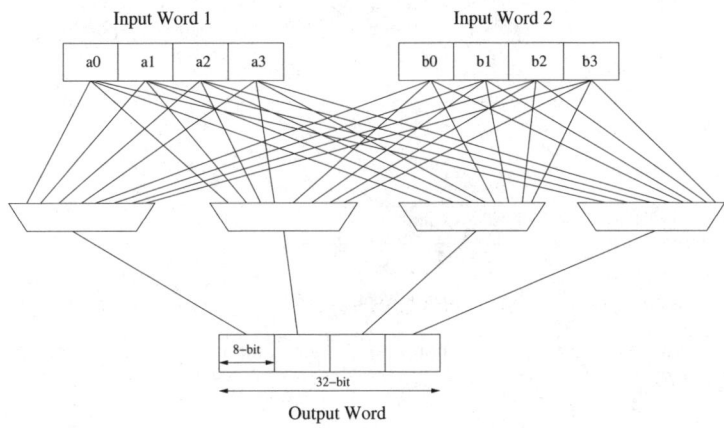

Fig. 1. Full Crossbar with two inputs and one output

If a shuffler is needed that can implement just those shuffle operations represented by the family $32\_m8\_O1\_F\_FFT$, which are shown in Table 2. From the table it is evident that in such a design not all inputs are required to be routed to each of the outputs. E.g., first sub-word output MUX requires inputs a_0, a_1, and a_2 only. Figure 2 shows the customized crossbar which can implement the shuffle operations of Table 2. Thus, given a set of shuffle operations/families that is required, corresponding customized crossbar can be instantiated by removing the unused input connections to each of the output muxes. This reduces both the MUX and the interconnect complexity. We still retain the encoding of MUX selection signals of the crossbar for design simplicity reasons. It should be noted that further energy savings can be achieved by choosing optimal encoding for selection lines (potentially different encoding across MUXes), but it is not explored in this work.

---

[3] In our experiments we observed that a shuffle operation on this implementation consumes nearly the same amount of dynamic energy as that of a 32-bit add operation.

Table 2. Different Shuffle Operations for the *32_m8_O1_F_FFT* family assuming the input words are $a_0a_1a_2a_3$ and $b_0b_1b_2b_3$

Family Name	Patterns
32_8_O1_F_FFT	$a_0b_0a_2b_2$
	$a_1b_1a_3b_3$
	$a_0a_1b_0b_1$
	$a_2a_3b_2b_3$

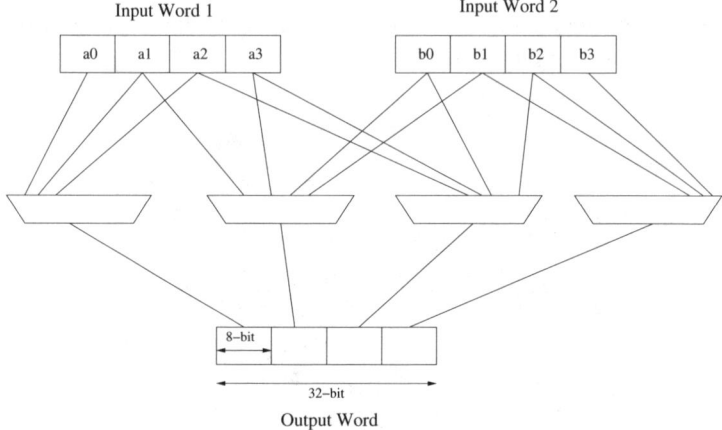

Fig. 2. Crossbar with two 32-bit inputs and one output customized for the family *32_m8_O1_F_FFT*

Another opportunity for optimization is in the implementation of broadcast-based shuffle operations. Since broadcast operations use only the first input, we propose that both inputs a_i and b_i are identical. This implies that implementing broadcast on a shuffler that implements other families will require much less extra connections. E.g., if the two inputs are not forced to be identical, implementing broadcast in the design shown in Figure 2 requires all a_is to be connected to all output MUXes and hence require 1, 3, 2, 4 extra connections to the output MUXes from left to right respectively. If we enforce that both the inputs are identical, to implement broadcast on the same design requires only 1, 2, 1, 1 extra connections to the MUXes.

It can be inferred that the more families a shuffler needs to implement, the larger the interconnect and MUX overhead are and the larger the power consumption will be. On the other hand, if a given customized shuffler only needs to implement a few families, less flexibility is needed and hence it will be less suitable to be used in a processor. To provide more insight on this trade-off, Figure 3 depicts the the average number of inputs to output MUXes in various implementations of customized crossbars for 64-bit datapath. O1 and O2 indicate the number of outputs of the shuffler. mX indicates the sub-word sizes that the shuffler can handle, namely 8-bit (m8), 16-bit (m16), and both 8-bit & 16-bit (mB). Each bar corresponds to one customized shuffler which can implement the indicated shuffle operation families - namely:

- both the filter and interleave families (Filter + Interleave)
- all families discussed in Section 3 and that belong to the wireless domain (WL)
- all multimedia families except broadcast (MM w/o BC)
- all multimedia families including broadcast but not applying aforementioned optimization (MM w/ unopt BC), i.e., broadcast operations are implemented as shown in Table 1
- all multimedia families with optimized broadcast implementation (MM w/ opt BC)
- both multimedia and wireless families and with optimized broadcast (MM+WL)

The figure also shows the number of MUX inputs of a full crossbar. It is clear from the figure that a customized shuffler which offers the flexibility required in embedded applications has a significantly reduced complexity compared to a full crossbar. The benefits of the proposed optimization for broadcast implementation are also explicit from the figure. For the rest of the paper the only the optimized version of the broadcast operation is taken for the multimedia domain.

5 Results

In this section we present the experimental setup and analyze the different crossbar customizations and the effect on power and flexibility.

5.1 Experimental Setup

The synthesis and power estimation flow shown in Figure 4 is used to study the benefits of the customized shufflers. Different shufflers are first coded in behavioral VHDL and implemented using Synopsys Physical Compiler [21] and a UMC130nm standard cell library. The post-synthesis gate-level netlist, including parasitic delays provided by Physical Compile, is used for simulation in ModelSim [22] to obtain the signal activity of the design. This activity information (in SAIF format) is then back-annotated in Physical Compiler/Power Compiler to estimate the average power consumption of the custom shuffler for the shuffle operations that the design is customized for.

To perform the exploration, we use datapath sizes of 32-bit, 64-bit and 128-bit and sub-word modes of 8-bit (*m8*), 16-bit (*m16*) and both 8-bit and 16-bit (*mB*). These datapath sizes and sub-word modes are chosen as they are quite representative of most embedded system processors and data-types [5]. All permutations and combinations of these sub-word sizes and datapath sizes are explored.

5.2 Results and Analysis

To customize the crossbar based shuffler, all the shuffle operations required for one application domain (wireless or multimedia) are used to make one architecture instead

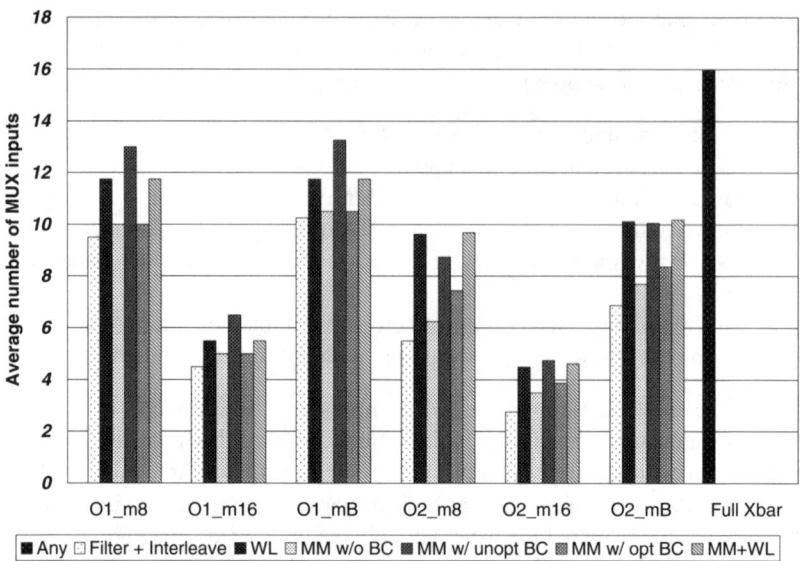

Fig. 3. Reduction in the number of MUX inputs for crossbars (for 64-bit datapath) customized for different sets of families

of making one architecture for every family. To observe the effect of added flexibility on the power consumption of the shuffler, we use another architecture which supports *both* wireless and multimedia shuffle families (MM+WL). Also shufflers are constructed such that they supports the following sub-word modes: only 8-bit sub-word, only 16-bit sub-word, both 8 and 16-bit sub-word modes. To see the effect of the complexity of the design we experiment with different datapath sizes.

Figure 5, 6 and 7 show the power consumption of a 32-bit[4], 64-bit and 128-bit shuffler datapath respectively, with architectures that generate both one and two outputs. Architectures based on sub-word modes 8-bit, 16-bit and both 8-bit and 16-bit based are also compared. All the power numbers are normalized w.r.t a two outputs full crossbar of corresponding datapath size. The *Full X bar* used as the baseline can handle both 8-bit as well as 16-bit sub-word sizes. The figures also show the comparison of the full crossbar (*Full X bar*) with respect to a customized crossbar for the multimedia (*MM*), wireless (*WL*) and both multimedia and wireless (*MM+WL*) domains. For the power estimates shown in Figures 5, 6 and 7, synthesis is performed for a *200MHz*[5] frequency target. It should be noted that each bar corresponds to different custom shuffler design as indicated by the labels.

---

[4] Note that in case of the 32-bit datapath, only sub-word mode of 8 is considered. Using 16-bit sub-word mode on a 32-bit datapath give only 8 possible shuffle operations and which cannot be categorized into the above mentioned families cleanly. Therefore modes *m16* and *mB* are dropped.

[5] All the presented designs are synthesized at various frequencies (100MHz, 200MHz, 333 MHz, 400MHz and 500MHz). It is observed that the presented trends are consistent across the frequencies.

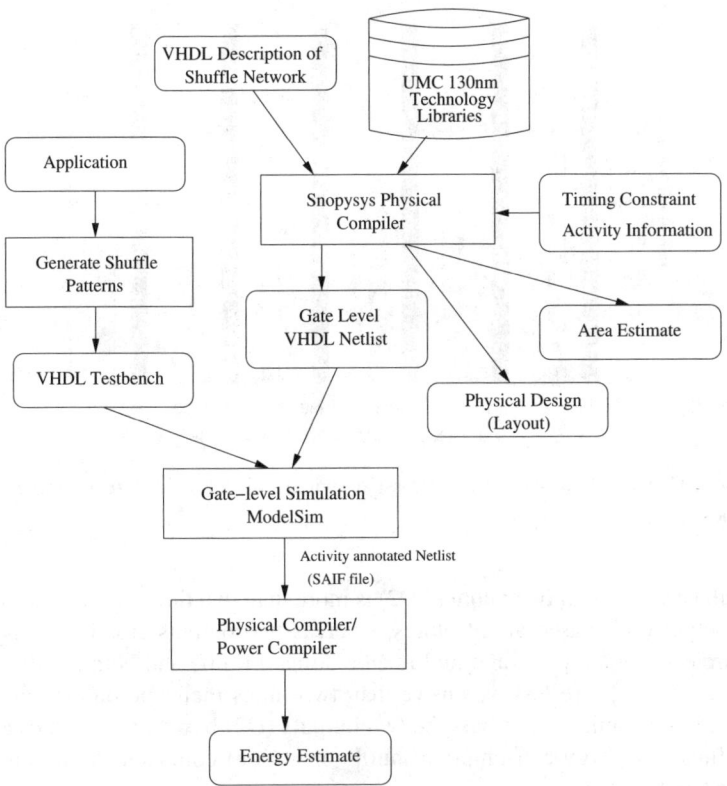

Fig. 4. Tool Flow Used for Assessing Power Efficiency of Custom Shufflers

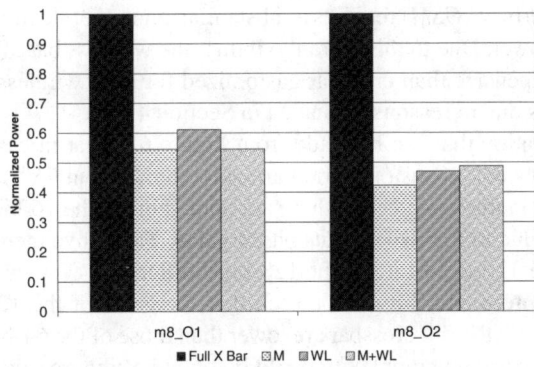

Fig. 5. Power Consumption of the 32 bit crossbar switch over all the different families and sub-word modes

Figure 6 shows that the 16-bit sub-word (*m16*) architecture is more energy efficient compared to the 8-bit sub-word architecture (*m8*) as the amount of routing and MUXing is lower. The overhead of the architecture with the flexibility of both 8-bit and 16-bit sub-words (*mB*) is quite low, compared to the 8-bit sub-word architecture.

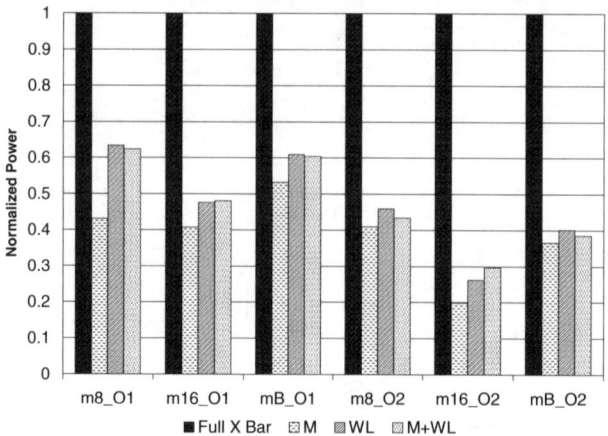

Fig. 6. Power Consumption of the 64 bit crossbar switch over all the different families and sub-word modes

The full crossbar with two outputs (*O2*) is more than two times more expensive than the one output (*O1*) based architectures, whereas two outputs crossbars, customized to the wireless domain (*WL*) and multimedia domain (*MM*) and both multimedia and wireless (*MM+WL*), are less expensive than two times their one output counterpart. Therefore, in customized crossbars the two outputs (*O2*) based architectures are more energy efficient (energy consumption/shuffle operation) compared to the one output (*O1*) based architecture.

It can also be inferred from Figure 6 that the power consumption of the crossbar customized for wireless (*WL*) is more than that of the multimedia (*MM*). This is because of the fact that Viterbi (*F_GSM*) requires a substantial amount of flexibility and therefore consumes more power. Due to this extra flexibility, the wireless based crossbar (*WL*) is not much more expensive than crossbar customized for both wireless and multimedia (*MM+WL*). This is due to reasons explained in Section 4.

Another observation that can be made from Figure 6 is that in case of *m16_O2* architecture, the gains due to customization are quite high (about 75%). These gains are due to the fact that in this architecture there are both gains of the 16-bit sub-word architecture as well as due to the two outputs based gains. The above mentioned trends are valid in case of the 32-bit, 64-bit and the 128-bit shufflers.

Comparing Figure 6 and 7 it can be seen that the gains of the 128-bit based customized crossbar over the full crossbar are lower than those of the 64-bit case. Although increased shuffle operation complexity could be one plausible reason, analysis has revealed that relative (to the full crossbar) the decrease in average number of MUX inputs for 128-bit case is of the same order for 64-bit and thus ruling out the possibility of increased shuffle operation complexity for reduced gain from customization. Further investigation revealed that the smaller gain is due to poor synthesis optimizations on the flat behavioral description of a large design[6]. We also observed that the synthesizer

---

[6] By large designs we mean a wider input bitwidth.

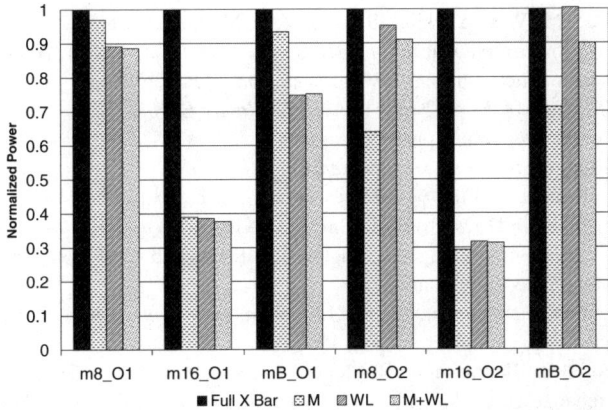

Fig. 7. Power Consumption of the 128 bit crossbar switch over all the different families and sub-word modes

is unable to fully exploit the don't care conditions for unused MUX selection lines on some (larger) designs, which means that the synthesized design still has some redundant logic. This is evident from the cases where custom shuffler that can implement MM+WL families consume less power than shuffler that can only implement WL families.

6 Conclusions

In this paper we presented the different shuffle operations that occur in the embedded systems domain and classified them into different families. The crossbar based SIMD shuffler was then customized to obtain domain specific instantiations of a shuffler which was shown to be power efficient compared to a conventional full-crossbar based implementation. A trade-off space between flexibility and energy efficiency of the shuffler was illustrated. Various datapath sizes as well as sub-word modes were also explored. It was shown that by customizing the crossbar, energy savings of up to 75% could be achieved. We are exploring the feasibility and benefits of using other non-crossbar based networks (such as Banyan, Benes, etc.) to implement the shuffle operations discussed in this paper.

References

1. Ruchira Sasanka. *Energy Efficient Support for All levels of Parallelism for Complex Media Applications*. PhD thesis, University of Illinois at Urbana-Champaign, June 2005.
2. Hyunseok Lee, Yuan Lin, Yoav Harel, Mark Woh, Scott Mahlke, Trevor Mudge, and Krisztian Flautner. Software defined radio - a high performance embedded challenge. In *Proc. 2005 Intl. Conference on High Performance Embedded Architectures and Compilers (HiPEAC)*, November 2005.
3. IBM, http://www.research.ibm.com/cell/. *The Cell Microprocessor*, 2005.
4. K. Van Berkel, F. Heinle, P. Meuwissen, K. Moerman, and M. Weiss. Vector processing as an enabler for software-defined radio in handsets from 3G+WLAN onwards. In *Proc. of Software Defined Radio Technical Conference*, pages 125–130, November 2004.

5. Y. Lin, H. Lee, M. Woh, Y. Harel, S. Mahlke, T. Mudge, C. Chakrabarti, and K. Flautner. SODA: A low-power architecture for software radio. In *Proc of ISCA*, 2006.
6. Freescale Semiconductor, http://www.freescale.com/files/32bit/doc/ ref_manual/MPC7400UM.pdf?srch=1. *Altivec Velocity Engine*.
7. Intel, http://www.intel.com/support/processors/ sb/cs-001650.htm. *Streaming SIMD Extension 2 (SSE2)*.
8. Freescle Semiconductor, http://www.freescale.com/webapp/sps/site/ overview.jsp?nodeId=0162468rH3bTdGmKqW5Nf2. *Altivec Engine Benchmarks*, 2006.
9. Hugo DeMan. Ambient intelligence: Giga-scale dreams and nano-scale realities. In *Proc of ISSCC, Keynote Speech*, February 2005.
10. Jose Duato, Sudhakar Yalamanchili, and Lionel Ni. *Interconnection Networks: an Engineering Approach*. IEEE Computer Society, 1997.
11. Nabanita Das, B.B. Bhattacharya, R. Menon, and S.L. Bezrukov. Permutation admissibility in shuffle-exchange networks with arbitrary number of stages. In *Intl Conference on High Performance Computing (HIPC)*, pages 270–276, 1998.
12. H. Cam and J.A.B. Fortes. Rearrangeability of shuffle-exchange networks. In *Proc. of Frontiers of Massively Parallel Computation*, pages 303 – 314, 1990.
13. I.D. Scherson, P.F. Corbett, and T. Lang. An analytical characterization of generalized shuffle-exchange networks. In *IEEE Proc of Computer and Communication Societies (INFOCOM)*, pages 409 – 414, 1990.
14. Krishnana Padmanabhan. Design and analysis of even-sized binary shuffle-exchange networks for multiprocessors. In *IEEE Transactions on Parallel and Distributed Systems*, pages 385–397, 1991.
15. S. Diana Smith and H.J. Siegel. An emulator network for SIMD machine interconnect networks. In *Computers*, pages 232–241, 1979.
16. Krishnan Padmanabhan. Cube structures for multiprocessors. *Commun. ACM*, 33(1):43–52, 1990.
17. J.P.McGregor and R.B. Lee. Architecture techniques for acclerating subword permutations with repetitions. In *Trans. on VLSI*, pages 325–335, 2003.
18. X. Yang, M. Vachharajani, and R.B. Lee. Fast subword permutation instructions based on butterfly networks. In *Proc of SPIE, Media Processor*, pages 80–86, 2000.
19. J.P. McGregor and R.B. Lee. Architectural enhancements for fast subword permutations with repetitions in cryptographic applications. In *Proc of ICCD*, 2001.
20. A. Elnaggar, M. Aboelaze, and A. Al-Naamany. A modified shuffle-free architecture for linear convolution. In *Trans on Circuits and Systems II*, pages 862–866, 2001.
21. Synopsys, Inc. *Physical Compiler User Guide*, 2006.
22. Mentor Graphics. *ModelSim SE User's Manual*, 2006.

Customized Placement for High Performance Embedded Processor Caches

Subramanian Ramaswamy and Sudhakar Yalamanchili

Center for Research on Embedded Systems and Technology
School of Electrical and Computer Engineering
Georgia Institute of Technology
Atlanta, GA 30332
ramaswamy@gatech.edu, sudha@ece.gatech.edu

Abstract. In this paper, we propose the use of compiler controlled customized placement policies for embedded processor data caches. Profile driven customized placement improves the sharing of cache resources across memory lines thereby reducing conflict misses and lowering the average memory access time (AMAT) and consequently execution time. Alternatively, customized placement policies can be used to reduce the cache size and associativity for a fixed AMAT with an attendant reduction in power and area. These advantages are achieved with a small increase in complexity of the address translation in indexing the cache. The consequent increase in critical path length is offset by lowered miss rates. Simulation experiments with embedded benchmark kernels show that caches with customized placement provide miss rates comparable to traditional caches with larger sizes and higher associativities.

1 Introduction

As processor and memory speeds continue to diverge, the memory hierarchy remains a critical component of modern embedded systems and the focus of significant optimization efforts [1]. With the adoption of larger and faster caches, cache memories have become major consumers of area and power, even in the embedded processor domain [2,3] where area and power constraints make large caches particularly undesirable. For example, the Intel IXP2400 network processor does not employ a cache, the ARM720T processor has an 8KB unified L1 cache, the TI TMS320C6414 has a 16KB data cache, NEC VR4121 has a 8KB data cache. However, at the same time, the cost of off-chip accesses is significant (e.g., the DRAM access time is approximately 300 cycles for the IXP [4]). This has led to the development of many techniques [5,6,7,8,9] for compiler controlled on-chip scratch-pad memory management as a low power, small footprint alternative to hardware managed caches. However, scratch-pad memories require explicit control of *all* data movement to/from the scratch-pad in the spirit of overlays. However, like overlays we would rather have the advantages of automated techniques if they are feasible at an acceptable cost. This paper proposes such a class of techniques - one that effectively combines static knowledge of memory usage when available with run-time simplicity of caching.

P. Lukowicz, L. Thiele, and G. Tröster (Eds.): ARCS 2007, LNCS 4415, pp. 69–82, 2007.

Consider the management of data transfers between other levels of the memory hierarchy. Register allocation is a static application-centric technique to effectively manage memory bandwidth demand to the register file. The page table/TLB contents reflect customized page placement to optimize memory usage and the bandwidth demand to secondary storage. However, caches still use a fixed hardware mapping between memory and the cache hierarchy primarily due to the fact that caches lay on the load/store instruction execution's critical path. Conventional cache placement polices map memory line L to cache set $L \bmod S$ where there are S sets. Caches are indexed with the $log_2 S$ address bits and additional lower order bits are used to access the target word in a line. Historically, compiler-based and hardware-based memory system optimizations have worked with this fixed placement constraint.

In this paper we argue for a simple, yet powerful change in managing caches - software synthesized placement. The mapping, or *placement* of main memory lines in the cache is derived from an analysis of application's memory reference profile. Many embedded applications are amenable to analysis where the memory behavior of application phases can be captured and characterized. This paper describes an approach to the reduction of conflict misses for these phases via better sharing of cache resources across the memory lines. The consequences of the significant reductions in the miss rate is a smaller application footprint in the cache. This can be leveraged to reduce the cache size with an attendant reduction in power, or alternatively improve the average memory access time (AMAT) and thereby reduce execution time. This advantage is achieved with a small increase in area to accommodate the increase in complexity of address translation, and potentially a slight drop in maximum processor clock speed that can be recovered via decreased miss rates. The implementation proposed here can be extended to include run-time optimizations that change placement policies as a function of program region and dynamic behavior to improve the overall miss rate. The synthesized placement is naturally performed by the compiler, but can also be under the control of dynamic optimizers to improve existing optimizations or explore new ones which were hitherto infeasible due to the fixed cache placement constraint in hardware caches.

The following section describes the overall model and the broader implications of software customized placement policies in embedded processor caches. Section 4 describes the algorithmic formulation and solution for the computation of a placement function followed by a description of the address translation mechanism. Section 5 presents the results of an analysis of the design space for a set of kernel benchmarks. The cache employing customized placement consistently outperformed traditional caches with AMAT as the metric, with little impact to area and energy costs. The paper concludes with a description of the relationship of this work with prior and ongoing efforts and concluding remarks that cover our ongoing efforts and extensions.

2   Caches with Customized Placement

Figure 1 conceptualizes the placement policy used in traditional cache architectures. A memory line at address L is placed in, or mapped to, the cache set $L \bmod S$ where the cache has S sets. The set of memory lines mapped to a cache set is referred to as a *conflict set* and this placement policy will be referred to as *modulo* placement. Associativity has a major impact on the number of conflict misses caused by accesses to lines in a conflict set. Increasing associativity reduces conflict misses at the expense of increased hit time, increased energy in tag matching logic, and reduction in the number of sets (for a fixed size cache). The proposed customized placement policy redefines the mapping from memory lines to sets.

In the most general case, a memory line can be mapped to any set in the cache. The difficulty with this approach lies in the problem size (mapping millions of lines) and the complexity of resulting address decoding. Therefore we propose a more structured approach as follows. Let us represent the total number of lines in the cache by C_l and define a *partition*, P_i as the set of memory lines where memory line $L_k \in P_i$ if $L_k \bmod C_l = i$. A traditional k-way set associative cache with S sets will define S conflict sets. The set of lines in each conflict set will be the union of k partitions. In a direct-mapped cache, partitions and conflict sets are equivalent. Figure 2 illustrates the concepts of partitions and conflict sets. The optimization problem can now be stated as follows: Given a memory reference profile, synthesize an assignment of partitions to conflict sets to minimize the number of conflict misses. Figure 3 illustrates the concept of customized placement.

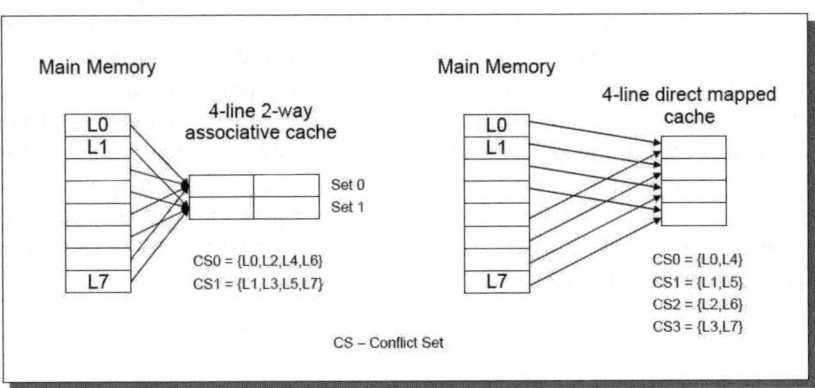

Fig. 1. Traditional cache placement

3   Program Phases and Interference Potential

Program execution evolves through *phases* [10] wherein each phase is characterized by a working set of memory references. Within a program phase, memory reference behavior exhibits spatial and temporal locality around a set of memory locations. The utilization of each partition in a program phase is measured and used to capture the potential conflicts between references to two partitions using the concept of *interference potential*.

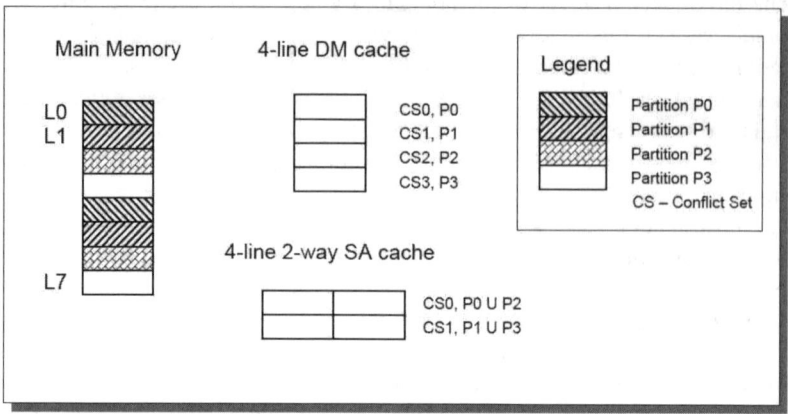

Fig. 2. Partitions and conflict sets in a traditional cache

A memory reference trace is partitioned into contiguous segments of references called *windows* where each window represents a program execution phase. For window w the number of references to partition i is defined as $r[w][i]$. In this study, the interference potential between partitions i and j in window w is $min(r[w][i], r[w][j])$ - representative of the average increase in the number of conflict misses in window w if partitions i and j are mapped to the same cache set (the maximum increase in the number of conflict misses is twice the interference potential). The interference potential between two partitions is the sum of the interference potential between the partitions across all windows. As associativity increases, the interference potential is an increasingly pessimistic measure of conflict misses as the merged partitions will share an increasing number of lines in the target set. As described in Section 4 customized placement will create new conflict sets by composing partitions based on their interference potential and assign these conflict sets to cache sets.

3.1 Architecture and Programming Model

Customized placement is conceived of as a compiler or software controlled optimization that is selectively invoked for specific program phases such as nested loops and functions, or even entire programs that have memory reference behavior that is amenable to analysis and characterization. The implementation is composed of a *base* conventional cache that selectively uses a programmable address translation mechanism for those program phases that employ customized placement as shown in Figure 6. Customized placement is invoked under compiler or software control for individual program regions or across programs in a thread as illustrated in Figures 4 and 5. In our analysis no special support is provided for multiple threads - each thread may individually customize program regions, but otherwise share the cache as in conventional embedded processors.

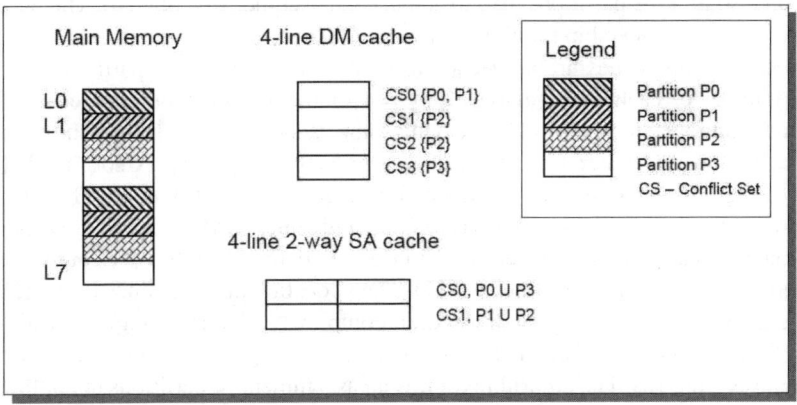

Fig. 3. Customized cache placement

1: placement(*kernel*1)
2: kernel1();
3: placement(*kernel*2)
4: kernel2();

Fig. 4. Example 1

1: placement(*kernel*1 − *phase*0)
2: ... {kernel1 execution begins}
3: placement(*kernel*1 − *phase*1)
4: ... {kernel1 execution continues}

Fig. 5. Example 2

4   Customized Placement

The placement algorithm computed over a trace and the accompanying address translation unit are described in the following sections.

4.1   Placement Algorithm

Algorithm 1 captures the pseudo-code for the computation of customized placement for a set-associative cache. For a k-way set-associative cache with C_s sets, there are $P = k*C_s$ partitions. These P partitions are merged to form C_s sets of partitions, which form the conflict sets for the cache. The primary inputs to the algorithm are the reference counts for each partition in each profile window ($r[P][W]$, where W represents the total number of windows, interference potential between partitions ($ip[P][P]$), and, the number of sets (C_s).

A greedy algorithm iteratively traverses the interference potential matrix to select the partition pair with the minimum interference potential (line 5) and merges them to form a new conflict set (line 6) and updates the reference counts (line 7) and the interference potential matrix (line 8) to reflect the removal of one partition. This process is iterated $P - C_s$ times, to ensure that the resulting number of sets of partitions (the new conflict sets) is equal to the number of cache sets. A partition can be allocated a maximum of one cache set, and multiple partitions may share a single cache set. Finally, the mapping is updated so that the partitions map to actual cache set indexes (line 10). The output is the conflict set membership for all partitions ($map[P]$).

Direct-mapped caches are a special case where the number of partitions is equal to the number of cache sets (associativity = 1) and each set is one cache line. Thus partitions cannot be merged without leaving some cache lines unassigned and therefore unused. Thus we treat direct mapped caches separately as shown in Algorithm 2. There are two stages to the algorithm - an allocation stage where partitions are allocated one or more cache lines based on their access demand (lines 3–10), followed by merging partition pairs based on interference potential and updating the reference counts and the interference potential matrix (lines 12–17). The allocation of cache lines is restricted to powers of two to reduce address translation complexity. The mapping of partitions to conflict sets (cache lines) is updated such that all mappings correspond to actual cache line indexes (line 18). The algorithm returns an assignment of partitions to conflict sets (cache lines), and an allocation count - the number of cache lines allocated to each partition.

4.2   Address Translation

This flexibility in placement is accompanied by a relatively complex translation of 32-bit physical addresses to access the cache set, tag, and word. In a conventional modulo placement, $log_2\ C_s$ bits of the address, the *set index*, are used to determine the set containing the referenced memory line. With custom placement a partition may be mapped to any of the sets in the cache, and therefore requires the re-mapping of the set index bits. This remapping is implemented using a look-up table as shown in Figure 6. There is no constraint on how conflict sets may be composed of partitions and thus two lines in memory with the same tag may share the same cache set. Therefore the conventional tag is extended with the original set index to ensure that all lines can be correctly differentiated in a cache set with a unique tag.

Algorithm 1. Set-associative cache placement

Input: $r[P][W]$, $ip[P][P]$, C_s
Output: $map[P]$

  1: **for** $i = 0$ to $P - 1$ **do**
  2:    $map[i] = i$ {Initialize}
  3: **end for**
  4: **for** $i = 1$ to $P\ -\ C_s$ **do**
  5:    find i_{min}, j_{min} s.t. $ip(i_{min}, j_{min}) = min(ip[P][P])$
  6:    $map[j_{min}] = map[i_{min}]$ {Merge i_{min}, j_{min}}
  7:    $update(r)$ {Update reference counts after merging}
  8:    $update(ip)$ {Update ip[P][P] after merging}
  9: **end for**
10: update(map[P]) {Update mappings to point to actual cache sets}
11: **return** $map[P]$

Address decoding in a direct mapped cache is a bit more complex because each conflict set corresponds to a single partition and is traditionally allocated to a single cache line. If partitions are coalesced based on interference potential to form new conflict sets, and if we wish to use all cache lines, then at least one partition will have to be allocated multiple cache lines. When this happens, the memory lines in a partition are mapped to the allocated cache lines using modulo placement. In this paper we limit the number of cache lines that can be allocated a to a single partition, to be 1, 2, or 4. Address translation now operates as illustrated in Figure 7 and implemented as shown in Figure 8.

Each entry in the lookup table is, again, indexed by the cache line index and the table entry contains the address of the first cache line in the block of cache lines allocated to the particular partition. When four lines are allocated to a partition, the lower order two bits of the tag determine which of the four cache lines is the target location (hence the addition operation). The mask operation (logical AND) helps enforce the correct offset of 1, 2, 3, or 4 (corresponding 0, 1, or 2 lower order bits of the tag). Note the maximum number of cache lines allocated to a partition can be 4. Hence, an additional two bits per entry is required to be stored along with the base cache line address in

---

Algorithm 2. Direct-mapped cache placement

Input: $r[P][W]$, $ip[P][P]$, C_s
Output: $map[P]$, $alloc[P]$

1: $avg = (\sum\limits_{i=0}^{P-1} \sum\limits_{j=0}^{W-1} r[i][j]) \ / \ P$

2: $partitions\_to\_merge = 0$

3: **for** $i = 0$ to $P - 1$ **do**

4:      $map[i] = i$ {Initialize}

5:      $alloc[i] = 1$ {Initialize one cache line per partition}

6:      $p\_ref = \sum\limits_{j=0}^{W-1} r[i][j]$ {Partition reference count}

7:      **if** $p\_ref > avg$ **then**

8:          $alloc[iter] = 2^{\lfloor log_2 \ val \rfloor}$ {Allocate lines to partition in powers of two}

9:          $partitions\_to\_merge$ += $alloc[i]$

10:     **end if**

11: **end for**

12: **for** $iter = 1$ to $partitions\_to\_merge$ **do**

13:     find i_{min}, j_{min} s.t. $ip(i_{min}, j_{min}) = min(ip[P][P])$

14:     $map[j_{min}] = map[i_{min}]$ {Merge i_{min}, j_{min}}

15:     $update(r)$ {Update reference counts after merging}

16:     $update(ip)$ {Update ip[P][P] after merging}

17: **end for**

18: update(map[P]) {Update mappings to point to actual cache sets}

19: **return** $map[P]$, $alloc[P]$

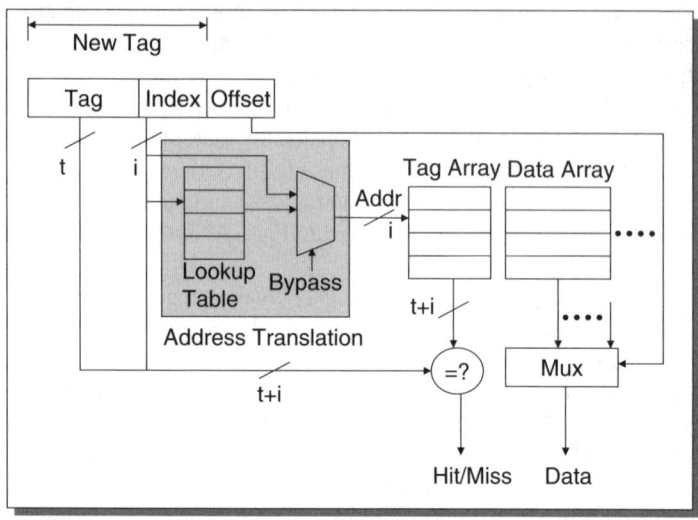

Fig. 6. Address decoding for set-associative caches (bypass path shown)

the lookup table. The direct-mapped configuration also possesses the bypass path (not shown in figure). The lookup table can be implemented using fast SRAM, and loaded using special instructions.

5   Performance Evaluation

The techniques described in this paper were evaluated by modifying the cachegrind [11] simulator to support alternative placement functions. Cachegrind has a write-allocate policy on write misses with LRU replacement in set-associative caches. The benchmarks studied included a subset of kernels from the *mibench* [12] benchmark suite including *basicmath, cjpeg, djpeg, fft, inverse fft, susan, tiff2bw, tiff2rgba, tiffmedian, tiffdither, patricia, ispell,* and, *ghostscript.* The area, power and cache latency estimates were generated using *cacti* [13] for 90nm technology. The placement function was computed using this window size and applied for each benchmark kernel.

As mentioned previously, the analysis does not implement phase detection, but rather, each phase (window) was chosen to be one million references. The window chosen has to be sufficiently large such that interference potential is not tied too closely to the profile (since any slight change in profile would affect performance), whereas it has to be sufficiently small to ensure the interference information captured is useful enough to drive miss rates down. The kernels had reference counts ranging from 40 million to 100 million references, and a value of one million references was chosen as the window size as a compromise between the conflicting arguments. A sensitivity analysis of the relationship of customized placement performance to window sizes is part of our ongoing investigation.

Figure 9 illustrates the AMAT (averaged over the *mibench* kernels) obtained for various cache configurations, with a miss penalty of 200 cycles. In this case, no access

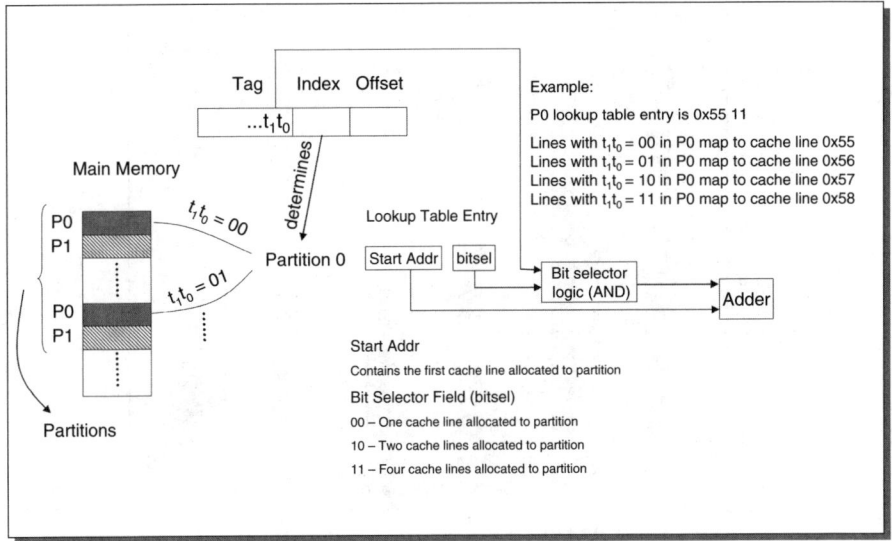

Fig. 7. Address translation of direct-mapped caches - concept

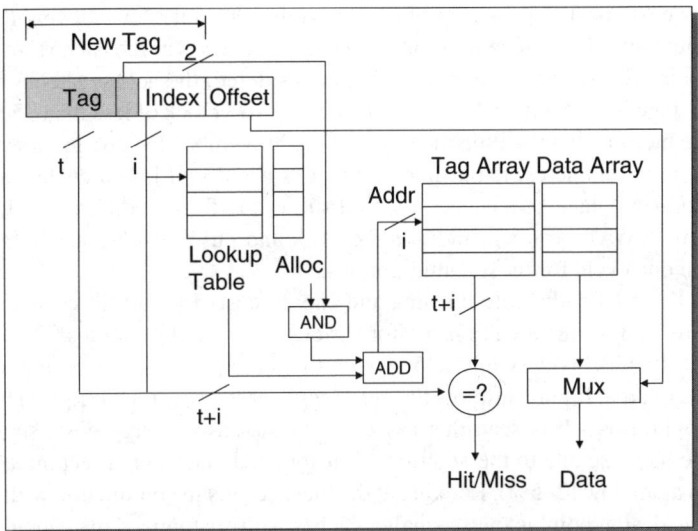

Fig. 8. Address decoding direct-mapped caches (bypass path not shown)

time penalty was assessed for the look-up table access. The AMAT using the CPC is consistently better, and can be seen to offer the same effect as increasing associativity in traditional caches. We conclude that the improved sharing of cache lines is responsible for improved performance. This improved sharing comes at the expense of the addition remapping but garners the effect of increased associativity.

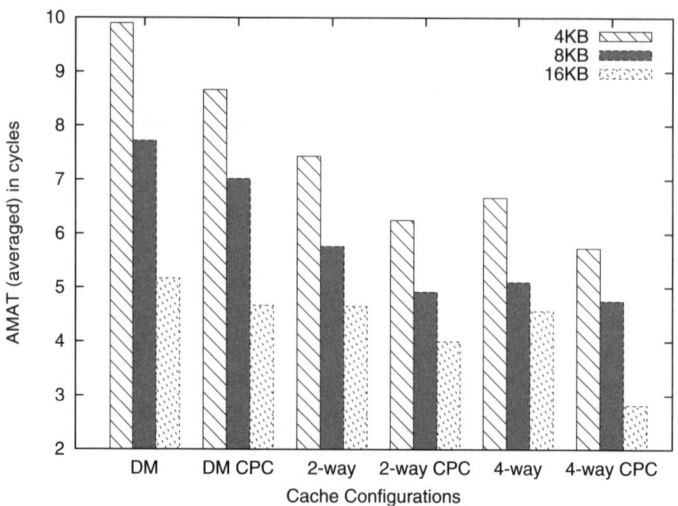

Fig. 9. AMAT for various cache configurations

The latency penalty for customized placement is approximately $0.3\ ns$, over the base latency of $0.6\ ns$ for a direct-mapped 4KB cache and $0.8\ ns$ for a 4-way 4KB cache, The latency cost consists primarily of the decoding necessary for indexing the SRAM based lookup table. The effect of the added latency decreases as the cache size increases or the associativity increases. For the caches considered, the access latency including the lookup stage is well within $1.2\ ns$ corresponding to a clock of less than 750 MHz for single cycle hit time - well within the scope of modern embedded processors. Therefore analysis assumes single cycle hit times. Even if a penalty of half a cycle was applied to the customized placement cache, it would still outperform traditional caches, as the difference in AMAT between traditional caches and customized placement cache is greater than one cycle for most configurations.

Figures 10 and 11 illustrate the area and energy costs for various cache configurations. Figure 10 plots the area in mm^2 for various cache configurations. The additional area cost of the CPC is very low(2–5%) with the lookup table contributing to most of the increase in area. Figure 11 plots the per access energy consumed (in nJ) by various cache configurations. It is seen that associativity increases energy costs significantly, whereas the increase due to the addition of customized placement (keeping associativity fixed) is again low (2–5%). Looking at the these results in conjunction with Figure 9, the desirable design point is to use smaller caches with customized placement, or lower associativity with customized placement to have miss rates comparable to larger caches or caches with higher associativity. The result is significant energy savings.

Figure 12 illustrates the (interpolated) relationship between the AMAT provided and the energy consumed by traditional and customized placement caches. For caches with the same configuration (consuming approximately the same energy - CPC consumes slightly higher energy than the modulo cache for the same configuration corresponding to two nearby points in the x - axis), customized placement caches can provide significantly lower AMAT (10–40%) compared to traditional placement caches. Alternatively,

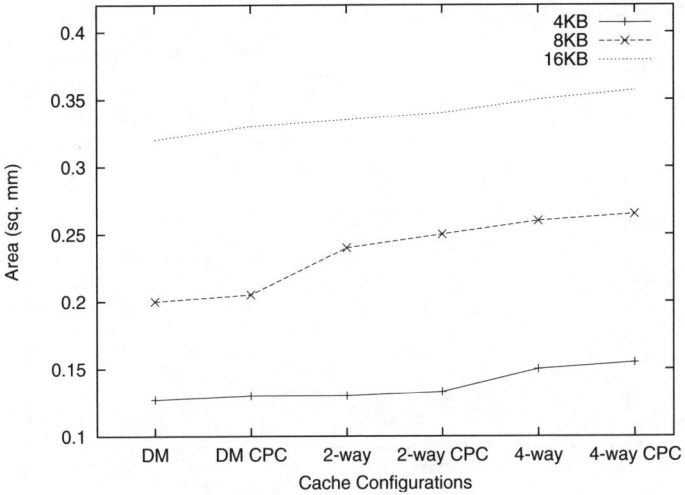

Fig. 10. Area costs of various cache configurations

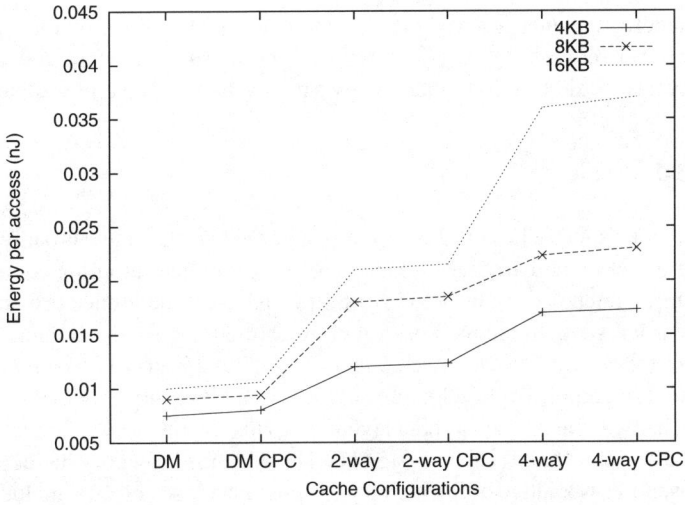

Fig. 11. Energy costs of various cache configurations

if one were to design caches for a specific AMAT, customized placement offers considerable energy savings (25% or more) over the traditional placement caches as caches of much lower size and associativity may be chosen. The energy savings increase as lower AMATs are desired - this is because using customized placement low AMATs are realized with low associativity and cache sizes, whereas with traditional caches energy hungry caches with higher associativity and sizes are required. This discussion

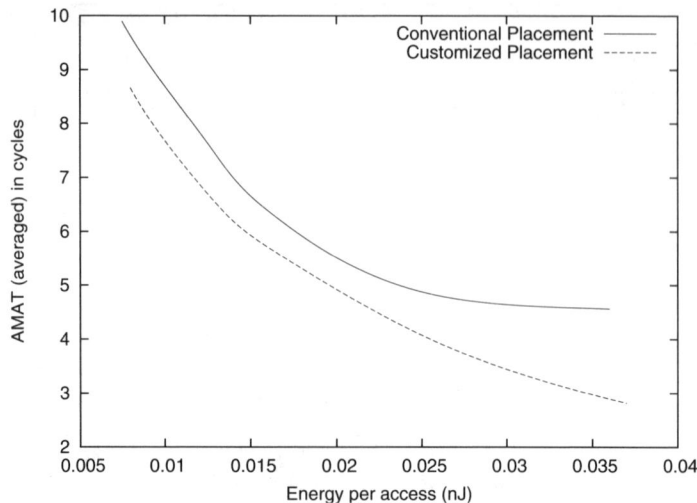

Fig. 12. Energy-AMAT curves compared for traditional and customized placement caches

did not considered the energy saved off-chip memory due to a lower number of misses resulting in fewer requests to off-chip memory. Considering the decreased number of off-chip requests would lead to higher energy savings than reported in this paper.

6 Related Work

Relevant previous approaches towards optimizing cache performance broadly falls into four categories. The first category includes several techniques adopting compiler controlled on-chip scratch pad memories [5,6,7,8,9] which were mentioned in Section 1. As mentioned earlier, scratch-pad memories require explicit control of all data movement between the scratch pad and the off-chip memory and this leads to an increase in code size and software complexity. Techniques in the second category propose adapting the application characteristics, such as data layout, such that its memory reference behavior is a better match for a (fixed) cache design [14,15]. The third category includes designs which are pseudo associative and which try to spread accesses evenly, including those with innovations in indexing and hashing such as [16,17,18,19,20,21,22,23,24]. Finally, the fourth category includes cache partitioning strategies to exploit reference locality in data structures [25,26,27].

The hardware approaches generally seek to find better *fixed* placements than traditional modulo placement. Our technique offers flexible placement policies that can be controlled by the programmer, and is optimized to the target application or kernel. We seek to share the cache resources better among main memory lines targeted to the specific application that is executing. This technique can also complement existing compiler optimizations.

7 Conclusion and Future Work

This work proposes software controlled and synthesized placement, whereby the mapping of a memory line to a cache set is customized to the application profile. This work augments existing techniques used for miss rate reduction by relaxing the constraint of fixed hardware placement. Our results indicate that customized placement produces performance typically achieved with increased associativity or larger sized caches. This behavior can be used to optimize embedded caches along the dimensions of cost (power, area) and performance.

Apart from the run-time placement optimizations we are pursuing, these insights can be exploited in several other ways. Data re-layout optimizations can be pursued in conjunction with selecting the placement policy by the compiler. We are focused on developing a framework for compiler-driven cache downsizing to optimize power and AMAT across phases of application execution. A special case of cache downsizing which maps memory lines to non-faulty cache lines is being extended into a design framework for robust caches in deep sub-micron technologies [28].

References

1. McKee, S.A.: Reflections on the memory wall. In: Conf. Computing Frontiers. (2004)
2. Zhang, M., Asanovi, K.: Fine-grain CAM-tag cache resizing using miss tags. In: ISLPED. (2002)
3. Hu, Z., Martonosi, M., Kaxiras, S.: Improving cache power efficiency with an asymmetric set-associative cache. In: Workshop on Memory Performance Issues. (2001)
4. Intel Corporation: Intel IXP2800 Network Processor Hardware Reference Manual. (2002)
5. Banakar, R., Steinke, S., Lee, B.S., Balakrishnan, M., Marwedel, P.: Scratchpad memory: design alternative for cache on-chip memory in embedded systems. In: CODES. (2002)
6. Steinke, S., Wehmeyer, L., Lee, B., Marwedel, P.: Assigning Program and Data Objects to Scratchpad for Energy Reduction. In: DATE. (2002)
7. Panda, P.R., Dutt, N.D., Nicolau, A.: Efficient Utilization of Scratch-Pad Memory in Embedded Processor Applications. In: EDTC '97. (1997)
8. Miller, J.E., Agarwal, A.: Software-based instruction caching for embedded processors. In: ASPLOS. (2006)
9. Udayakumaran, S., Dominguez, A., Barua, R.: Dynamic allocation for scratch-pad memory using compile-time decisions. Trans. on Embedded Computing Sys. 5(2) (2006)
10. Sherwood, T., Varghese, G., Calder, B.: A pipelined memory architecture for high throughput network processors. In: ISCA. (2003)
11. Nethercote, N., Seward, J.: Valgrind: A Program Supervision Framework. Electr. Notes Theor. Comput. Sci. 89(2) (2003)
12. Guthaus, M., Ringenberg, J., Ernst, D., T. Austin, T.M., , Brown, R.: MiBench: A free, commercially representative embedded benchmark suite. In: 4th IEEE International Workshop on Workload Characteristics. (2001)
13. Tarjan, D., Thoziyoor, S., Jouppi, N.P.: CACTI 4.0: An Integrated Cache Timing, Power,and Area Model (2006)
14. Rabbah, R.M., Palem, K.V.: Data remapping for design space optimization of embedded memory systems. ACM Transactions in Embedded Computing Systems 2(2) (2003) 186–218
15. Chilimbi, T.M., Hill, M.D., Larus, J.R.: Cache-Conscious Structure Layout. In: PLDI. (1999)

16. Qureshi, M.K., Thompson, D., Patt, Y.N.: The V-Way Cache: Demand Based Associativity via Global Replacement. In: ISCA. (2005)
17. Chiou, D., Jain, P., Rudolph, L., Devadas, S.: Application-specific memory management for embedded systems using software-controlled caches. In: DAC. (2000)
18. Zhang, C.: Balanced cache: Reducing conflict misses of direct-mapped caches. In: ISCA. (2006)
19. Hallnor, E.G., Reinhardt, S.K.: A fully associative software-managed cache design. In: ISCA. (2000)
20. Peir, J.K., Lee, Y., Hsu, W.W.: Capturing dynamic memory reference behavior with adaptive cache topology. In: ASPLOS. (1998)
21. Seznec, A.: A Case for Two-Way Skewed-Associative Caches. In: ISCA. (1993)
22. Calder, B., G, D., Emer, J.: Predictive sequential associative cache. In: HPCA. (1996)
23. Agarwal, A., Pudar, S.D.: Column-associative caches: A technique for reducing the miss rate of direct-mapped caches. In: ISCA. (1993)
24. Jouppi, N.P.: Improving direct-mapped cache performance by the addition of a small fully-associative cache and prefetch buffers. In: ISCA. (1990)
25. Petrov, P., Orailoglu, A.: Towards effective embedded processors in codesigns: customizable partitioned caches. In: CODES. (2001)
26. Ramaswamy, S., Sreeram, J., Yalamanchili, S., Palem, K.: Data Trace Cache: An Application Specific Cache Architecture. In: Workshop on Memory Dealing with Performance and Applications (MEDEA). (2005)
27. Dahlgren, F., Stenstrom, P.: On reconfigurable on-chip data caches. In: ISCA. (1991)
28. Ramaswamy, S., Yalamanchili, S.: Customizable Fault Tolerant Embedded Processor Caches. In: ICCD. (2006)

A Multiprocessor Cache for Massively Parallel SoC Architectures

Jörg-Christian Niemann, Christian Liß, Mario Porrmann, and Ulrich Rückert

Heinz Nixdorf Institute, University of Paderborn, Germany
{niemann, liss, porrmann, rueckert}@hni.upb.de

Abstract. In this paper, we present an advanced multiprocessor cache architecture for chip multiprocessors (CMPs). It is designed for the scalable GigaNetIC CMP, which is based on massively parallel on-chip computing clusters. Our write-through multiprocessor cache is configurable in respect to the most relevant design options. It is supposed to be used in universal co-processors as well as in network processing units. For an early verification of the software and an early exploration of various hardware configurations, we have developed a SystemC-based simulation model for the complete chip multiprocessor. For detailed hardware-software co-verification, we use our FPGA-based rapid prototyping system RAPTOR2000 to emulate our architecture with near-ASIC performance. Finally, we demonstrate the performance gains for different application scenarios enabled by the usage of our multiprocessor cache.

1 Introduction

Responding to the emerging performance demands on modern processor systems, we have developed an architecture for single-chip multiprocessors (CMPs) that can be deployed in network processing nodes and universal coprocessors, the GigaNetIC architecture (cf. Fig. 1).

Following our architectural concept [1], the system on chip (SoC) may comprise a multitude of processing elements (PEs), which are locally connected in clusters via on-chip buses like Wishbone or AMBA AHB, and these clusters are globally interconnected by using a network on chip, the GigaNoC [2].

On cluster level, we focus on a shared memory approach with a shared memory bus, which represents one potential bottleneck of our architecture. As we can connect up to 8 PEs to one local bus, the congestion of the bus becomes evident. Another problem is the growing gap between CPU performance and SDRAM memory performance. One solution that trades off between high speed and low costs is the usage of caches. By the usage of caches, the memory bus bottleneck can be widened thus leading to an increase of system performance.

We have developed a multiprocessor cache, which suits the specific requirements of our GigaNetIC architecture, and which is also usable for other multiprocessor architectures due to its standardized AMBA interface and generic CPU interface, respectively. We provide a MOESI-based multiprocessor cache coherency scheme as well as many enhanced cache features (e. g., compiler initiated prefetching, run-time

P. Lukowicz, L. Thiele, and G. Tröster (Eds.): ARCS 2007, LNCS 4415, pp. 83–97, 2007.

partitioning of the *data cache* in cache coherent cache lines and private scratchpad memory on cache line level, and early continuation) for raising the system performance. Actually, the clock rate of the instantiated PEs does not exceed the capabilities of the on-chip memories yet, but the problem of the congestion in accessing the shared memory is tackled.

As the development of modern cache hardware consumes a considerable amount of time, we provide a SystemC simulation model (SiMPLE [2]) of the GigaNetIC CMP that also integrates the multiprocessor cache. This transaction-level-based model allowed us to shorten the development process by doing hardware and software development in parallel and therefore being able to debug new hardware configurations in an early design phase. The coarse-grained SystemC model is finally modeled in a fine grained RTL (Register Transfer Level) description in VHDL. The debugging can be sped up by using a mixed-level simulation: VHDL models for system modules already implemented and SystemC modules for the system modules still under development. After completing the VHDL implementation it can be used for more accurate performance measurements, debugging, verification and profiling of the software. We use our FPGA-based rapid prototyping system RAPTOR2000 [4] for speeding up these software-related tasks and for prototyping the CMP, which is the final step before fabricating the CMP as an ASIC. Both, the SystemC model and the rapid prototyping system allow us a fast verification of the system specification and an early testing of finished parts of the VHDL description.

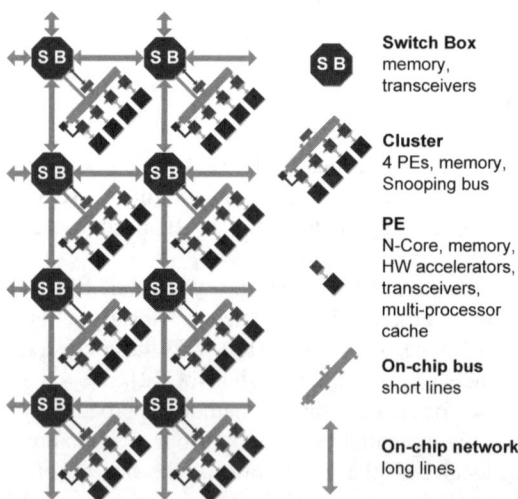

Fig. 1. System architecture based on massively parallel embedded processing clusters

In section 2, the GigaNetIC CMP is discussed in more detail. Section 3 gives an overview over similar chip-multiprocessor caches. We describe the specific features of our cache architecture in section 4. Section 5 presents analysis results in respect to application-specific and implementation-specific performance, power and area requirements concerning standard cell libraries in CMOS technology as well as FPGAs. We conclude this paper with section 6.

2 GigaNetIC Multiprocessor Architecture

The backbone of the GigaNetIC CMP architecture [5] (cf. Fig. 1) is the GigaNoC [2], a hierarchical hybrid network on chip (NoC). It provides high scalability due to its modularity and its regular interconnect scheme. Local on-chip buses are used to connect small numbers of processing elements (PEs) [6][7] or peripherals. These clusters are interconnected by switch boxes (SBs) on the higher level [8]. These IP blocks transparently handle and terminate our on-chip communication protocol similar to [9]. The parallel and redundant structure of our switch box approach offers a high potential for fault tolerance: in case of a failure of one or more components, other parts can take over the respective operations. This can cause a significant increase in production yield.

The interconnection between SBs can form arbitrary topologies, like meshes, tori or butterfly networks. A central aim of our approach is that the resulting CMP is supposed to be parameterizable in respect to the number of clusters, the processors instantiated per cluster, memory size, available bandwidth of the on-chip communication channels, as well as the number and position of specialized hardware accelerators and I/O interfaces. By this, a high reusability of our architecture can be guaranteed, providing a concept for scalable system architectures.

The central processing element is the 32 bit RISC processor N-Core [6]. As hardware accelerators can be added to each cluster or to any SB port to trade run-time functional flexibility for throughput, they help to reduce energy consumption and increase the performance for the system for dedicated applications [10]. The CPUs are relieved and have free resources for control tasks.

Further advantages of our uniform system architecture lie in the homogeneous programming model and in the simplified testability and verification of the circuit. The on-chip communication is based on a packet switched network-on-chip [9]. Data is transmitted by means of packet fragments called flits (flow control digits), which represent the atomic on-chip data transmission units. We have chosen a mesh topology, due to efficient hardware integration.

A close cooperation with the compiler group of our computer science department provides us with an efficient C compiler tailored to our system [11][12]. This compiler will enable the use of the advanced features of our cache (cf. section 4), such as cache line locking, prefetching, etc. The parallel development of the compiler and of the target system allows an optimization of the CMP in both directions. The advices of the compiler developers contributed to important decisions concerning the structure of the multiprocessor system and vice versa.

3 Chip-Multiprocessor Caches

A lot of research has been done on caches and on the optimization of caches in the last four decades, but little research has been done using very flexible, synthesizable caches. In current publications often simulators like the single-processor simulator SimpleScalar [13] or the cache simulator Cacti [14] are used to determine and compare the performance of various system variants (e.g., in [15]). These benchmarks determine either the application-specific performance or the physical characteristics of a cache. Not supporting both, they do not allow for consistent application specific performance and resource usage determination. Configurable systems like Tensilica's Xtensa [16] and ARC's configurable cores [17] are synthesizable and can be used for accurate evaluation of application-specific performance, but they are either configurable in just some of the important cache parameters or they are not easily included in an automated flow that generates many different variants of a system. Therefore, we report on a multiprocessor cache that allows for the determination of all the mentioned characteristics.

4   GigaNetIC Multiprocessor Cache Architecture

Our cache (cf. Fig. 2) can be operated as a unified cache or as a split cache, consisting of a simple cache for instructions (*instruction cache*) and a more complex cache for data (*data cache*). Besides these sub-caches our cache includes a *communication buffer*, which is used for message passing between the CPUs of the same cluster, and a buffer for uncached accesses (e.g., accesses to peripherals, DMA controllers or to the SBs).

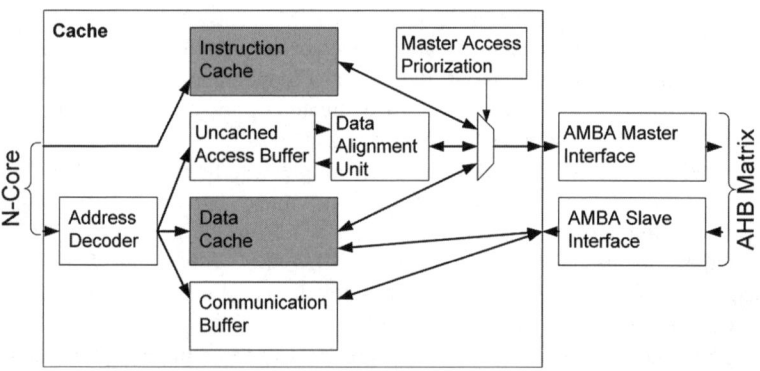

Fig. 2. Structure of the cache architecture

We decided to implement our cache with a write-back architecture to increase scalability by decoupling processor bus traffic from system bus traffic. The cache uses write-allocation to simplify coherency preservation. Our cache has a set-associative architecture with a parameterizable amount of ways. This allows for trade-offs between area and hit-rate. The cache features a true-LRU replacement policy to maximize the hit rate and easy cache command implementation. Coherency is maintained by using a snooping bus (MOESI coherency-protocol [18]) that is accessed through the use of a snooping slave on the system bus and uses direct data intervention to shorten cache miss penalties. The used programming model for the complete SoC

(BSP [1]) is based on weak consistency; therefore the cache supports software-initiated synchronization barriers, which are automatically inserted by the compiler. This offers a high degree of programmability at low complexity. Efficient message passing is done by using the *communication buffer* mentioned above as a "mailbox".

The parameterization of memory depth (8 lines - 2^{20} lines), cache line size (4 Byte – 128 Byte), associativity (2 - 32) and AMBA bus width (32 Bit – 1024 Bit) increases the flexibility and therefore decreases the adaptation costs in case of use of our cache architecture in various system environments. Cache commands to trigger cache actions by software are: *prefetching* of a cache line for the *data cache* or the *instruction cache*, *prefetching of an exclusive copy* of a cache line for the *data cache*, updating of the main memory with a copy of a valid cache line of the *data cache* (*write back*), removing a cache line (*invalidation*) from the *data cache* (and update the main memory's copy, if necessary), *locking* or *unlocking* a cache line in the *data cache* or in the *instruction cache*, *making a cache line private* in the *data cache* (disabling coherency instruction; excluding a cache line from the coherency scheme to keep it local and uninfluenced by other processors). The cache offers early-out on fetches and is capable to generate a hit-under-miss after an early continuation or after a write miss.

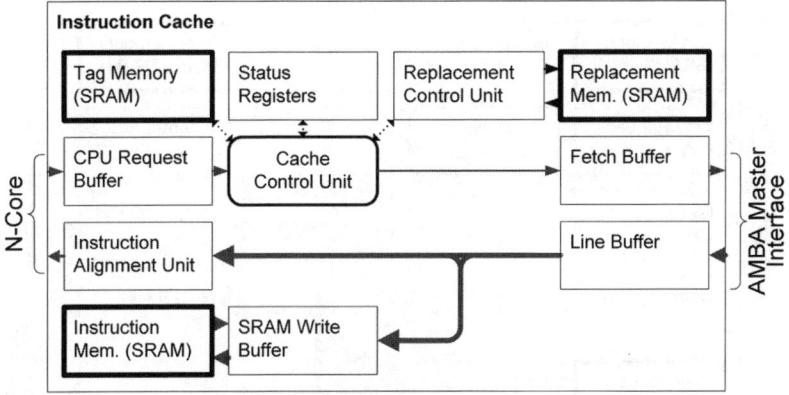

Fig. 3. Structure of the instruction cache

The *instruction cache* (cf. Fig. 3) is controlled by a central *cache control unit*. It initializes the cache memory (SRAM), checks the tag memory and the status bits for cache hits and initiates the fetching in case of a cache miss. A read request is issued by the CPU to the *CPU request buffer* where it is stored until the read is completed. The *cache control unit* checks the availability of the requested instructions and gets the instructions from the *instruction memory* if found. If a miss occurs, a fetch is initiated by sending a request to the *fetch buffer*, which sends it through the *AMBA master interface* to the memory subsystem. In addition, a location in the *instruction cache's* memory is selected by the *replacement control unit* to store the instructions that are to be fetched. The fetched instructions are temporarily stored in the *line buffer* until they are forwarded to the CPU and the *SRAM write buffer*. The CPU preceding *instruction alignment unit* arranges the instructions according to the CPU's bus protocol. The

SRAM write buffer stores fetched instructions that are to be written into the *instruction cache's instruction memory*.

The *data cache* (cf. Fig. 4) provides a *barrier control unit*, a *data combination unit*, a *coherent access control unit* and two extra buffers: a *write back buffer* and a *line sharing buffer*. The *barrier control unit* is used to hold information about synchronization barriers, which are issued by the CPU.

The *data combination unit* stores write requests issued by the CPU in case of write misses until they can be executed. With this, hit-under-misses are possible. The *coherent access control unit* initiates and controls accesses to the *snooping slave* (section 4.2) as well as the main memory in case of data fetch requests or accesses for maintaining coherency. The *write back buffer* is used for storing cache lines that will be written back into the main memory. This can happen either through ordinary line replacement or through the use of the write back or invalidation cache instructions by the CPU. The *line sharing buffer* holds cache lines that will be picked up by other caches. The *cache control unit* can move or copy a cache line to this buffer if the system's snooping slave requests to do this.

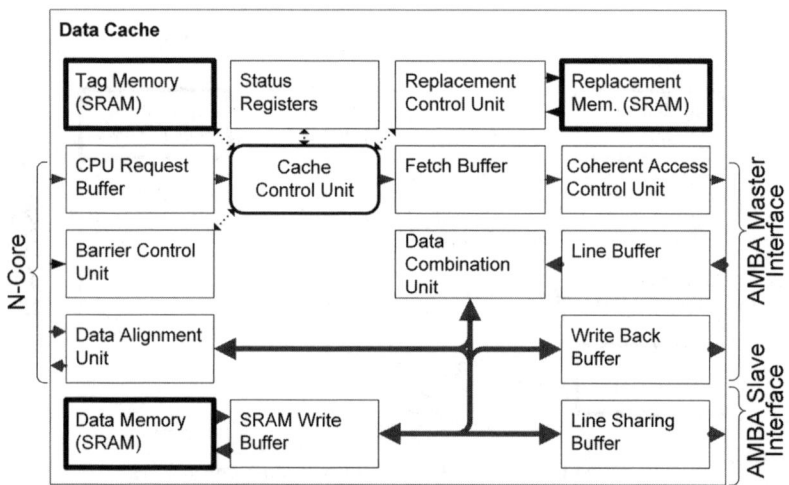

Fig. 4. Structure of the data cache

In both sub-caches the Status Bits of each cache line (the *valid* bit in both, and additionally the *dirty* bit, the *exclusive* bit, and the *coherent* bit in the *data cache*) are stored in registers instead of SRAM memory to allow an easy and multi-ported access to that information. The first three bits are used for controlling the state in the coherency maintenance protocol (MOESI), the fourth bit indicates if this protocol should be applied to the respective cache lines.

4.1 Reuse and Application-Specific Optimizations

We implemented most of the settings related to the behavior and structure of the cache in a parameterizable way, to allow a high degree of flexibility. Anyway, the cache implementation has been influenced by some characteristics of the system environment: e.g., the N-Core processor's architecture, the ability of the memory bus to perform parallel transfers, the absence of a level-2 cache, and the programming model. To allow for new system configurations some extra parameters were added, e.g., in case of a uniprocessor cache the *coherency control unit* and all snooping related parts are removed from the *data cache*. In this case the amount of status bits in the *status registers* is reduced as well: the multiprocessor encoding needs four bits (valid, dirty, exclusive, and coherent), while the uniprocessor encoding uses only two of them (valid, dirty). The valid bit shows if the according cache line is empty or is a valid cache line. The dirty bit indicates a modification by the local CPU so it has to be written back to the memory subsystem on cache-line replacement. The exclusive bit signals if this is the only cache holding a valid copy of the line, which is relevant in case of a write to this cache line. Finally, the coherent bit is used for distinguishing between ordinary cache lines that are coherent within the whole system and not-coherent/ private cache lines. These will not be replaced by the line replacement scheme and are not influenced by snooping activities and these cache lines can be accessed only by the local CPU. The coherency bit can be toggled by the *disabling coherency* cache instruction.

The cache implements the MOESI coherency protocol. Its main advantages become evident if the main memory access is much slower than the fetching of data from other caches in the system. In this case the direct data intervention feature of the MOESI protocol (the direct data exchange between caches) may strongly reduce the miss penalty. In our system architecture each cluster contains a local memory, but the direct data intervention feature of MOESI increases the effective size of memory available to the caches at a low latency and therefore allows the local memory to be small-sized.

4.2 Snooping

The cluster level communication infrastructure is based on our implementation of an AMBA interconnection matrix [3] that fully complies with the AMBA 2.0 specification. It supports concurrent parallel transfers, but prevents "native snooping" by not supporting broadcast transfers. To compensate for this, supplementary snooping buses are usually used. We implemented our snooping bus architecture in a new way by reusing the arbiter of the AMBA matrix: We created a snooping slave, which links the snooping bus to the AMBA matrix, so snoop requests can be initiated in parallel to normal data transfers using the standard bus interface of the caches. The snooping slave is an ordinary slave at the AMBA matrix that controls the snooping bus. Snooping requests, sent to the slave by using the AMBA matrix, are forwarded to the snooping bus, which distributes them to all of the caches. The answers of the slaves are preprocessed and the resulting snoop response is then sent through the AMBA matrix to the snoop issuing cache.

In this system the snooping slave is no bottleneck: instruction fetches do not need snooping slave accesses, so a cache can fetch instructions while another one is using the snooping slave. Data fetches use the snooping slave only for investigating the location of the data. The actual fetching of the data can exploit the parallel transfer capabilities of the AMBA matrix. An example for a scenario of concurrent transfers is shown in Fig. 5 (the arrows show the directions of the read data, write data, and snoop responses within the AMBA matrix): a snooping access to the snooping bus (1), a write access to status registers within the switch box (2), a sharing access based on direct data intervention (3), and a read access (4).

We have chosen the combination of a snooping architecture instead of a directory-based architecture and a non-broadcasting AMBA matrix instead of a broadcasting bus, because of the communication requirements of our system:

First, the main memory is shared to all clusters of the mesh, but consistency has to be maintained only within the clusters. Consistency checks between clusters are not necessary, because in our system a central system component called load balancer assures that data is never shared between different clusters at the same time. In addition, assuming that each cluster consists of a small number of processors (up to eight) a snooping bus protocol promises less overhead than a distributed directory. Another advantage is the easier replaceability of the AMBA matrix by less area-intensive designs.

Second, the non-broadcasting AMBA matrix offers an enormous bandwidth that scales with the amount of active components, like, e.g., CPUs. The high bandwidth enables low latency data fetches, as far as the data is contained within reach (in other caches of the same cluster or in the local memory).

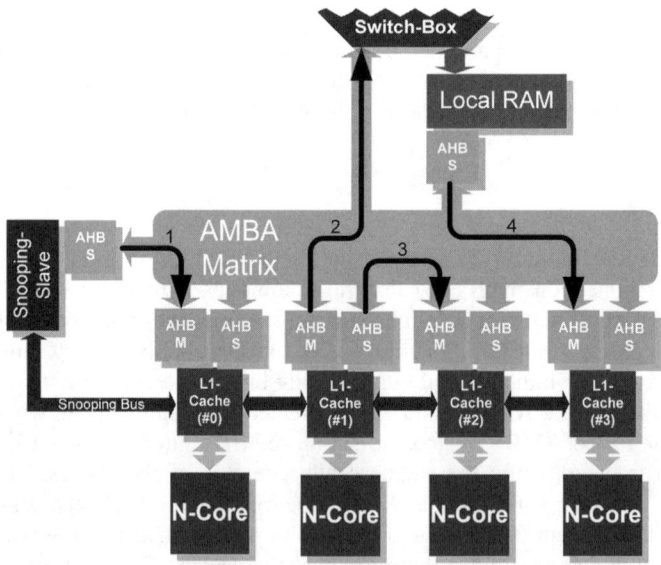

Fig. 5. Parallel transfers through an AMBA-Matrix on cluster level

Our *data cache* architecture allows for just one snooping request each clock cycle to assure a high utilization of CPU while keeping cache complexity at a reasonable level. The *instruction cache* does not need coherency checks, since the programming model does not allow self-modifying code.

To allow more than two concurrent requests (a snoop request and a CPU request) to the *data cache*, the tag ram would have to have more than two ports. This would be very expensive in terms of area.

5   Analysis Results

In the following, we show how our SystemC model of the whole CMP and our rapid prototyping system RAPTOR2000 help us to speed up the test and verification of our VHDL implementation. Additionally, the test and verification of the software is accelerated. Afterwards, we show the synthesis results for a state-of-the-art 90 nm standard cell CMOS process and combine this with benchmark results to illustrate the advantages and disadvantages of several typical configuration variants for the benchmarks.

5.1   Simulation Models and the RAPTOR2000 Rapid Prototyping System

All of the four options for simulation or emulation of our system have their advantages and handicaps. Before the implementation of a SystemC model, we had just a simple C-based prototype that simulates cycle-accurate only one processing cluster of any number of N-Cores with caches of a random size. This model is designed for very fast simulation and is less accurate in respect to bus accesses and memory accesses, like, e.g., cache behavior. For these accesses only the average access duration could be configured. The model considers if the needed data is found in the cache, in the main memory or in another cache. So the effect of direct data intervention is accurately modeled by the cluster level simulator. Based on this first simulator, a SystemC model was developed, which gave us opportunities for more detailed design space exploration and the quick comparison of different design variants. During the implementation phase of the final VHDL model it served as a host system for mixed level simulation to verify and debug parts of the VHDL implementation. This enabled an early test, verification and profiling of the application software. After finalizing the VHDL implementation we had two additional options: the simulation of the complete system using a VHDL simulator, which is the slowest option, but offers the widest opportunities to monitor and trigger every signal, and the emulation of the system using our rapid prototyping system RAPTOR2000. The rapid prototyping system offers near-ASIC speed at the same level of detail that is offered by the VHDL simulator, but in contrast to, e.g., the SystemC model, the amount of complexity that can be emulated directly influences the utilization of the limited FPGA resources. However, the rapid prototyping system's main advantages are its speed and its physical properties: it is possible to use it with "real-world hardware", e.g., Ethernet interfaces to communicate with other network devices.

We measured the performance of our system using four different benchmarks: the Dhrystone 2.1 benchmark, a bubblesort implementation, a quicksort implementation and a packet processing task (IP header check), which is, e.g., a computational intensive part of a DSL Access Multiplexer (DSLAM) application scenario. Given below (in Table 1) is an overview over the speeds that can be reached using the four different implementation options for our four benchmarks:

Table 1. The four simulation / emulation models

	C model	SystemC simulation	VHDL simulation	RAPTOR2000
Execution speed	10 MHz	100 kHz	100 Hz	20 MHz
Level of detail	Poor	Medium	High	High
Constraints	Cache is fixed to unified architecture. Memory and bus simulation limited to an average miss penalty for direct data intervention, an average miss penalty for main memory accesses, and the average latency of cache hits.	Cache is fixed to unified architecture. Associativity (2) and cache line size (16 Bytes) fixed.	None	None

5.2 Performance Analysis

In this section, we analyze the results of our performance measurements. The performance was measured using the four benchmarks described above. All of our measurements were done independently, so they were done using a cold-start cache. To simplify the comparison of the results, we assumed a clock frequency of 250 MHz. The upper bound of our measurements was a system with large local memories attached to each processor. Without caches and local memories, each memory access would cost at least 10 cycles due to the bus access pipelines, the global routing and the off-chip access duration. In reality this value would be much higher due to slower off-chip memory and bus contention. Bus contention is caused by the multiprocessor environment in conjunction with the arbiter of the different communication stages.

To determine the performance of our system we measured the runtime of the mentioned benchmarks using our toolchain. Our toolchain provides a comfortable and cycle accurate profiling and run time analysis of software blocks. Even so, as the number of simulations grows very fast with the number of variables and their values, we limited our measurements to the associativity (2, 4, or 8), the number of cache lines per sub-cache (32, 64, 128, or 256), the cache line width (32, 64, 128, or 256), the main memory latency (0, 5, 10, or 20 cycles), the sub-cache architecture (split or unified) and the availability of a cache (use a cache or do not use a cache). In total a workload of 2716 Benchmark runs were performed. To automatically execute the benchmarking process, we created a comfortable simulation automation tool called MultiSim for ModelSim and SystemC. This enabled us to receive "Excel-ready" preprocessed analysis data of all these simulations.

2716 benchmark runs result in a lot of data. This is why we utilized statistics to determine the dependencies between cache architecture and application performance. We computed the correlation coefficients between those two to discover linear dependencies. The correlation coefficients are shown in Table 2.

Please note that positive values correspond to a lower performance for the first three benchmarks, but a higher performance for the last one. The main memory latency given is the minimal latency. During the benchmark runs we sometimes experienced a higher latency because of bus congestion as well as interference between the CPUs due to bad regular access patterns.

Table 2. Correlation coefficients of our simulations

Correlation coefficients	Higher associativity	Higher number of lines	Higher line width	Higher memory latency	Split cache	Do not use cache
IP Header Check						
Small packets (46 Bytes) processing duration	-0.05	-0.01	-0.39	0.62	0.05	0.35
Middle sized packets (552 Bytes) processing duration	-0.08	-0.01	-0.20	0.32	-0.06	0.65
Large packets (1550 Bytes) processing duration	-0.08	0.00	-0.16	0.26	-0.07	0.68
Quicksort						
Processing duration	-0.11	-0.02	-0.03	0.10	-0.08	0.72
Bubblesort						
Processing duration	-0.12	-0.12	-0.13	0.11	-0.08	0.71
Dhrystone						
DMIPS	0.41	0.38	0.16	0.14	0.26	-0.21

Based on our results we are able to draw some conclusions:

- The larger the IP packets processed by the IP Header Check algorithm are, the more important the use of a cache is.
- The smaller the IP packets are, the more important is low main memory latency.
- The performance of the IP Header Check usually can be improved by increasing the size of the cache lines. Processing of smaller packets is accelerated much more than processing of larger packets.
- The availability of a cache in most cases has a strong positive influence on sorting performance.
- A split cache architecture often is more suitable for executing the Dhrystone benchmark than a unified cache architecture.
- The Dhrystone benchmark is usually accelerated by using a cache and by either a higher associativity or more cache lines.

Conclusions based on statistics are a good direction for optimizing a SoC for a certain application, but to gain maximum performance the actual performance values have to be checked in depth.

5.3 Synthesis Results

We synthesized a processing element consisting of our N-Core processor, our cache and both, our AHB master interface and our AHB slave interface using a state-of-the-art 90 nm process. We used a front-end flow for a 90nm standard cell process considering worst case conditions. Setting place and route aside, our synthesis results can be seen as a good indicator for the area of the final product.

While a simulation of a complex SoC can be performed in a reasonable amount of time, synthesis is still a problem. Taking into account the benchmarks used, our system's size and the target hardware architecture (standard cells), synthesis takes a multiple of the simulation duration. In addition, running a series of syntheses incorporates more effort for the engineer, due to the lack of flexible memory. For simulation highly parameterized memory models can be used, for synthesis no analogous IP can be used. A memory generator was used to generate an instance of low power SRAM to be used in a specific variant of our system. The memory generation was initiated by hand. Therefore we limited our synthesis effort to seven cases: starting with a "standard configuration" with 256 lines of 128 bit, an associativity of 2, a unified cache architecture and a 32 bit wide system bus interface, we changed just one of these variables to get another 5 configurations. Additionally, we synthesized a configuration without caches. The results of the synthesis are shown in Table 3.

Table 3. Resources used by the cache in five different configurations

Configuration	Line size [Bit]	Depth [lines]	Associativity	Split cache	Area [mm²]	Speed [ns]	Power at 250 MHz [mW]
Short cache lines	64	256	2	no	0.61	4.16	136.79
Split cache	128	256	2	yes	1.11	4.08	276.13
High associativity	128	256	4	no	1.31	4.48	376.18
Less cache lines	128	128	2	no	0.58	3.92	164.87
Standard	128	256	2	no	0.73	4.10	180.99

For the standard configuration this leads to the following performance for the whole system, including 32 processing element (Table 4):

Table 4. Resources used by the whole system

SoC main components	Amount	Area [mm²]	Total area [mm²]	Frequency [MHz]	Power at 250 MHz [mW]	Total power at 250 MHz [mW]
Caches	32	0.7293	23.34	243.90	180.99	5791.68
N-Cores	32	0.1273	4.07	158.73	11.74	375.68
AMBA master interfaces	32	0.0075	0.24	354.61	1.11	35.52
AMBA slave interfaces	56	0.0008	0.04	465.12	0.15	8.40
AMBA matrixes	8	0.0969	0.78	265.96	25.16	201.28
Snooping slaves	8	0.0020	0.02	420.17	0.55	4.40
Packet buffer	8	0.9519	7.62	250.00	35.50	284.00
Switchboxes (NoC)	8	0.5748	4.60	434.78	53,00	424.00
Total			40.70	158.73		7124.96

Figure 6 shows the distribution of area and power in the system, based on the table above. As shown, in our system the cache has the highest impact on both. For doing a cost-benefit analysis we calculated the performance values of the complete SoC. These are shown in Table 5 in relation to our standard configuration.

All benchmarks show a huge increase in performance by using caches, although the systems area grows at high rates. In addition, the table shows approaches for optimization. For example, instead of using the standard configuration the configuration with half of the cache lines would result in a reduction of area (decrease by 11%) as well as huge speed-up (speed-up factor: 4.51) for a system executing the bubblesort algorithm. The table's values are in line with the conclusions drawn in the last chapter, except the one stating that a higher associativity improves Dhrystone performance.

Since the systems clock frequency is limited by the N-Core processors architecture and all cache syntheses lead to clock frequencies higher than the processors speed, the cache options have no influence on the frequency of the whole system using this processing element. But as our cache architecture can be adapted to a broad variety of processing cores, we will benefit by using high-speed core variants.

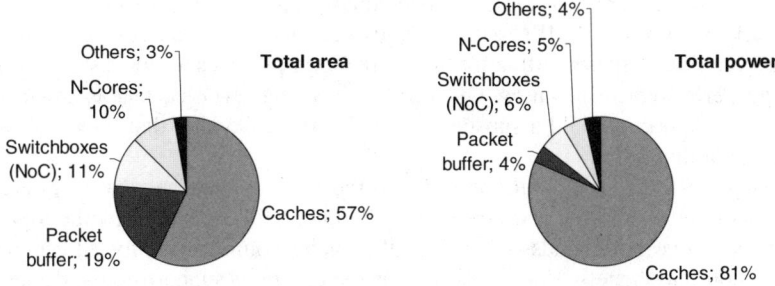

Fig. 6. The influence of the system's components on total system area and power consumption

Table 5. Cost and benefit of various cache configurations in relation to the standard configuration

Configuration	Area [mm²]	IPHC Small	IPHC Medium	IPHC Large	Performance Quicksort	Bubblesort	Dhrystone
Standard	**100%**	**100%**	**100%**	**100%**	**100%**	**100%**	**100%**
Short cache lines	91%	64%	69%	71%	98%	451%	108%
Split cache	130%	76%	91%	97%	104%	469%	114%
High associativity	146%	110%	101%	101%	100%	451%	98%
Less cache lines	89%	100%	100%	99%	100%	451%	95%
Without cache	43%	22%	8%	7%	5%	21%	5%

Increased energy efficiency is another outcome of using our cache. This is due to a much shorter runtime of the benchmarks in relation to the power consumption of the cache. In standard configuration the system needs only between 11.4% (Dhrystone) and 51.6% (IPHC small) of the energy compared to the system without cache.

6 Conclusion

In this paper, we have shown both, the cost and the benefits of a cache in an on-chip MPSoC evaluating four benchmarks and using a state-of-the-art 90 nm standard cell CMOS process. Despite high costs in area and power consumption we observed high performance increases and remarkable energy reduction (up to 88.6%) concerning the presented benchmarks. Even better results could be achieved using a cache configuration tailored to the application.

Our research was enabled by our MultiSim tool that allows for accurate and extensive design space exploration. Further research has to be done in automating the synthesis workflow to integrate synthesis into the benchmarking process. This way an automatic cost-benefit analysis would be possible. In addition, the application specific power consumption may be determined, because switching activities based on statistical models are not very accurate.

In future experiments measurements with warm-start caches will be performed to do research with the more realistic scenarios of processing elements executing programs for longer runtimes instead of one time execution runs. This will be done, in particular, using a complete DSLAM benchmark that is mapped to our system and benchmarks similar to EEMBC using various network traffic patterns. Especially for tasks as complex as these, with a lot of computation per packet, our results let us expect high performance increases. Our results show that even caches as small as our standard configuration yield a substantial acceleration, showing that they are suitable for highly parallel SoCs.

Looking at the different configurations synthesized, one can see that there are high fixed costs in terms of area and power. Beside others, this is caused by the logic supporting the cache commands. Currently, the cache commands only add additional costs, without any benefit, due to the lack of the compiler supporting the cache commands. Further experiments with a compiler supporting these commands are expected to exhibit noticeable higher system performance.

Besides design space exploration in regard to various cache configurations, the high flexibility of our cache allows it also to be used in other multiprocessor systems that are not based on our GigaNetIC architecture. As a subset even single processor environments are supported. The only requirements for the processor bus interface and the system bus interface to work efficient is the support of the widespread address pipelining and 32-bit-wide address and data buses from and to the processing elements.

Acknowledgements. The research described in this paper was funded by the Federal Ministry of Education and Research (Bundesministerium für Bildung und Forschung), registered there under grant number 01AK065F (NGN-PlaNetS), and by the Collaborative Research Center 614.

References

1. O. Bonorden, N. Brüls, D. K. Le, U. Kastens, F. Meyer auf der Heide, J.-C. Niemann, M. Porrmann, U. Rückert, A. Slowik and M. Thies. A holistic methodology for network processor design. In Proceedings of the Workshop on High-Speed Local Networks held in conjunction with the 28th Annual IEEE Conference on Local Computer Networks, pages 583-592, October 20-24 2003.
2. J.-C. Niemann, M. Porrmann and U. Rückert. A Scalable Parallel SoC Architecture for Network Processors. In IEEE Computer Society Annual Symposium on VLSI (ISVLSI), Tampa, FL., USA, 2005.
3. Christian Liß. Implementation of an AMBA AHB Interconnection Matrix. Technical Report. University of Paderborn, Paderborn, Germany. May 2004.
4. H. Kalte, M. Porrmann and U. Rückert. A Prototyping Platform for Dynamically Reconfigurable System on Chip Designs. In Proceedings of the IEEE Workshop Heterogeneous reconfigurable Systems on Chip (SoC), Hamburg, Germany, 2002.

5. J.-C. Niemann, C. Puttmann, M. Porrmann and U. Rückert. GigaNetIC - A Scalable Embedded On-Chip Multiprocessor Architecture for Network Applications. In ARCS'06 Architecture of Computing Systems, pages 268-282, 13-16 March 2006.

6. D. Langen, J.-C. Niemann, M. Porrmann, H. Kalte and U. Rückert. Implementation of a RISC Processor Core for SoC Designs FPGA Prototype vs. ASIC Implementation. In Proc. of the IEEE-Workshop: Heterogeneous reconfigurable Systems on Chip (SoC), Hamburg, Germany, 2002.

7. M. Grünewald, U. Kastens, D. K. Le, J.-C. Niemann, M. Porrmann, U. Rückert, M. Thies and A. Slowik. Network Application Driven Instruction Set Extensions for Embedded Processing Clusters. In PARELEC 2004, International Conference on Parallel Computing in Electrical Engineering, Dresden, Germany, pages 209-214, 2004.

8. R. Eickhoff, J.-C. Niemann, M. Porrmann and U. Rückert. Adaptable Switch boxes as on-chip routing nodes for networks-on-chip. In From Specification to Embedded Systems Application, International Embedded Systems Symposium (IESS), A. Rettberg , M. C. Zanella and F. J. Rammig Ed., pages 201-210, Manaus, Brazil, 15-17 August 2005.

9. W. J. Dally and B. Towles. Route Packets, Not Wires: On-Chip Interconnection Networks. In Proceedings of the Design Automation Conference, pages 684-689, Las Vegas, Nevada, USA, June 18-22 2001.

10. J.-C. Niemann, M. Porrmann, C. Sauer and U. Rückert. An Evaluation of the Scalable GigaNetIC Architecture for Access Networks. In Advanced Networking and Communications Hardware Workshop (ANCHOR), held in conjunction with the ISCA 2005, 2005.

11. E. Stümpel, M. Thies and U. Kastens. VLIW Compilation Techniques for Superscalar Architectures. In Proc. of 7th International Conference on Compiler Construction CC'98, K. Koskimies Ed., 1998.

12. U. Kastens, D. K. Le, A. Slowik and M. Thies. Feedback Driven Instruction-Set Extension. In Proceedings of ACM SIGPLAN/SIGBED 2004 Conference on Languages, Compilers, and Tools for Embedded Systems (LCTES'04), Washington, D.C., USA, June 2004.

13. D. Burger and Todd M. Austin. The SimpleScalar tool set, version 2.0. SIGARCH Computer Architecture News. Vol. 25, No. 3, p. 13-25. ACM Press. New York, NY, USA, 1997.

14. D. Tarjan, S. Thoziyoor and N. P. Jouppi. CACTI 4.0. Technical Report. HP Laboratories Palo Alto, Palo Alto, CA, USA, June 2006 .

15. J. Mudigonda, H. Vin, R. Yavatkar. Managing Memory Access Latency in Packet Processing In SIGMETRICS '05: Proceedings of the 2005 ACM SIGMETRICS international conference on Measurement and modeling of computer systems, pages 396-397, Banff, Alberta, Canada. June 2005.

16. Tensilica. Xtensa LX Microprocessor, Overview Handbook. Internet publication, Santa Clara, CA, USA, 2004. Source: http://tensilica.com/pdf/xtensalx_overview_handbook.pdf, Seen online: 05.10.2006.

17. ARC International. ARC 700 configurable core family. Internet publication. San Jose, CA, USA, 2005. Source: http://arc.com/evaluations/ARC_700_Family.pdf, Seen online: 05.10.2006

18. P. Sweazey and A. J. Smith. A class of compatible cache consistency protocols and their support by the IEEE futurebus. In 13th Annual International Symposium on Computer Architecture, ISCA, Japan, 1986.

Improving Resource Discovery in the Arigatoni Overlay Network*

Raphaël Chand[1,**], Luigi Liquori[2], and Michel Cosnard[2]

[1] University of Geneva, Switzerland
Raphael.Chand@cui.unige.ch
[2] INRIA Sophia Antipolis, France
{Michel.Cosnard,Luigi.Liquori}@inria.fr

Abstract. Arigatoni is a structured multi-layer overlay network providing various services with variable guarantees, and promoting an intermittent participation to the virtual organization where peers can appear, disappear and organize themselves dynamically. Arigatoni mainly concerns with how resources are declared and discovered in the overlay, allowing global computers to make a secure, PKI-based, use of global aggregated computational power, storage, information resources, etc. Arigatoni provides fully decentralized, asynchronous and scalable resource discovery, and provides mechanisms for dealing with dynamic virtual organizations. This paper introduces a non trivial improvement of the original resource discovery protocol by allowing to register and to ask for *multiple instances*. Simulations show that it is efficient and scalable.

1 Introduction

The explosive growth of the Internet gives rise to the possibility of designing large *overlay networks* and *virtual organizations* consisting of Internet-connected *global computers*, able to provide a rich functionality of services that makes use of its aggregated computational power, storage, information resources, etc. Arigatoni [3] is a structured multi-layer overlay network which provides resource discovery with variable guarantees in a virtual organization where peers can appear, disappear and organize themselves dynamically.

The virtual organization is structured in *colonies*, governed by *global brokers*, GB. A GB (un)registers global computers, GCs, receives service queries from clients GCs, contacts potential servants GCs, trusts clients and servers and allows the clients GC and the servants GCs to communicate. Registrations and requests are performed via a simple query language *à la* SQL and a simple *orchestration language à la* LINDA. Communication intra-colony is initiated via only one GB, while communication inter-colonies is initiated through a chain of GB-2-GB message exchanges. Once the resource offered by a global computer has been found in the overlay network, the real resource exchange is performed out of the overlay itself, in a peer-to-peer fashion.

---

* This work is supported by Aeolus FP6-2004-IST-FET Proactive.
** This work was done while the author was at INRIA Sophia Antipolis, France.

P. Lukowicz, L. Thiele, and G. Tröster (Eds.): ARCS 2007, LNCS 4415, pp. 98–111, 2007.
© Springer-Verlag Berlin Heidelberg 2007

The main challenges in Arigatoni lie in the management of an overlay network with a dynamic topology, the routing of queries and the discovery of resources in the overlay. In particular, resource discovery is a non trivial problem for large distributed systems featuring a discontinuous amount of resources offered by global computers and an intermittent participation in the overlay. Thus, Arigatoni features two protocols: the *virtual intermittence protocols*, VIP, and the *resource discovery protocol*, RDP. The VIP protocol deals with the *dynamic topology* of the overlay, by allowing individuals to login/logout to/from a colony. This implies that the process of routing may lead to some failures, because some individuals have logged out, or are temporarily unavailable, or because they have been *manu militari* logged out by the broker because of their poor performance or avidity (see [9]).

The total decoupling between GCs in *space* (GCs do not know each other), *time* (GCs do not participate in the interaction at the same time), and *synchronization* (GCs can issue service requests and do something else, or may be doing something else when being asked for services) is a major feature of Arigatoni overlay network. Another important property is the encapsulation of resources in colonies. All those properties play a major role in the scalability of Arigatoni's RDP.

The version V1 of the RDP protocol [6] enabled to ask for *one service* at the time, like, *e.g.* CPU or a particular file. The version V2, presented in this paper, allows *multiple instances of the same service*. Adding multiple instances is a non trivial task because the broker must keep track (when routing requests) of how many resource instances were found in its own colony before delegating the rest of the instances to be found in the surrounding supercolonies.

As defined above, GBs are organized in a dynamic tree structure. Each GB, leader of its own subcolony, is a node of the overlay network, and a root of the subtree corresponding to its colony. It is then natural to address scalability issues that arise from that tree structure. In [6], we showed that, under reasonable assumptions, the Arigatoni overlay network is scalable. The technical contributions of this paper can be summarized as follows:

- A new version of the resource discovery protocol, called RDP V2, that allows for multiple instances; for example, a GC may ask for 3 CPUs, or 4 chunks of 1GB of RAM, or one compiler gcc. Multiple services requests can be also sent to a GB; each service will be processed sequentially and independently of others. If a request succeeds, then via the orchestration language of Arigatoni (not described in this paper), the GC client can synchronize all resources offered by the GC's servants.
- A new version of the simulator taking into account the non trivial improvements in the service discovery.
- Some simulation results that shows that our enhanced protocol is still scalable.

The rest of the paper is structured as follows: after Section 2 describing the main machinery underneath the new service request, Section 3 introduces the pseudocode of the protocol. Then, Section 4 shows our simulation results and Section 5 provides related work analysis and concluding remarks. An Appendix conclude with some auxiliary algorithms. For obvious lack of space, we refers to http:// www-sop.inria.fr/mascotte/Luigi.Liquori/ARIGATONI for an extended version of this paper.

2   Resource Discovery Protocol RDP V2

Suppose a GC X registers to its GB and declares its availability to offer a service S, while another GC Y issues a request for a service S'. Then, the GB looks in its *routing table* and *filters* S' against S. If there exists a solution to this matching equation, then X can provide a resource to Y. For example, S ≜ [CPU=Intel, Time<10sec] and S' ≜ [CPU=Intel, Time>5sec] match, with attribute values Intel and Time between 5 and 10 seconds. When a global computer asks for a service S, it also demands a certain number of instances of S. In RDP V2 this is denoted by "SREQ:[(S, n)]".

Each GB maintains a table $\mathcal{T}$ representing the *services* that are registered in its colony. The table is updated according to the *dynamic registration and unregistration* of GC in the overlay. For a given S, the table has the form $\mathcal{T}[S] = [(P_j, m_j)]^{j=1...k}$, where $(P_j)^{j=1...k}$ are the address of the direct children in the GB's colony, and $(m_j)^{j=1...k}$ are the instances of S available at P_j.

For a service request SREQ:[(S, n)], the steps are:

- Look for q *distinct* GCs capable of serving S in the local colony.
- If $q<n$, then search r remaining instances $(n-q)$ in local subcolonies.
- If $r<(n-q)$, then delegate remaining instances $(n-q-r)$ to the leader of the colony.

A GC receiving a service request chooses the services that it *accepts/rejects* to serve. It then generates a SRESP message containing the lists of services accepted/rejected, and sends it to its GB. The response messages are then propagated back in the overlay, following the reverse path.

A Service Request. SREQ:[(S, n)] may arrive bottom-up to the GB directly from its colony, or top-down from its own leader. In both cases, the leader tries to find n distinct GC that can serve S. More precisely, the list $[(P_j, m_j)]^{j=1...k}$ contains all the direct children in GB's colony that can serve S (child P_j with m_j instances of S). The discovery protocol features two search modes, *selective* and *exhaustive*. The selective mode is resource conservative at the price of important delays in case of low acceptance rates. The exhaustive mode is resource eager, but is independent of the acceptance rate. Let SREQ:[(S, n)], and $\mathcal{T}[S] = [(P_j, m_j)]^{j=1...k}$. The selective mode consist in:

- If $\sum_{i=1}^{k} m_i \geq n$, then there are enough resources in the GB's colony to serve S. Let $y \leq k$ be the smallest index such that $\sum_{i=1}^{y} m_i \geq n$, and $\sum_{i=1}^{y-1} m_i < n$. Then, SREQ:[(S, m_i)] is sent to all P_i with $(i \leq y-1)$, and SREQ:[(S, $n - \sum_{i=1}^{y-1} m_i$)] is sent to P_y.
- If $\sum_{i=1}^{k} m_i < n$, then there are not enough GCs in the GB's colony that can serve S. Then, SREQ:[(S, m_i)] is sent to all P_i $(i \leq k)$, and SREQ:(S, $n - \sum_{i=i}^{k} m_i$)]

is delegated to the GB's leader. The rationale is that we first try to ask for *as many resources* in GB's colony, and then ask GB's leader for the *remaining resources*.

The exhaustive search mode consists in sending SREQ:$[(S, min(m_i, n))]$ to all P_i $(1 \leq i \leq k)$, and to delegate SREQ:$[(S, n - \sum_{i=1}^{k} min(m_i, n)]$ to the GB's leader. The rationale is to first ask for *all* resources in the GB's colony, and then ask the GB's leader for the remaining resources.

A Service Response. SRESP:ACC:$[(S, a)]$, or SRESP:REJ:$[(S, d)]$, may follow service requests for services S. That is, "*a*" GCs accepted to serve S, and "*d*" denied. Due to the asynchrony of Arigatoni, more replies can arrive to the colony's leader (*i.e.* $a+d \geq n$). As for requests, there exists two modes that determine the way those acceptances are propagated back to the leader. In the *selective reply* mode, we return at most the number of instances of S that were asked by the leader whereas in the *exhaustive reply* mode, we return *all* acceptances.

As for acceptances there exists two modes that determine the way those acceptances are propagated back to the leader. In the *selective search* mode, the *whole colony* was asked for n instances of S, at most. This implies that exactly d instances of S must now be looked for to fulfill the original request. Hence, we first try to find d instances of S in other subcolonies. We then delegate the instances that could not be found to the leader. Finally, the remaining instances are reported back as rejected. In the *exhaustive search* mode, each *sub-colony* was asked for n instances of S, at most. Hence, there may be other sub-colonies that have not replied yet, and which may reply with enough acceptations to fulfill the request. The remaining instances must be delegated to the leader.

3 RDP Pseudo-code

In this section, we detail the pseudo-code of the RDP V2. Five variables are used for each Arigatoni's interaction "ask-route-reply-route-back": *Path, asked, downstream, upstream,* and *SendList*. Each message (SREQ or SRESP) contains a unique identifier id, that is initially set by the GC that sends the initial SREQ message. Variable *Path* is a simple hash keyed by the identifier of the message. The other variables are double hashes which first key is the identifier of the message, and second key a given service S. The intuitive meaning of those variables is listed below.

- *Path*{id}: Peer address: identifies the child from which the original SREQ message came from.
- *asked*{id}{S}: Integer: number of instances of S asked and not replied.
- *downstream*{id}{S}: Integer: instances of S asked in colony and not replied.
- *upstream*{id}{S}: Integer: instances of S delegated but not replied.
- *SendList*{id}{S}: (Peer address,Integer): the list of direct children that are potentially capable of serving S.

Algorithm 1. Receiving $SREQ_{id}$:$[(S, n)]$ from P_{from} (executed by P)

```
 1: Path{id} ← P_from                                    // To trace back the reverse route
 2: if SendList{id}{S} = ∅ then
 3:     SendList{id}{S} ← Filter(S, P_from)               // Filter S in P's routing table
 4: end if
 5: (RoutingList, remaining) ← Route(P_from, S, n, search_mode)      // Build a routing list
 6: asked{id}{S} ← asked{id}{S} + n
 7: if remaining ≠ 0 then                                 // Remaining instances to find
 8:     if L ≠ ∅ and L ≠ P_from then                      // L exists and is different from P_from
 9:         Insert L : (S, remaining) in RoutingList
10:         upstream{id}{S} ← upstream{id}{S} + remaining
11:     else                                             // P's colony is isolated
12:         Send SRESP_id:REJ:[(S, remaining)] to P_from
13:         asked{id}{S} ← asked{id}{S} − remaining
14:     end if
15: end if
16: for each Q : (S, m) ∈ RoutingList do
17:     Send SREQ_id:[(S, m)] to Q                        // Send SREQ_id to every element in RoutingList
18: end for
```

The pseudo-code of RDP V2 is showed in Algorithms $[1 - 5]$. For obvious lack of space, we details only Algorithms 1, and 2, and 3. The Appendix presents succintely the remaining auxiliary algorithms.

Case of Service Request (Alg. 1). Consider an individual P receiving a reply message $SREQ_{id}$ from a neighbor P_{from}, and let L be P's leader.

- In line 1, the originator of the request is first recorded in *Path*, so as to allow reply messages to follow the reverse path.
- In line 3, the *Filter* function (Alg. 4) determines the *SendList* corresponding to service S, *i.e.*, the list of direct children of P potentially able of serving S.
- In line 5, the *Route* function (Alg. 5) builds (*RoutingList, remaining*), *i.e.*, the list of children that will be sent a particular service request, according to the selected search mode, and the positive number of the remaining instances for which no servant has been found. The *RoutingList* contains a list of mappings of the form Q:$[(S, m)]$ which means that neighbor Q is to be sent a service request SREQ:$[(S, m)]$.
- In line 8, if L exists and is not the originator of the request (to avoid routing loops), then the entry L:(S, *remaining*) is appended to the *RoutingList* (line 9), and the *upstream* counter is incremented accordingly (line 10); else (line 11, L exists and it is the originator of the request), since servants can be found for *remaining* instances of service S, a rejection reply is sent back to the originator of the request (line 12), and the *asked* counter is decremented accordingly (line 13).
- In line 17, a service request is sent to each neighbor Q having an entry in the *RoutingList*.

Case of Service Response (Alg. 2,3). Consider an individual P receiving a reply message $SRESP_{id}$ from a neighbor P_{from}. The operation of the resource discovery

---

Algorithm 2. Receiving $SRESP_{id}$:ACC:$[(S, a)]$ from P_{from} (executed by P)

```
1:  case search_mode is
       "selective" :
2:       Send SRESP_id:ACC:[(S, a)] to Path{id}              // Forward the SRESP
3:     "exhaustive" :
4:       if P_from = L then                                   // Top-down request
5:         upstream{id}{S} ← max(upstream{id}{S} − a; 0)
6:       else                                                 // Bottom-up request
7:         downstream{id}{S} ← max(downstream{id}{S} − a; 0)
8:       end if
9:       if asked{id}{S} ≥ a then              // More instances asked than accepted
10:        asked{id}{S} ← asked{id}{S} − a
11:        acc_return ← a
12:      else                                  // More instances accepted than asked
13:        acc_return ← asked{id}{S} − a
14:        asked{id}{S} ← 0
15:      end if
16:      case reply_mode is
           "selective" :
17:          Send SRESP_id:ACC:(S, a) to Path{id}             // Accepted "a" instances
18:        "hexaustive" :
19:          Send SRESP_id:ACC:(S, acc_return) to Path{id}    // Accepted "acc_return" instances
20:      end case
21: end case
```

---

algorithm is detailed in pseudo-code in Algorithms 2 and 3 and explained hereafter.

- *Acceptance (Alg. 2).* Let $SRESP_{id}$:ACC:$[(S, a)]$ arrive from P_{from} at P, *i.e.*, "a" global computers in P's colony accepted to serve service S.

  If the *selective search* mode was used to route the original service request $SREQ_{id} : (S, n)$ (issued by $Path\{id\}$), then the *whole colony* was asked for at most n instances of S. Hence, no more than n acceptances may arrive from P's colony. Thus, the reply message is simply forwarded back to $Path\{id\}$ (line 2).

  If the *exhaustive search* mode was used, then *each child* was asked for at most n instances of S. Hence, it is possible that a number of acceptances higher than n arrives from L's colony. To do this, counters *asked, upstream, downstream*, and *acc_return* are updated accordingly (lines $5 - 14$).

  The *selective reply* mode simply reply back to $Path\{id\}$ with "a" acceptation instances (line 17), while the *exhaustive reply* reply with "*acc_return*" instances (line 19).

- *Rejections (Alg. 3).* Let $SRESP_{id} : REJ:[(S, d)]$ arrive from P_{from} at P, *i.e.*, "d" global computers in P's colony refused to serve S. This implies that *all* global computers in P's colony have been sent a request for S.

  If the sender of the message is the leader L, then no other potential servants for the d instances of S can be found. Consequently, the rejection message is simply forwarded back (line 2), and counters *asked* and *upstream* updated accordingly (lines 3 and 4).

  If L is not the sender of the rejected message, then there may be other potential servants in the colony or in other surrounding colonies. The operation of the protocol depends on the search mode that was used.

Algorithm 3. Receiving $SRESP_{id}:REJ:[(S, d)]$ from P_{from} (executed by P)

1: **if** $P_{from} = L$ **then**   // *Return rejections*
2:     Send $SRESP_{id}:REJ:[(S, d)]$ to $Path\{id\}$
3:     $asked\{id\}\{S\} \leftarrow asked\{id\}\{S\} - d$
4:     $upstream\{id\}\{S\} \leftarrow upstream\{id\}\{S\} - d$
5: **else**   // *Retry at other children or delegate*
6:     **case** search_mode **is**
       "*exhaustive*" :   // *Try to delegate or reject*
7:         $downstream\{id\}\{S\} \leftarrow max(downstream\{id\}\{S\} - d; 0)$
8:         **if** $asked\{id\}\{S\} \leq downstream\{id\}\{S\} + upstream\{id\}\{S\}$ **then**
9:                                    // *Less instances asked than down/upstream'ed*
10:            Wait for more replies from other children
11:        **else**   // *More instances asked than down/upstream'ed*
12:            $remaining \leftarrow asked\{id\}\{S\} - downstream\{id\}\{S\} - upstream\{id\}\{S\}$
13:            **if** $L \neq \emptyset$ and $L \neq Path\{id\}$ **then**
14:                $upstream\{id\}\{S\} \leftarrow upstream\{id\}\{S\} + remaining$
15:                Send $SREQ_{id}:(S, remaining)$ to L
16:            **else**
17:                $asked\{id\}\{S\} \leftarrow asked\{id\}\{S\} - remaining$
18:                Send $SRESP_{id}:REJ:(S, remaining)$ to $Path\{id\}$
19:            **end if**
20:        **end if**
21:        Remove P_{from} from $SendList\{id\}\{S\}$
22:     "*selective*" :   // *Try other children, delete, or reject*
23:        Remove P_{from} from $SendList\{id\}\{S\}$ // *Don't send requests to P_{from} anymore*
24:        $(RoutingList, remaining) \leftarrow Route(P_{from}, S, d, search\_mode)$
25:        **if** $remaining \neq 0$ **then**   // *Still some remaining instances to treat*
26:            **if** $L \neq \emptyset$ and $L \neq P_{from}$ **then**   // *L exists and is different from P_{from}*
27:                Insert $L : (S, remaining)$ in $RoutingList$
28:                $upstream\{id\}\{S\} \leftarrow upstream\{id\}\{S\} + remaining$
29:            **else**   // *P's colony is isolated*
30:                Send $SRESP_{id}:REJ:(S, remaining)$ to $Path\{id\}$
31:                $asked\{id\}\{S\} \leftarrow asked\{id\}\{S\} - remaining$
32:            **end if**
33:        **end if**
34:        **for each** $Q : \{(S, e)\} \in RoutingList$ **do**
35:            Send $SREQ_{id}:[(S, e)]$ to Q // *Send an SREQ for every element in RoutingList*
36:        **end for**
37:     **end case**
38: **end if**

If the *exhaustive search* mode was used, then there are no other potential servants in L's colony but there may be some in other surrounding colonies. Hence, we first determine the number of instances of S that need to be found to fulfill the request.

If $asked \leq downstream + upstream$ (line 8), then there are enough potential servants in the colony or in surrounding colonies that have not replied yet, to fulfill the request. Consequently, we simply wait for more replies (line 10).

In contrast, if $asked \geq downstream + upstream$, then we must look for more potential servants in order to fulfill the request. There are ($asked - downstream - upstream$) of them to be found (line 12). As said before, those may only be found via a delegation to the leader L. Hence, the latter is sent a request for the remaining instances of S, if possible, (line 15), or a rejection is sent back to the original sender of the request (line 18). The *upstream* or *asked* counters are updated accordingly (lines 14 and 17).

If the *selective* search mode is used, then there may be other potential servants in P's colony. The process is the same as in Algorithm 1, except that we do not consider children that have already been sent a request (line 21, 23). For that purpose, we use the *SendList* that was originally created by the *Filter* function (during the processing of the original service request message), and produce another *RoutingList* with the *Route* function (line 24).

Finally, we proceed as in Algorithm 1 (lines 25 − 36).

4   Protocol Evaluation

The actual Arigatoni's topology is tree-based with routing complexity of $O(logN)$ (N being the number of nodes). However, in each GB, an extra complexity is due to solve the matching equation between the service request and the routing table T containing the mapping between peers and resources (this complexity is usually linear in the size of S).

To assess the effectiveness and the scalability of the protocol, we have conducted simulations using large numbers of units and service requests. For lack of space, we only present the results that correspond to the new features of the protocol, namely, the possibility to specify multiple instances of a service.

We have generated a network topology of 103 GBs, using the transit-stub model of the Georgia Tech Internetwork Topology Models package [18], on top of which we added the Arigatoni overlay network.

We take 120 distinct services, and we define the *overlap interval* $1 \leq L \leq 120$, as the interval of indices inside which services match each other. That is, for all $(i, j) \in L^2$, S_i and S_j match. If L=120, then all services match each other; if L=1, then each service only matches itself. At each GB, we added a random number of GCs chosen randomly between 0 and 100.

To simulate subscription load, we then randomly registered at each GC each service with a probability ρ denoting the *global availability of services*. We then randomly raised 50, 000 service requests per GC. Each request contained either a certain number of instances I of a service, chosen uniformly at random.

Each service request was then handled by the new RDP V2, described in this paper. We used a service acceptance probability of α=75%, which corresponds to the probability that a GC receiving a request for a service S (and offering S), accepts to serve it.

Upon completion of all the requests, we measured for each GB its load as the number of requests (messages) it received. We then computed the average load as the average value over the population of GB s in the system. We also computed the maximum load as the maximum value of the load over all the GBs in the system.

Similarly, we computed the average and maximum load fractions as the average and maximum loads divided by the number of requests. The average load represents the average load of a GB due to the completion of all the requests. The average load fraction represents the fraction of requests that a GB served, on average. The maximum fraction represents the maximum fraction of the requests that a GB served.

We computed the average service acceptance ratio as follows. For each GC, we computed the local acceptance ratio as the number of service requests that yielded a positive response (*i.e.* the system found at least one GC), over the number of service requests issued at that GC agent. A service request that contained multiple instances of a service counts as a positive response only if the system found as many GCs as the number of instances specified in the request.

We then computed the average acceptance ratio as the average value over the number of GC (that issued at least one service request). We repeated the experiments for different values of ρ, L and I. Results are illustrated in Figures 1 and 2. The algorithm V2 was implemented in C++.

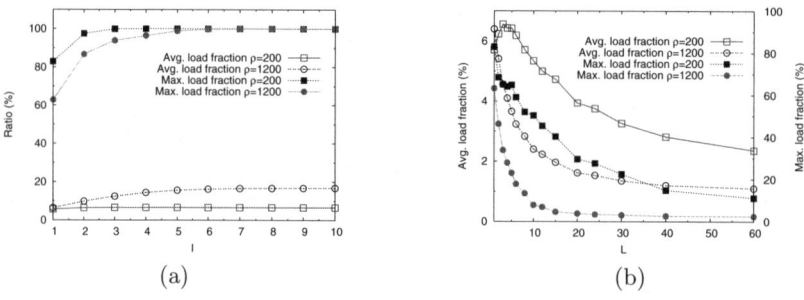

Fig. 1. Average and maximum load fraction w.r.t. (a) number of instances of a service in service requests (b) overlap interval

Figure 1(a) shows the evolution of the average and maximum load fraction w.r.t. the number of instances of a service in service requests. Unsurprisingly, we observe that asking more instances of a service in a service request requires much more resource from the system. Indeed, for each instance, the system tries to find a different GC capable of providing the service. We observe that low-level GBs participate more, since there are more delegations. For values of I of circa 3 for ρ=0.02, and circa 6 for ρ=0.12%, the average and maximum load fractions stabilize. For values of I higher than those values, there are not enough resources in the system to completely fulfill the request (*i.e.*, not enough GCs capable of providing the requested service).

Figure 1(b) illustrates the evolution of the average and maximum load fractions w.r.t. the overlap interval L. Unsurprisingly, we observe that the more services match, the smaller the load imposed to the system. Indeed, for a given requested service, there are more potential candidates capable of providing a resource that satisfies it. For high enough values of L, the load stabilizes, and the resources are found very quickly (most often at the nearest GB).

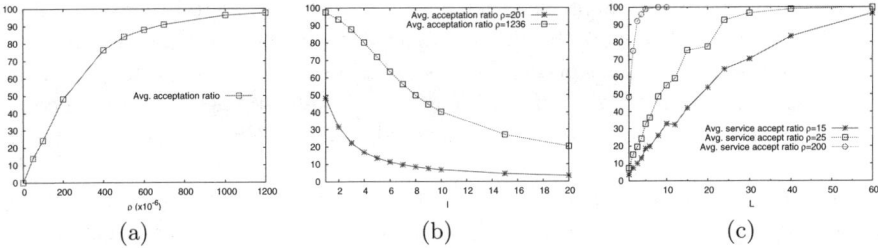

(a)   (b)   (c)

Fig. 2. Average success rate (shown in the Y Axis in %) w.r.t. (a) service availability (b) number of instances (c) overlap interval

Figure 2(a) shows the average success rate w.r.t. the service availability ρ. Unsurprisingly, the average service acceptance ratio increases exponentially with the availability of services. This shows that **Arigatoni** is *efficient* in searching individuals for requested services. Indeed, a service availability of $\rho=0.06\%$ enables the system to achieve an acceptance rate of 90%.

Figure 2(b) shows the evolution of the success rate w.r.t. the number I of instances of S in a service requests. We observe that the average success rate decreases with the increasing of I and eventually stabilizes to 0%. This is due to the fact that the more instances we ask, the less **GCs** can be found to fulfill the request.

Figure 2(c) shows the average success rate w.r.t. the overlap interval L. We observe that the success rate increases with L, and eventually stabilizes to 100% at for higher values of L. This is due to the fact that the more services match, the higher the number of **GCs** capable of providing a resource that satisfies a given request.

5 Related Work and Conclusions

Many technologies, algorithms, and protocols have been proposed recently on resource discovery. Some of them focus on **Grid** or **P2P** oriented applications, but none of those targets the full generality of **Arigatoni** which only deals with generic resource discovery for building an overlay network of global computers, structured via a virtual organization of variable topology and clear distinct roles between leader and individuals (**GCs** or subcolonies).

Discussion on Closest Overlay Architectures (from [1]). The main challenges of pervasive computing are *how to build* an overlay network with respect to its topology, and *how to route queries* and *discover resources*.

There are essentially in the literature many basic types of overlays: structured (tree, ring or grid), unstructured, hybrid overlays (a combination of the two above), and multi-layer (or n-layer) overlays. **Arigatoni** falls in the latter category that is widely used in many **P2P** systems.

In a nutshell, in a *n-layer* overlay network, the responsibility assigned to Individuals differs (think of the different roles and responsibilities of GBs and GCs), since there are super-peers (GBs) serving as a server for a subset of all peers. Ordinary peers (GCs) submit queries to their super-peers and receive results from it. Super-peers are also connected to each others as ordinary peers (Individuals), routing messages over the overlay network, submitting, delegating, and answering queries on behalf of their peers (in their colony) and themselves. This structure is replicated *recursively*, creating a *n-layer topology*, where some peers become super-peers with decreasing responsibilities.

Typical issues of n-layer overlays are the size of each colony, together with the interests and the resources offered and demanded in each colony. Typical bottlenecks of n-layers are reliability and service availability (related to few points of failure) and load balancing. Classical solutions to cope with these problems are adding redundancy at the broker-layer. Historically, the n-layer topology generalizes the *two-layer topology*, such as the one we can find in the *hierarchical* DHT of Canon [12] and Coral [11].

Discussion on Closest Technologies. The Globus toolkit [13], is an open-source set of technology, protocols and middleware, used for building Grid systems and applications. Possible applications range from sharing computing power to distributed databases in a heterogeneous overlay network, where security is seriously taken into account. The toolkit includes stand-alone software for security, information infrastructure, resource management, data management, communication, fault detection, and portability.

The analogies with the Arigatoni model are in the *Community Scheduler Framework* component and the *Web Service Grid Resource Allocation and Management* of the toolkit concerning the resource discovery, and the *Globus Teleoperations Control Protocol* to allow units to cooperate (analogy with our *ad hoc* protocol). However, Globus does not target the full generality of Arigatoni, thanks to its generic, resource discovery algorithm that can be also suitable for pervasive compiting in addition to pure Grid-oriented applications.

Promoted by Sun, the JXTA [15] technology is a set of open peer-to-peer protocols that enable any device to communicate, collaborate and share resources. After a peer discovery process, any peer can interact *directly* with other peers. Hence, the overlay network of peers induced by the JXTA technology is *flat*.

Moreover, the main concern of the Arigatoni is the design of protocols for generic resource discovery, and intermittent participation, while the main concern of the JXTA technology is to offer some tools to implement a P2P model.

In addition, the Arigatoni focuses on the evolution/devolution of colonies and the mechanism of resource discovery, while JXTA technology allows peers to communicate using an already existing overlay network of peers. Arigatoni's aim is the dynamicity of the overlay network while JXTA's is the freedom of connectivity between peers. Finally, peers in the JXTA architecture come with their proper JXTA-ID (logical JXTA peers addressing) while Arigatoni relies on the more conventional IP addresses.

Pub/sub [10] is a communication paradigm for asynchronous dissemination of information. Consumers subscribe to the system (typically called the *Notification Service*) to specify the type of information that they are interested in. Producers publish data to the system. The notification service disseminates the data to all (if possible) the consumers that are interested in receiving it, according to the data *and* the interests declared by the consumers.

Many pub/sub systems have been developed recently, such as XNet [8,7], Siena [4] or IBM Gryphon [2]. In [14], the authors propose to adapt the Siena publish/subscribe system to achieve Gnutella-like resource discovery. Their work resembles ours in the sense that Arigatoni is also inspired by the pub/sub paradigm. However, in [14], resource discovery is achieved by publishing queries to the notification service. In contrast, Arigatoni implements its own resource discovery algorithm, especially designed for generic and scalable resource lookup.

Conclusions. In this paper, we describe the V2 of the Arigatoni's generic resource discovery protocol. The first version RDP V1 permitted to ask for *one service* at the time. The new improved protocol RDP V2 presented in this paper allows for *multiple instances*, the latter point being a non-trivial improvement. Other main achievements are the complete decoupling between the different units in the system, and the encapsulation of resources in local colonies, which enable Arigatoni to be potentially scalable to very large and heterogeneous populations.

The reliability of the RDP V2 itself, although desirable, is of lesser importance, given the fact that service provision is not guaranteed at all in Arigatoni (indeed it is not a requirement). In other words, when a GC issues a service request, it is possible that no individuals are found for some of the services included in the request. This happens, for example, if those services have not been declared by any GCs in the system, or if all the GCs that have declared themselves as potential individual refuse to serve them.

However, at the cost of memory and bandwidth requirements, it is still possible (future work) to implement *reliable* resource discovery by using a reliable transmission protocol (*e.g.* TCP), an applicative *acknowledgment scheme* in combination with a retransmission buffer, and persistent data storage, and leader's replication.

The subscription mechanisms of classical tree-based pub/sub systems [7,5,4] can be used for the maintenance and update of consistent routing tables. Furthermore, as for the reliability of subscription advertisement, we can adapt the reliability mechanisms described in [8] to allow Arigatoni to be fault-tolerant or to adapt to dynamic topology changes due to the intermittent participation of individuals [9].

We are currently still improving Arigatoni with several new features, such as the possibility to embed services in strong conjunctions (*i.e.*, the services in a strong conjunction should be provided by the *same* GC). We are also working on the implementation of a real prototype and the subsequent deployment on the PlanetLab experimental platform [16], and/or on GRID5000, the experimental platform available at the INRIA [17].

As part of our ongoing research, we are also working on a more complete statistical study of our system, based on more elaborate statistical models and realistic assumptions, as well as the possibility to include some hierarchical DHT in addition to the routing tables. The possibility to change the Arigatoni topology from a hierarchical tree to a graph is also intriguing.

Acknowledgment. We would like to thank Luc Hohwiller for a careful reading of the paper, and the anonymous referees for the very useful comments.

References

1. AEOLUS. Deliverable D2.1.1: Resource Discovery: State of the art survey and Algorithmic Solutions, 2006. http://aeolus.ceid.upatras.gr.
2. G. Banavar, T. Chandra, B. Mukherjee, J. Nagarajarao, R.E. Strom, and D.C. Sturman. An efficient multicast protocol for content-based publish-subscribe systems. In *Proc. of ICDCS*, 1999.
3. D. Benza, M. Cosnard, L. Liquori, and M. Vesin. Arigatoni: A Simple Programmable Overlay Network. In *Proc. of John Vincent Atanasoff International Symposium on Modern Computing*, pages 82–91. IEEE, 2006.
4. A. Carzaniga, D.S. Rosenblum, and A.L. Wolf. Design and Evaluation of a Wide-Area Event Notification Service. *ACM TOCS*, 19(3), 2001.
5. R. Chand. *Large scale diffusion of information in Publish/Subscribe systems*. PhD thesis, University of Nice-Sophia Antipolis and Institut Eurecom, 2005.
6. R. Chand, M. Cosnard, and L. Liquori. Resource Discovery in the Arigatoni Overlay Network. In *I2CS: International Workshop on Innovative Internet Community Systems*, volume LNCS. Springer, 2006. To appear. Also available as RR INRIA 5928.
7. R. Chand and P. Felber. A scalable protocol for content-based routing in overlay networks. In *Proc. of NCA*, 2003.
8. R. Chand and P. Felber. XNet: A Reliable Content-Based Publish/Subscribe System. In *SRDS 2004, 23rd Symposium on Reliable Distributed Systems*, 2004.
9. M. Cosnard, L. Liquori, and R. Chand. Virtual Organizations in Arigatoni. *DCM: International Workshop on Developpment in Computational Models. Electr. Notes Theor. Comput. Sci.*, 2006. To appear.
10. P. Th. Eugster, P. Felber, R. Guerraoui, and A.M. Kermarrec. The many faces of publish/subscribe. *Computing Survey*, 35(2):114–131, 2003.
11. M. J. Freedman and D. Mazières. Sloppy Hashing and Self-Organizing Clusters. In *Proc. of IPTPS*, pages 45–55, 2003.
12. P. Ganesan, P. Krishna, and H. Garcia-Molina. Canon in g major: Designing DHTS with Hierarchical Structure. In *Proc. of ICDCS*, 2004.
13. Globus Alliance. Globus Home Page. http://www.globus.org/.
14. D. Heimbigner. Adapting publish/subscribe middleware to achieve gnutella-like functionality. In *Proc. of SAC*, pages 176–181, 2001.
15. JXTA Community. JXTA Home Page. http://www.jxta.org/.
16. Planet Lab Consortium. Planet Lab Home Page, 2006. http://www.planet-lab.org/.
17. The Grid 5000 Consortium. Grid 5000 Home Page, 2006. http://www.grid5000.org/.
18. E.W. Zegura, K. Calvert, and S. Bhattacharjee. How to Model an Internetwork. In *Proc. of INFOCOM*, 1996.

A The *Filter* and *Route* Algorithms

Algorithm 4. *Filter*(S, P_{from})

1: **for each** entry $T[S'] = [(P_k, n_k)]^{k=1\ldots C}$ in T **do**
2:    **if** S' *matches* S **then**
3:       **for each** $j \le C$ such that $P_j \ne P_{\text{from}}$ **do**
4:          $SendList\{\text{id}\}\{S\}\{P_j\} \leftarrow SendList\{\text{id}\}\{S\}\{P_j\} + n_j$ // (P_j, m) becomes $(P_j, m + n_j)$
5:       **end for**
6:    **end if**
7: **end for**
8: **return** $SendList\{\text{id}\}\{S\}$

Filter builds the *SendList* corresponding to service S (and interaction with identifier id), *i.e.*, the list of P's children that are potentially capable of serving S. The function parses all the services in the routing table accordingly.

Algorithm 5. *Route*$(P_{\text{from}}, S, n, search\_mode)$

1: $remaining \leftarrow n$
2: $RoutingList \leftarrow \emptyset$
3: **for each** $(Q, f) \in SendList\{\text{id}\}\{S\}$ **do**
4:    **if** $Q = P_{\text{from}}$ or $Q = Path\{\text{id}\}$ **then**
5:       continue   // *Go to next iteration in loop*
6:    **end if**
7:    **case** search_mode **is**
       "*exhaustive*" :
8:       **if** $n \ge f$ **then**   // *More instances asked than offered*
9:          Insert $Q : (S, f)$ in $RoutingList$
10:         $remaining \leftarrow remaining - f$
11:         $downstream\{\text{id}\}\{S\} \leftarrow downstream\{\text{id}\}\{S\} + f$
12:         Remove (Q, f) from $SendList\{\text{id}\}\{S\}$
13:       **else**   // *More instances offered than asked*
14:         Insert $Q : (S, n)$ in $RoutingList$
15:         $remaining \leftarrow 0$
16:         $downstream\{\text{id}\}\{S\} \leftarrow downstream\{\text{id}\}\{S\} + n$
17:         $f \leftarrow f - n$
18:       **end if**
       "*selective*" :
20:       **if** $remaining \ge f$ **then**   // *More instances asked than offered*
21:         Insert $Q : (S, f)$ in $RoutingList$
22:         $remaining \leftarrow remaining - f$
23:         Remove (P, f) from $SendList\{\text{id}\}\{S\}$
24:       **else**   // *More instances to offer than asked*
25:         Insert $Q : (S, remaining)$ in $RoutingList$
26:         $f \leftarrow f - remaining$
27:         $remaining \leftarrow 0$
28:       **end if**
29:       **if** $remaining = 0$ **then**   // *No more instances to treat*
30:         break   // *Break loop*
31:       **end if**
32:    **end case**
33: **end for**
34: **return** $(RoutingList, remaining)$

Route builds *RoutingList*, *i.e.*, the list of neighbors that will be sent a particular service, according to the selected search mode; it has the form $\{(P_i : (S, n_i))\}^{i=1\ldots h}$ that is neighbors P_i will receive a request for n_i instances of S. The function also returns the remaining instances for which no servant has been found.

An Effective Multi-hop Broadcast in Vehicular Ad-Hoc Network

Tae-Hwan Kim, Won-Kee Hong*, and Hie-Cheol Kim

Department of Information and Communication Engineering,
Daegu University, Gyeong-San, Gyeong-Buk, 712-714, Korea
{thkim76, wkhong, hckim}@daegu.ac.kr

Abstract. Multi-hop broadcast protocols in vehicular ad-hoc network (VANET) require more prompt message dissemination than traditional broadcast protocols because they mainly deal with vital data involved in driver safety. In this paper, a time reservation-based relay node selection algorithm is proposed in order to achieve immediate message dissemination. All nodes in the communication range of a relay node randomly choose their waiting time within a given time-window. The time-window range is determined by a distance from a previous relay node and a reservation ratio of the time-window. A node with the shortest waiting time is selected as a new relay node. The experimental results show that the proposed algorithm has a shorter end-to-end delay time than the distance-based relay node selection algorithm no matter how node density varies. In particular, when the node density is low, the proposed algorithm has a 25.7% shorter end-to-end time and a 46% better performance in terms of the compound metric than the distance-based relay node selection algorithm.

Keywords: Vehicular Ad-hoc Network, Multi-hop Broadcast,Inter-vehicle Communication, Data Dissemination.

1   Introduction

Vehicular Ad-hoc Network (VANET) is temporarily established through a wireless connection between moving vehicles without any additional infrastructures like MANET[1]. However, when compared to MANET, VANET has several characteristics such as frequent changes of network topology, node[1] density, high mobility and frequent network fragmentation[2]. Therefore, the traditional network protocols cannot be directly applied to VANET since these characteristics are not considered. In VANET, it requires directional message dissemination because the nodes move along the road. In addition, rapid message dissemination should be guaranteed by broadcast protocols in VANET since it mainly deals with vital data involved with driver safety.

---

* Corresponding author.
[1] Node stands for vehicles in this paper.

P. Lukowicz, L. Thiele, and G. Tröster (Eds.): ARCS 2007, LNCS 4415, pp. 112–125, 2007.

The message dissemination schemes for VANET can be categorized into flooding-based[3], cluster-based[4], table-based[5] and distance-based[6] broadcast methods. The flooding-based broadcast scheme has excellent message arrival rates even with a high mobility of nodes, but when the node density grows, the bandwidth consumption is dramatically increased. The table-based and the cluster-based broadcast schemes can lead to reduced level of performance if there are increasing control message exchanges between the nodes. The distance-based broadcast scheme has better performance than other schemes in VANET as it has less network traffic and an end-to-end delay. In this scheme, the relay node that has the role of delivering messages is selected by the distance-based relay node selection (DBRS) algorithm. The chosen relay node with the shortest waiting time is located at the border of the transmission range of the previous relay node. However, the relay node takes a longer waiting time, if it is selected among nodes that are not placed at the border of the transmission range in the low node density network. Consequently, the end-to-end delay is lengthened because the relay nodes have to waste some time untill the waiting time expires.

In this paper, we propose a time reservation-based relay node selection (TRRS) algorithm. It has shorter end-to-end delay time and is desirable regardless of the node density. TRRS uses a time-window. The size of the maximum time-window is inversely proportional to the distance from the previous relay node. In TRRS, a part of the time-window is reserved so that the farthest node from the previous relay node is to be selected as the next relay node with the shorter waiting time than itself. Each node within transmission range of a relay node randomly chooses its waiting time within the given time-window range. The time-window is longer than the reserved time range and shorter than the maximum time-window. To avoid multiple reception of broadcasting messages of nodes in the broadcast region, TRRS prevents a node, which received many duplicate broadcast messages from previous relay nodes, to be selected as the next relay node. The nodes that have duplicate broadcast messages have a higher reservation ratio of the time-window. Therefore, the farther the node from the previous node is, the narrower the time-window. This means that a node farther from the previous relay node is selected as the next relay node because it can take a shorter waiting time within a narrowed time-window than the closer nodes. TRRS can minimize end-to-end delay time because it takes a shorter waiting time than DBRS even though the nodes are located far from the border of a transmission range with a low node density. The experimental results show that the broadcast protocols using TRRS have a shorter end-to-end delay time than the distance-based broadcast protocol using DBRS regardless of the node density. In particular, when the node density is low, TRRS has a 25% shorter end-to-end delay time and a 46% better performance than DBRS in terms of the compound metric.

This paper is organized as follows. Section 2 introduces the related work. The TRRS algorithm is explained in section 3. The result of the experiments and performance evaluation is described in section 4. Finally, section 5 contains concluding remarks.

2   Related Work

As shown in Table 1, VANET has several characteristics which are different from MANET. There are frequent changes of node density and network topology, high mobility and frequent network fragmentation. Due to these characteristics, the communication link between nodes is frequently broken and it is difficult to disseminate messages to neighboring nodes effectively. Hence, we cannot expect that the network protocols proposed for MANET will operate well for VANET[1][2]. Moreover, the performance of the multi-hop routing algorithm for MANET declines over 3 4 hop in VANET[7]. That is because the lifetime of the communication link between nodes is very short and the end-to-end delay and the network traffic are increased by increasing control message exchange for a new routing path discovery. The routing path discovery is mainly established by the flooding scheme. However, the flooding causes high bandwidth consumption and a long end-to-end delay like broadcast storm problem[8] if the node density is high. To improve the performance of the network protocols proposed for VANET, it is necessary to equip the effective message dissemination protocol with a short end-to-end delay and low network traffic.

Table 1. Comparison of MANET and VANET

Item	MANET	VANET
Mobility	Low (walking speed)	High(up to $200km/h$)
Cost of Production	Cheap	Expensive
Network topology change	slow	Fast and frequent
Node density	sparse	dense and frequently variable
Transmission range of nodes	up to $100m$	up to $500m$
Bandwidth	several hundreds kps	several thousands kbps
Life time of nodes	Depending on power resource	depending on vehicle's life time
Computation capability	8~16 bit CPU	Over 32 bit CPU
Addressing scheme	Attribution-based addressing	Location-based addressing or Unique ID
Ability of multi-hop routing	Available (Depending on the scale of local network)	Weakly available (Depending On node density and limited to 3~4 hop)
Reliability	medium	Very high
Position acquisition	Triangulation using RSSI and ultra sonic	GPS, RADAR and Vision
Moving patter of nodes (speed and direction)	Too random	Mostly regular

Lots of broadcast protocols for VANET are found in the literature as shown in Figure 1. They can be classified into four categories based on the roles of nodes and the relay node selection methods such as flooding-based, cluster-based, table-based and distance-based broadcast schemes. Most of these protocols assume that vehicles can ascertain their positions from the GPS.

The flooding-based scheme tries to suppress the number of re-broadcasting nodes but still has heavy network traffic and long delivery latency because of

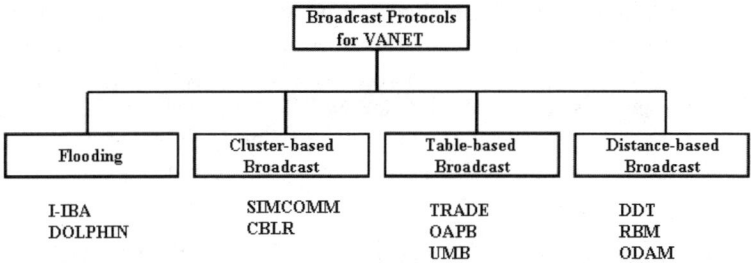

Fig. 1. Broadcast protocols for VANET

the broadcast storm problem. The representative flooding schemes are I-IBA[3], DOLPHIN[9].

The cluster-based broadcast scheme divides the road into several clusters and selects a cluster head among nodes within a cluster. Only the cluster head is eligible to disseminate messages. This scheme performs well when changes in network topology is minimal. However. it suffers from heavy network traffic and a long latency time in high node mobility since it is necessary to reorganize the member of a cluster and to reelect a cluster head more frequently. SIMCOMM[4], CBLR[10] are the cluster-based broadcast schemes.

In the table-based broadcast scheme, every node in the network maintains a list of neighbor nodes that is periodically updated through a query-reply mechanism. The relay node is determined by the previous relay node. This scheme provides better performance in terms of network traffic and end-to-end delay when the nodes are moving slowly. However, if the network topology changes frequently, the performance declines sharply and thus, it is not suitable for a network which exhibits high node mobility. When node mobility increases, the period of control message exchange between the nodes gets shorter which wastes network bandwidth and increase end-to-end delay. The table-based broadcast schemes are TRADE[5], OAPB[11], and UMB[12].

The distance-based broadcast scheme allows only one node to be involved at a time in relaying a broadcast message to decrease network traffic and end-to-end delay. The DBRS is a representative relay node selecting algorithm in distance-based broadcast scheme. The relay node is determined by the distance from the previous relay node. In other words, every intermediate node in transmission range of the previous relay node is eligible to re-broadcast a message and has to hold the messages for a given waiting time before re-broadcasting it. The waiting time is computed by the distance from the previous relay node. Because the waiting time of a node is different from one another, only the node with the shortest waiting time is assured of re-broadcasting the message. The farther the node from the previous relay node, the shorter is the waiting time. There-fore, the closest node to the border of the transmission range of the previous relay node is selected as the next relay node. There are DDT[6], RBM[13] and ODAM[14] in the distance-based broadcast protocol. If a relay node is a border node, the shortest waiting time will be spent so that the lowest network traffic and the shortest end-to-end delay time can be guaranteed in the distance-based broadcast protocol. The border node means a node that is placed at the border of transmission range of a previous relay node. However, that is not always the case. If a relay node is not at the border, it will hold back the message unnecessarily for a longer time because the waiting time is determined only by the distance. Especially, in case of low node density, this often happens.

3   Time Reservation-Based Relay Node Selecting Algorithm

As mentioned in section 2, the relay node in DBRS is determined only by the distance from the previous relay node. Figure 2 illustrates the DBRS, where n_i stands for an intermediate node within transmission range of the previous relay node. If n_1, n_2 and n_3 are located at a distance d_1, d_2 and d_3 from the previous relay node, each of them has to spend its given waiting time inversely proportional to the distance from the previous relay node before disseminating a message. The waiting time, denoted by DWT_i, of an intermediate node is represented by the following equation:

$$DWT_i = DWT_{max} \times (1 - d_i/R) \tag{1}$$

where d_i is a distance between an intermediate node and the previous relay node, DWT_{max} is the maximum predetermined waiting time and R is the transmission distance of the relay node. According to the equation (1), the best performance is given only when all relay nodes are located at the border of the transmission range of the previous relay node. However, each relay node cannot be guaranteed to be at the border in the VANET, if it is not in the very high node density. If a relay node is not at the border, it has to spend a given waiting time wastefully. Consequently, this algorithm is very simple, but the overall end-to-end delay time is increased because the waiting time is blindly determined by the distance.

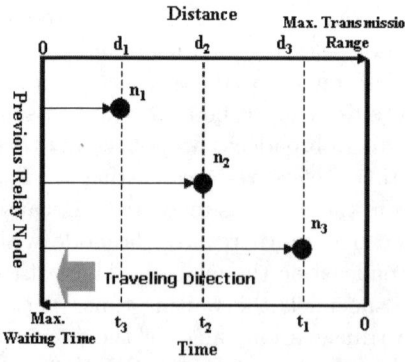
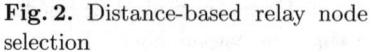

Fig. 2. Distance-based relay node selection

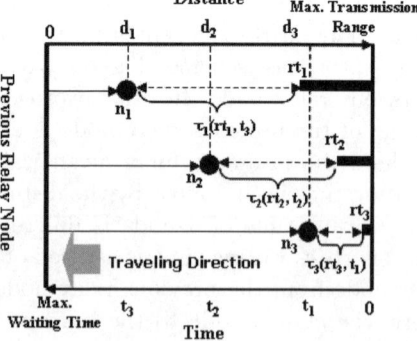

Fig. 3. Time reservation-based relay node selection

We propose a new relay node selection algorithm, called TRRS, to achieve low end-to-end delay time as well as low network traffic in the VANET. This approach is similar to the DBRS in that the waiting time for re-broadcasting is based on the distance from the previous relay node. In the TRRS, however, a relay node can choose its waiting time in a given time-window so that a relay node which is not at the border is allowed to have the waiting time of the border node. As shown in Figure 3, the intermediate nodes, n_1, n_2, and n_3 have maximum time-windows that are inversely proportional to the distance from the previous relay node t_1, t_2 and t_3, respectively. Each node reserves a proportional time period rt_1, rt_2 and rt_3 in the given time-windows in order to let the relay node that is farther from the previous node with a shorter waiting time be selected next. Therefore, the range of the time-window τi is greater than rt_i and lower than or equal to t_i. Each node has a time-window τi with a different lower limit and upper limit and it randomly takes the waiting time within the given time-window range. Thus, the waiting time of an intermediate node in TRRS is always lower than or equal to the one in DBRS. It means that the TRRS can reduce the unnecessary time spent by a relay node. In the worst case, it may happen that a node close to the broadcasting node has a shorter waiting time than the remote one. However, the rate is very low because the closer nodes have a wider range of time-window than the remote ones. Therefore, closer node's possibility of selecting a shorter waiting time is relatively low in comparison to the remote one.

Figure 4 illustrates the worst case scenario that may happen in TRRS and its prevention scheme. To avoid the worst case scenario in TRRS, the node that has many duplicated message receptions has a higher reservation ratio of the time-window. In other words, when the node is closer to the previous relay node, it will receive more duplicated message from previous relay nodes than other intemperate nodes. Therefore, the node that is closer to the border of the transmission range and has a few duplicated message reception is selected as the next relay node in TRRS. As shown in Figure 4, let's assume that the original dissemination node RN_1 has to broadcast a message to n_4. It requires

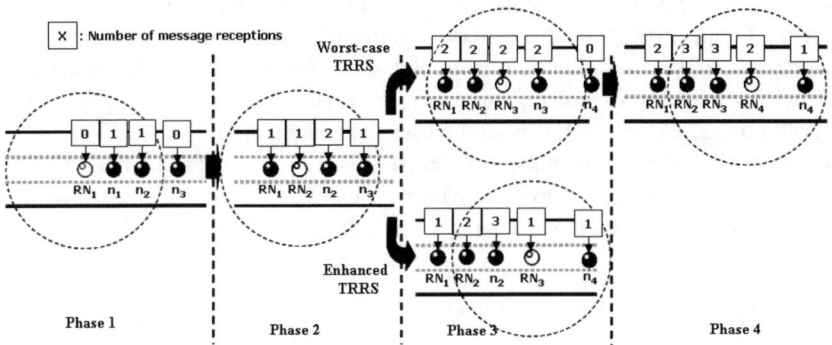

Fig. 4. Prevention of worst-case scenario in TRRS

multi-hop broadcasting because it cannot reach n_4 by a single-hop and it needs to disseminate messages by several relay nodes. In the worst case of TRRS, n_1, n_2 and n_3 are sequentially selected as the relay node like domino effect. In the worst case, the cost to deliver a message to n_4 includes 4 relay nodes and 11 packets. The enhanced TRRS (ETRRS), on the other hand, prevents a node that has many duplicated message to be selected as a new next relay node. In phase 2 in Figure 4, n_2, in which the number of duplicated message receptions is two, cannot achieve a shorter waiting time than n_3 which has just one message reception because the reservation ratio of the time-window of n_2 is higher than n_3. Consequently, n_2 cannot be selected as a next relay node of RN_2. In this case, 3 relay nodes and 8 packets are consumed and the costs are less than the worst case. In this paper, if a node received a message only one time, it caused the reservation ratio of the time-window to be under fifty percent and if over one time, it caused it to be over fifty percent. In TRRS, the waiting time, DWT_i of an intermediate node can be defined as follows:

$$DWT_i = \{DWT : rt_i \leq \tau_i \leq tw_{max}\}$$

$$rt_i = tw_{max} \times \rho, \qquad \rho(c_i) = \begin{cases} 0 \leq \rho \leq 0.5 & c = 1 \\ 0.5 < \rho \leq 1 & c > 1 \end{cases} \qquad (2)$$

$$tw_{max} = DWT_{max} \times (1 - d_i/R)$$

where rt_i denotes the maximum reserved time-window which is a lower limit of time-window τ_i, the reservation ratio of the time-window is ρ, the number of the duplicated message receptions is c and a upper limit of time-window τ_i is tw_{max}.

The TRRS requires every node to know the position of the original broadcasting node as well as the previous relay node in order to compute its waiting time. The information is contained in the broadcast message. The position of the original broadcasting node is used to determine when the message relay should be finished. As soon as it receives the broadcast message from the previous relay node, an intermediate node within communication range checks if the message is new. If not, the message will be discarded. Otherwise, intermediate node pulls information about positions of the original broadcasting node and the previous relay node out of a new broadcast message and obtain its positions from the GPS receiver. Then, an intermediate node calculates the distance from the original broadcast node and the distance from the previous relay node. An intermediate node with the shortest waiting time based on the equation (2) is selected as a new relay node. A new relay node changes the field of the previous relay node position in the broadcast message with its position and then disseminates it. A re-broadcasting message is no longer made when the broadcast zone exceeds the predetermined dissemination range.

4   Performance Evaluation

In this section, to evaluate and analyze the performance of TRRS in VANET, we conducted a simulation of the emergency warning service[14] which is one of the applications in VANET. The emergency warning service has to promptly offer emergency warning messages (EWM) such as icy roads, falling rocks, fog, construction, blind spots or vehicle accidents to vehicles within a risk zone. The simulation parameters are presented in Table 2. It is assumed that all of the nodes in the network can obtain their positions from GPS and use IEEE 802.11 DCF(Distribution Control Function) MAC. The length of EWM is 250 bytes and it consists of four fields such as an original broadcasting node position, a previous relay node position, a delivery range and emergency contents. We do not consider lane change and the overtaking of nodes. The evaluation metric for TRRS is the end-to-end (ETE) delay and the network traffic. The simulation was performed one hundred times and all of the simulation results are average values.

Table 2. Simulation paremeters

Network Environment		Road Environment	
Parameter	Value	Parameter	Value
Transmission range	150 m	Length of road	7 km
Packet length	250 $bytes$	Width of a lane	3.6 m
Channel Bandwidth	2 $Mbps$	Road Direction	One way
propagation Delay of a packet	0.125 μs	Number of lanes	3
Computation time	1 ms	Average speed of node	100 km/h
DWT_{max}	10 ms	Traffic density	13.33 $vehicles/lane/km$
EWM dessemination range	5 km	Length of a vehicles	4m

Table 3 shows the network traffic and the end-to-end delay time of TRRS when the reservation ratio of the time-window varies. The experimental result shows that when the reservation ratio of the time-window is 50%, it had the lowest network traffic and when the ratio is 0%, it shows the shortest end-to-end delay time. The Compound metric is a function of the end-to-end delay time and the network traffic. However, when the reservation ratio of the time-window is 10%, it had better performance in terms of the compound metric than any others. Because of the higher reservation ratio of the time-window, the network traffic rate becomes lower; but it cannot have the shortest waiting time due to the narrowed time-window range. In this section, when the reservation ratio of the time-window in TRRS is zero percent, it is denoted as RBRS(Range Based Relay node Selection). The the reservation ratio of time-window in TRRS and ETRRS is ten percent.

We assumed that network fragmentation does not occur in this experiment. When the informed rate reached over 97%, fragmentation increased only slightly. The results of experiments at DBRS, RBRS, TRRS and ETRRS show that the relationship between the informed rate and the node density is almost the same. Consequently, we conducted an experiment with node densities and transmission ranges that shows 97% of the informed rate.

Table 3. Efficiency of TRRS when the reservation ratio of the time window varies

Reservation ratio	0%	10%	20%	30%	40%	50%
Network traffic (Packets)	672	605	594	585	577	575
ETE Delay (ms)	61.421	62.061	65.438	68.495	70.596	73.992
Compound Metric	0.856	0.941	0.909	0.882	0.867	0.830

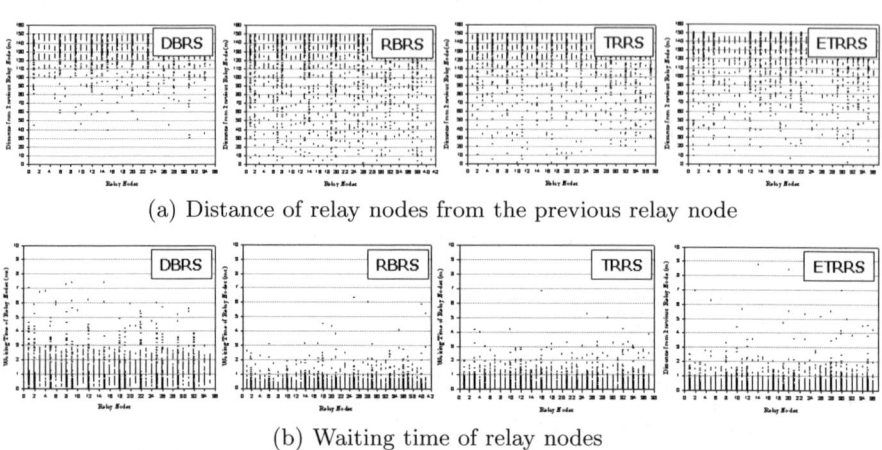

(a) Distance of relay nodes from the previous relay node

(b) Waiting time of relay nodes

Fig. 5. Distance and waiting time distribution of relay nodes

The distance from the previous relay node of each relay node and the waiting time are presented in Figure 5, when the informed rate is 97% and the node density is 18.33 vehicles/lane/km. The X-axis denotes the series of the relay nodes in Figure 5 and Y-axis denotes the distance from the previous relay node in Figure 5(a) and the waiting time of each relay node in Figure 5(b). As shown in Figure 5. the relay nodes in DBRS are distributed closer to the border of the transmission range of its previous relay node than other schemes because the node that is located the farthest from the previous relay node is only selected as the relay node. The waiting time of the relay nodes in DBRS indicates that it is inversely proportional to the distance.

The average results of the experiments in Figure 5 are presented in Table 4. DBRS has the longest waiting time although it indicates the lowest number and the farthest distance of the relay nodes. RBRS shows the shortest waiting time

Table 4. Comparison of the number of relay nodes, waiting time and distance from the previous relay node

	DBRS	RBRS	TRRS	ETRRS
Number of relay nodes	35	41	37	36.5
Waiting time(ms)	1.26	0.348	0.531	0.535
Distance(m)	131.09	112.95	123.64	125.32

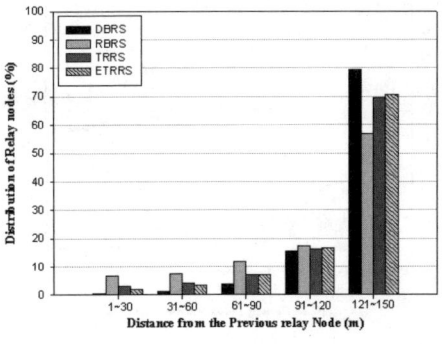
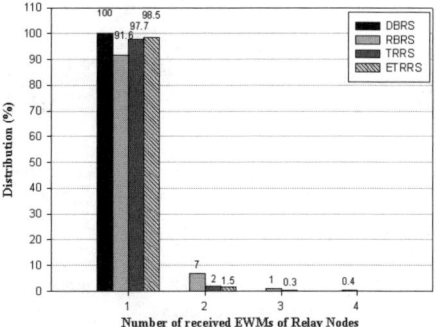

Fig. 6. Distribution of relay nodes by distance

Fig. 7. Distribution of relay nodes by the number of received EWMs

and distance than any other node but has more relay nodes because the relay nodes are widely distributed within the transmission range of the previous relay node. On the other hand, TRRS and ETRRS have short waiting times as RBRS and farther distance as DBRS. The ETRRS results are similar to those of TRRS but ETRRS showed some better results in terms of the number of relay nodes and the distance than TRRS because ETRRS prevents the worst case scenario.

Figure 6 describes the distribution of the relay nodes based on the distance from each previous relay node respectively. In DBRS, 79.4% of relay nodes are over 120m far from the previous relay node. On the other hand, 43.1% of the relay nodes in RBRS are located within 120m because the node can be selected as the next relay node even though it is not the farthest node from the previous relay node. In TRRS and ETRRS, however, 69.5% and 70.8% of the relay nodes are distributed over 120m because both of them can select the relay nodes that are farther from the previous relay node by means of the time-window reservation. In particular, the relay nodes in ETRRS are distributed more distantly than TRRS. Since the nodes that have many duplicated message receptions are distributed near to the previous relay node, ETRRS tries not to select these nodes as the next relay node.

Figure 7 shows the distribution of the received EWMs of the relay nodes. Each relay node in DBRS has only one message reception. Among the nodes that have more than two duplicated message receptions, 8.4% of the relay nodes are selected in RBRS and 2.3% in TRRS. ETRRS, however, shows 98.5% of the relay nodes which are similar to those of DBRS and just 1.5% of the relay nodes are selected among the nodes that have twice the number of duplicated message receptions.

The number of the received EWMs of nodes in the network is presented in Figure 8 as the node density varies. Ideally, the minimum number of receptions is two because one is from the following node and the other is from the ahead node. The experimental result shows that the number of the received EWMs decreased scarcely when the informed rate is higher than 97%. This is because the probability of its being located in the margin of the transmission range

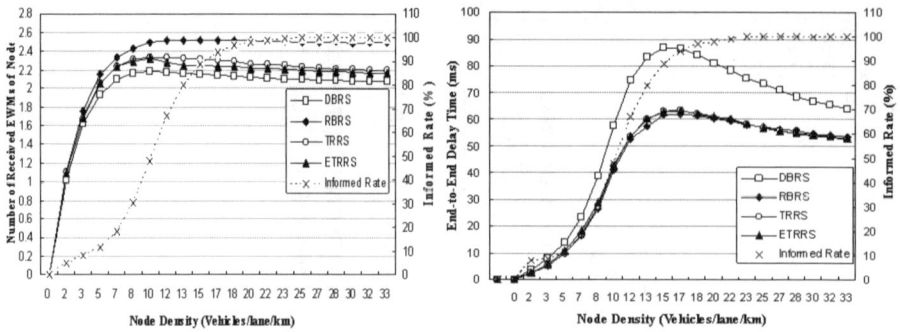

Fig. 8. The number of received EWMs of nodes in the network

Fig. 9. The end-to-end delay time

increases. Consequently, the number of relay nodes is minimized when node density is high, followed by lower number of duplicated EWM receptions. DBRS indicates the lowest number of duplicated EWM receptions. RBRS is 17.8% higher than DBRS. On the other hand, ETRRS is 12% lower than RBRS and 2.6% lower than TRRS, but 4.5% higher than DBRS.

Figure 9 illustrates the end-to-end delay time when the node density varies. The end-to-end delay time slowly decreased when the node density increased because it required fewer relay nodes. The experimental result indicated that DBRS showed the longest end-to-end delay time. RBRS, TRRS and ETRRS, on the other hand, indicated a much shorter end-to-end delay time than DBRS regardless of the node density. When the informed rate was 97%, ETRRS had 25.7% of the shorter end-to-end delay time than DBRS and was 1.7% and 0.7% longer than RBRS and TRRS, respectively. For over 97% of the informed rate, however,

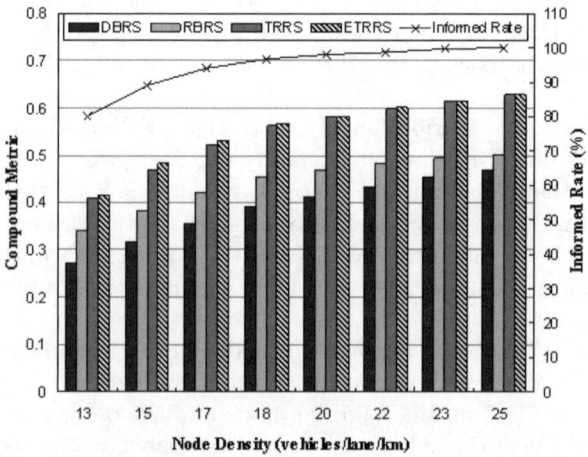

Fig. 10. Compound metric

it had the shortest end-to-end delay time. The RBRS shows almost the same result compared to TRRS and ETRRS even though it has the shortest waiting time. The reason is that RBRS requires more relay nodes than the others do.

The result of the compound metric is presented in Figure 10 when the informed rate is from 80% to 100%. ETRRS shows the highest performance regardless of the node density and DBRS indicates the lowest performance. ETRRS is 47% better than DBRS and 26% better than RBRS. The performance of TRRS and ETRRS is almost the same, but ETRRS is $1 \sim 1.7\%$ better than TRRS under the 97% informed rate.

In conclusion, DBRS shows the lowest network traffic, but a longer end-to-end delay time than the others do. RBRS has a shorter waiting time than the others, but higher network traffic because it requires more relay nodes than the others. TRRS and ETRRS have better performance than the others because they have the characteristic of having low network traffic which is similar to DBRS and an end-to-end delay time as RBRS. ETRRS performs better than the TRRS because it prevents the worst case scenario.

5 Conclusion

VANET is a temporarily established network through a wireless connection of moving vehicles without the aid of infrastructure. It provides drivers with useful traffic information as well as an extended range of awareness for safe driving. Therefore, the network protocol for VANET should be designed with all the characteristics because it has several challenges compared to MANET, such as frequent changes of network topology and node density, high mobility, and frequent network fragmentation. In particular, VANET requires a prompt message dissemination protocol since it mainly deals with vital data involved in driver safety.

In this paper, a TRRS algorithm is proposed to minimize the end-to-end delay time for a prompt message broadcasting in VANET. In TRRS, an intermediate node is allowed to have the maximum size of a time-window which is inversely proportional to the distance from the previous relay node. A part of the time-window is reserved for a node that is close to the border of the transmission range of the previous relay node to be selected as the next relay node. The waiting time of intermediate nodes within a communication range is randomly selected within a given time-window range that is longer than the reserved time range and shorter than the maximum time-window. To minimize multiple receptions of a broadcast message, TRRS prevents the worst case scenario. A node that has many duplicate broadcast messages has higher reservation ratio of the time-window. It means that the node cannot take the shortest waiting time. Therefore, the farthest node from the previous node has a higher probability to be selected as the relay node than the closer nodes because it takes a shorter waiting time. In addition, TRRS can minimize the an end-to-end delay even though the relay nodes are not located at the border of the transmission range of a previous relay node. The experimental results show that TRRS always has shorter end-to-end delay time than DBRS regardless of the node density, and it like DBRS has low network traffic. Especially, when the node density is low, TRRS has a 25.7% shorter end-to-end delay time and a 46% better performance than DBRS in terms of the compound metric.

We believe that the proposed algorithm is also useful for a rapid routing path discovery protocol as well as for a prompt emergency message dissemination protocol in VANET.

Acknowledgment

This research was supported by the Daegu University Research Grant, 2006.

References

1. J.J. Blum, A. Eskandarian,L.J. Hoffman, "Challenges of inter-vehicle ad hoc networks", Intelligent Transportation Systems, IEEE Tran. on Vol. 5, Issue 4, pp. 347-351, Dec. 2004.
2. M. Torrent-Moreno, M. Killat, H. Hartenstein, "The challenges of robust inter-vehicle communications", Vehicular Technology Conf. 2005. VTC-2005-Fall. 2005 IEEE 62nd Vol. 1, pp. 319-323, Sept. 2005.
3. S. Biswas, R. Tatchikou, F. Dion, "Vehicle-to-vehicle wireless communication protocols for enhancing highway traffic safety", Communications Magazine, IEEE Vol. 44, Issue 1, pp. 74-82, Jan. 2006.
4. M. Durresi, A. Durresi, L. Barolli, "Sensor inter-vehicle communication for safer highways", Advanced Information Networking and Applications, 2005. AINA 2005. 19th Int. Conf. on Vol. 2, pp. 599-604, Mar. 2005.
5. S. Min-Te, F. Wu-Chi, L. Ten-Hwang, K. Yamada, H Okada, K. Fujimura, "GPS-Based Message Broadcasting for Inter-vehicle Communication", Parallel Processing, 2000. Int. Conf., pp. 279-286, Aug. 2000,
6. S. Min-Te, F. Wu-Chi, L. Ten-Hwang, K. Yamada, H. Okada, K. Fujimura, "GPS-based message broadcast for adaptive inter-vehicle communications", Vehicular Technology Conference, 2000. IEEE VTS-Fall VTC 2000. 52nd Vol. 6, pp. 2685-2692, Sept. 2000.
7. M.M. Artimy, W. Robertson, W.J.Phillips, "Connectivity in inter-vehicle ad hoc networks", Electrical and Computer Engineering, 2004. Canadian Conference on Vol. 1, pp. 293-298, May 2004.
8. S. Ni, Y. Tseng, Y. Chen, J. Sheu., "The Broadcast Storm Problem in a Mobile Ad Hoc Network", In ACM MOBICOM '99, pp. 151-162, Aug. 1999.
9. K. Tokuda, M. Akiyama, H. Fujii, "DOLPHIN for inter-vehicle communications system", Intelligent Vehicles Symposium, 2000. IVS 2000. Proceedings of the IEEE, pp. 504-509, Oct. 2000.
10. R.A. Santos, R.M. Edwards, A. Edwards, "Cluster-based location routing algorithm for vehicle to vehicle communication", Radio and Wireless Conference, 2004 IEEE, pp. 39-42, Sept. 2004.
11. H. Alshaer, E. Horlait, "An optimized adaptive broadcast scheme for inter-vehicle communication", Vehicular Technology Conf., VTC 2005-Spring. 2005 IEEE 61st Vol. 5, pp. 2840-2844, May 2005.

12. K. Gokhan, E. Eylem, O. Fusun, O. Umit, "Urban Multi-Hop Broadcast Protocol for Inter-Vehicle Communication Systems", Proceedings of First ACM Workshop on Vehicular Ad-Hoc Networks (VANET 2004), pp. 76-85, Oct. 2004.
13. L. Briesemeister, G. Hommel, "Role-based multicast in highly mobile but sparsely connected ad hoc networks", Mobile and Ad-Hoc Networking and Computing, 2000. MobiHOC. 2000 First Annual Workshop, pp. 45-50, Aug. 2000.
14. B. Abderrahim, "Optimized Dissemination of Alarm Messages in Vehicular Ad-Hoc Networks (VANET)", High Speed Networks and Multimedia Communications 7th IEEE Int. Conf., HSNMC 2004, LNCS Vol. 3079, pp. 655-666, 2004.

Functional Knowledge Exchange
Within an Intelligent Distributed System

Oliver Buchtala and Bernhard Sick

Faculty of Computer Science and Mathematics – Institute of Computer Architectures
University of Passau, Innstrasse 33, 94032 Passau, Germany
oliver.buchtala@uni-passau.de, bernhard.sick@uni-passau.de

Abstract. Humans learn from other humans – and intelligent nodes of a distributed system operating in a dynamic environment (e.g., robots, smart sensors, or software agents) should do the same! Humans do not only learn by communicating facts but also by exchanging rules. The latter can be seen as a more generic, abstract kind of knowledge. We refer to these two kinds of knowledge as "descriptive" and "functional" knowledge, respectively. In a dynamic environment, where new knowledge arises or old knowledge becomes obsolete, intelligent nodes must adapt on-line to their local environment by means of self-learning mechanisms. If they exchange functional knowledge in addition to descriptive knowledge, they will efficiently be enabled to cope with a particular phenomenon before they observe this phenomenon in their local environment, for instance. In this article, we present an architecture of so-called organic nodes that face a classification problem. We show how a need for new functional knowledge is detected, how new rules are determined, and how the exchange of locally acquired rules within a network of organic nodes leads to a certain kind of self-optimization of the overall system. We show the potential of our methods using an artificial scenario and a real-world scenario from the field of intrusion detection in computer networks.

1 Introduction

Organic Computing [1] deals with the self-organization of complex technical systems by means of self-optimization, self-configuration, self-healing, self-protection, or self-learning capabilities. An example for such a complex technical system is an intelligent distributed system, e.g., a team of robots, a smart sensor network, or a multi-agent system. Often, the nodes of such a system have to perform the same or similar tasks, or they even have to cooperate to solve a given problem. Typically, these nodes know *how to observe* their local environment and *how to react* on certain observations, for instance, and this knowledge is represented by (symbolic) rules. However, many environments are dynamic. That is, new rules become necessary, old rules become obsolete, or rules change slightly over time (*concept drift*). That implies that really *intelligent* nodes (robots, smart sensors, agents, etc.) should adapt on-line to their environment by means of certain machine learning techniques.

Typically, nodes exchange information about what they observe in their environment in order to collaborate. We refer to this kind of knowledge as *descriptive knowledge*. We

P. Lukowicz, L. Thiele, and G. Tröster (Eds.): ARCS 2007, LNCS 4415, pp. 126–141, 2007.

claim that the rules that are adapted or learned on-line (we call this *functional knowledge*) are more abstract and often more valuable than descriptive knowledge! Furthermore, in the case of a dynamic environment descriptive knowledge may be inadequate or even wrong to describe novel or changing phenomena. That is, a phenomenon in the input space of one node might be misinterpreted by another node when the models that represent functional knowledge are different, for instance. Therefore, *organic nodes* should exchange functional knowledge instead of or in addition to descriptive knowledge (cf. Figure 1). Other advantages are obvious:

- Techniques and ontologies needed for functional knowledge exchange are independent from a particular application domain,
- The communication effort needed for functional knowledge exchange may be significantly lower than the effort needed for descriptive knowledge exchange, and
- Organic nodes may behave proactively: Before certain situations come up in their local environment, they will already be enabled to handle them.

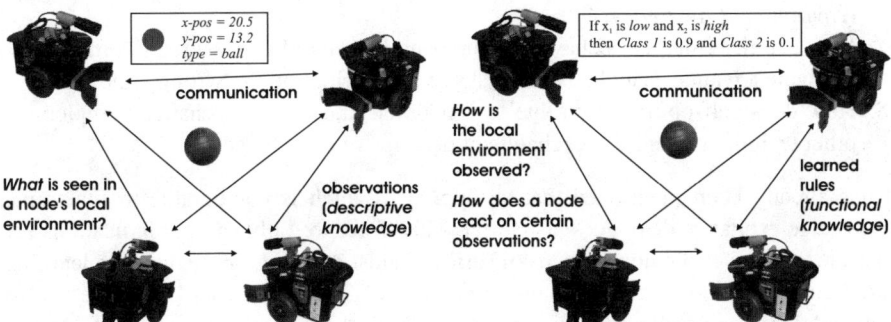

(a) Robots exchange descriptive knowledge that describes what is seen in their local environment (conventional approach).

(b) Robots exchange functional knowledge that describes how to interpret these observations and how to react (novel idea).

Fig. 1. A team of robots as an example for knowledge exchange applications: Robots decide autonomously when, with whom, and what kind of knowledge they exchange

Functional knowledge exchange is a completely novel idea and this article introduces an architecture that realizes this kind of behavior in a distributed system for the very first time. It should be seen as a kind of *proof of concept*, where many of the components will be further improved in the future. Section 2 describes an architecture of an organic node for functional knowledge exchange and the components we realized up to now, Section 3 provides some experimental results, and Section 4 summarizes the major findings and gives an outlook to our future work.

2   Architecture for Functional Knowledge Exchange

In this section we discuss the main questions that have to be answered if an organic node is going to be constructed. We introduce a generic architecture for such an organic node and describe the components we realized so far.

2.1 Research Issues and Related Work

If we think about how an organic node can be realized, we are facing many challenges:

1. Which machine learning paradigms for classification or regression, for instance, can be utilized and how can these be trained on-line?
2. How can functional knowledge (in form of rules) be extracted, fused, and inserted (i.e., exchanged)?
3. How can this exchange process be assessed (e.g., in terms of temporal effort or loss of information)?
4. How can the quality, the novelty, the reliability, and the understandability (and interpretability) of functional knowledge be assessed or even assured?
5. How do nodes assess their own knowledge and how is a need for functional knowledge exchange created (*self-awareness*)?
6. How do nodes assess the competence of other nodes and how do they decide when and with which other nodes they communicate to exchange functional knowledge (*environment-awareness*)?
7. How can the emergent behavior of the overall system be measured in order to investigate advantages and possible risks (e.g., distribution of "wrong" rules)?
8. How does self-optimization due to functional knowledge exchange complement other techniques, e.g., for exchange of descriptive knowledge?

It has already been mentioned that an approach which is comparable to functional knowledge exchange does not exist so far. Therefore, we only can get indications for answers to some of the questions from various fields. An appropriate machine learning paradigm, for example, can be found in the area of *Soft Computing* [2]. There is also some work on knowledge extraction in this field (see [3] for some references). Clustering methods can be taken from the field of *Pattern Recognition* [4]. The area of *Data Mining* [5] aims at measuring the interestingness of knowledge (validity, novelty, interpretability, etc.) [6]. However, new answers have to be found for most of the questions.

2.2 Architecture Overview

In our work, we are focusing on classification problems. That is, the rules we want to exchange assign a certain area of the input space of a classifier to a certain class. Depending on the classification result, certain reactions may be initiated.

The generic architecture of an organic node for functional knowledge exchange we propose consists of three layers that reflect certain aspects of human behavior (see Figure 2): reaction, cognition, and social behavior (cf. [7]):

– The **reaction layer** works quite autonomously. Sensor data are acquired from the environment, the data are classified, responses are triggered depending on the classification result, and the environment is controlled by means of actuators.
– The **cognition layer** realizes a certain kind of consciousness: The organic node must be aware of what he does know (or does not). This and other properties are realized by the *self-awareness* component in cooperation with a rule management component. Rules and classifiers are rated, e.g., with respect to validity or novelty.

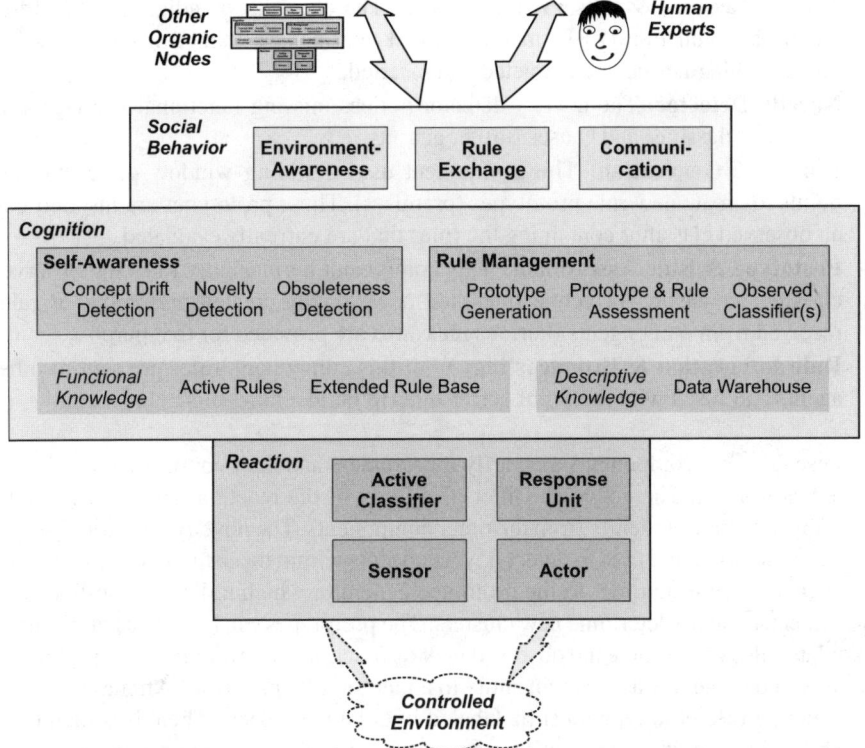

Fig. 2. Generic architecture of an organic node for functional knowledge exchange

These rules are active rules as well as rules self-created by the organic node or received from other nodes (extended rule base). Also, rules can be adapted on-line by means of training mechanisms. A database of specifically selected data is provided for that purpose (data warehouse for active learning). In contrast to the active classifier at the reaction layer, the observed classifiers at the cognition layer comprise rules that are not actively used but only evaluated at the moment.

– The **social behavior layer** enables the organic node to interact with other nodes and with human experts. Human experts are needed to label novel rules, if necessary. The *environment-awareness* component is aware of the competence and needs of other organic nodes. Based on information from the self-awareness component, the rule exchange component decides together with the environment-awareness component when, with whom, and which functional knowledge the node exchanges. Appropriate mechanisms are provided by the communication component.

It should be mentioned, that this architecture resembles existing approaches to the design of adaptive systems such as the approaches described in [8,9], for instance.

In our current prototypical implementation of the organic node, the following components are realized (detailed descriptions follow in the next section):

- **Active Classifier & Observed Classifier:** The classifier paradigm is a modified radial basis function (RBF) neural network that can be trained from data, Also, interpretable rules can be extracted and inserted.
- **Novelty Detector:** The novelty detector notices missing functional knowledge in the active classifier and causes further actions.
- **Prototype Generation:** This component uses a sliding window of recent data points to create new rule prototypes (premises). These prototypes are inserted into an observed classifier containing the rules that are currently evaluated.
- **Prototype & Rule Assessment:** This component permanently rates the observed classifier consisting of active rules and occasionally created prototypes or rules received from other agents. Various measures are provided for that purpose.
- **Rule Integration & Broadcasting:** With this component, rules are sent to other agents and fused with the set of active rules to build a new active classifier.

These different components are partly independent and their cooperation can be described as follows: The active classifier being part of the reaction layer processes the incoming data independently from the other components. The novelty detector observes the active classifier and tries to detect novel concepts within the data. In the case novelty is detected, it emits an order to the prototype generator which utilizes an on-line clustering mechanism to determine new clusters. The prototypes can be seen as premises of candidate rules that resolve the observed lack of functional knowledge. New prototypes are delivered to the rule assessment entity to undergo further analysis. After a successful evaluation, a rule prototype must be labeled by a human expert. Then, it is integrated into the active classifier and sent to other agents. To allow a stable processing in the reaction layer, the process of finding new rules is decoupled by introducing a secondary model (observed classifier). This model includes the active rules as well as evaluated rules. Several measures can then be applied to evaluate the observed classifier in order to predict the influence of the intended changes to the active classifier. It should also be emphasized that the human expert (who is simulated in our experiments) is involved in a very efficient way: He or she does not label many data points but only a few new rule prototypes.

2.3   Components of the Organic Node

In this section we describe the various components of our realization of an organic node.

Active Classifier & Observed Classifier: We define the radial basis function classifier RBFS (cf. [3,10]) as a hybrid system that can be seen as both, an RBF neural network (NN) and a Mamdani-type fuzzy system (FS). With the following definition we gain the advantages of two worlds: Trainability of NN and interpretability of FS. From the viewpoint of a NN, the RBFS may be defined as follows (cf. Figure 3):

1. The RBFS has three layers of neurons: input U_I, hidden U_H, and output layer U_O. Feed-forward connections exist between U_I and U_H as well as between U_H and U_O. A scalar weight ($w_{(i,j)}^{(I,H)}$ or $w_{(j,l)}^{(H,O)}$) is associated with each connection.

2. The activation of each hidden neuron $j \in U_H$ is determined using a multivariate Gaussian function:

$$a_j^{(H)}(k) \overset{\text{def}}{=} \frac{a_j'(k)}{\sum_{m=1}^{|U_H|} a_m'(k) + \max\{\varepsilon - \sum_{m=1}^{|U_H|} a_m'(k), 0\}}, \tag{1}$$

with

$$a_j'(k) \overset{\text{def}}{=} e^{\left(-\sum_{i=1}^{|U_I|} \frac{(w_{(i,j)}^{(I,H)} - x_i(k))^2}{r_{(i,j)}^2}\right)} = \prod_{i=1}^{|U_I|} e^{\left(-\frac{(w_{(i,j)}^{(I,H)} - x_i(k))^2}{r_{(i,j)}^2}\right)} \tag{2}$$

and a user-defined parameter ε (typically with $\varepsilon = \frac{1}{e}$, e being the base of the natural logarithm), where $k = 1, 2, \ldots$ denotes the number of the pattern and $\mathbf{x}(k) \overset{\text{def}}{=} (x_1(k), \ldots, x_{|U_I|}(k))$ is the external input. The activation function is parameterized by the weight vector $\mathbf{w}_j^{(I,H)} \overset{\text{def}}{=} (w_{(1,j)}^{(I,H)}, \ldots, w_{(|U_I|,j)}^{(I,H)})$ and a parameter vector $\mathbf{r}_j \overset{\text{def}}{=} (r_{(1,j)}, \ldots, r_{(|U_I|,j)})$.

3. Each output neuron $l \in U_O$ computes its activation as a weighted sum:

$$a_l^{(O)}(k) \overset{\text{def}}{=} \sum_{j=1}^{|U_H|} w_{(j,l)}^{(H,O)} \cdot a_j^{(H)}(k). \tag{3}$$

The external output vector of the network, $\mathbf{y}(k) \overset{\text{def}}{=} (y_1(k), \ldots, y_{|U_O|}(k))$, consists of the activations of output neurons, i.e. $y_l(k) \overset{\text{def}}{=} a_l^{(O)}(k)$.

Note that with an abbreviation for univariate Gaussians $a_j'(k) \overset{\text{def}}{=} \prod_{i=1}^{|U_I|} \varphi_{(i,j)}(k)$. In the following, the $\varphi_{(i,j)}$ are called basis functions; $w_{(i,j)}^{(I,H)}$ is the center of such a basis function and $r_{(i,j)}$ is its radius. The vectors $\mathbf{w}_j^{(I,H)}$ and $\mathbf{r}_j$ describe an axes-oriented hyperellipsiod in the input space of the RBFS. Thus, $\mathbf{w}_j^{(I,H)}$ can be regarded as a center of a hyperellipsoidal cluster – big x in Figure 3(b) – and $\mathbf{r}_j$ defines the shape of the cluster – axes-oriented ellipses in Figure 3(b). The activation of a hidden neuron describes the similarity between an input pattern $\mathbf{x}(k)$ and a center based on a matrix norm (Mahalanobis distance measure with diagonal covariance matrix).

The parameters of an RBFS must be determined by means of training algorithms such as gradient-based techniques or clustering techniques in combination with methods for the solution of linear least-squares (LLS) problems (see, e.g. [11]). For an iterative training we use penalty terms (*regularization technique*) to enforce small radii (*weight decay*) and to enforce normalized output weights in the interval $[0, 1]$.

For a classification problem, each class is typically assigned its own output neuron using an orthogonal representation of classes for training. A winner-takes-all approach is used for the final decision on class membership.

From the viewpoint of FS we can say that we have defined an FS with $|U_I|$ input variables, $|U_H|$ rules, and $|U_O|$ output variables (here: classes). The membership functions of the input variables correspond to the Gaussian basis functions of the hidden neurons, singletons are used for the output variables. That is, a fuzzy rule j ($j = 1, \ldots, |U_H|$) has the form

if x_1 is $\varphi_{(1,j)}$ $\ldots$ and $x_{|U_I|}$ is $\varphi_{(|U_I|,j)}$ then y_1 is $w_{(j,1)}^{(H,O)}$ $\ldots$ and $y_{|U_O|}$ is $w_{(j,|U_O|)}^{(H,O)}$.

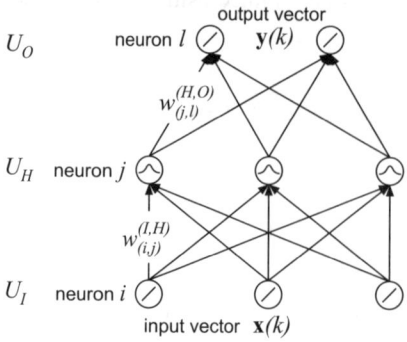
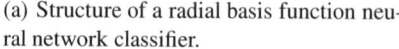
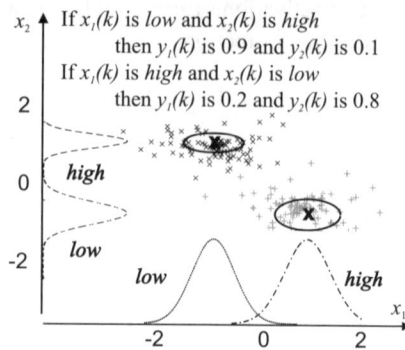

(a) Structure of a radial basis function neural network classifier.

(b) Example of a classifier consisting of two rules ($|U_I| = 2$ and $|U_H| = 2$).

Fig. 3. Radial basis function classifier

The conjunction of variables in the premise of a rule as well as the implications are realized by the product operator. The sum operator is taken to combine the rules (i.e., we use sum-prod-inference). For defuzzification, the height method is applied. We like to mention that the usage of other paradigms could be considered as well. Though, we prefer the RBFS due to the following reasons: Firstly, neural network training methods can be applied for on-line adaptation (e.g., reinforcement learning). Secondly, interpretable and even comprehensible rules can be extracted (without loss of performance). The usage of rules with Gaussian premises is motivated by the *generalized central limit theorem*: Processes with multi-causal origination tend to be normally distributed.

Novelty Detector: The task of this component is to detect possibly novel phenomena within the incoming data and to decide whether new rule prototypes should be created. We defined the RBFS paradigm in a way such that we can use it for novelty detection, too. The additional normalization term in equation 1 indicates missing activation in the active classifier. The following measure defines the so-called *recognition factor* $m_{\text{recognition}}$ for a time step k on a sliding data window of length l:

$$m_{\text{recognition}}(k) \overset{\text{def}}{=} \frac{1}{l} \sum_{k-l+1}^{k} 1 - \max\{\varepsilon - \sum_{m=1}^{|U_H|} a'_m(k), 0\}. \tag{4}$$

A novel concept is assumed to exist if the value of this measure sinks under a predefined threshold $\vartheta_{\text{novelty}}$. To avoid creating new prototypes while there exist prototypes that are already being evaluated, the decision is based on the novelty measure for a secondary (observed) classifier.

Prototype Generation: To determine rules in a dynamically changing environment, on-line mechanisms are necessary. Under the assumption that the measured data originate from Gaussian processes, this task can be solved by simple clustering algorithms. Conventional algorithms such as c-means have to be parametrized with the number of clusters [4]. Here we utilize a simple strategy that tends to produce too many prototypes.

More appropriate methods will be applied in the future, e.g., techniques from information theory that determine an appropriate number of clusters automatically (cf. [12]).

The following algorithm is executed on demand using a window of the most recent l data points $X_l(k) \overset{\text{def}}{=} \{\mathbf{x}(j) | j \in k - l + 1..k\}$. Thus, it can be seen as an on-line clustering algorithm.

1. Randomly choose an initial set of centers $C^{(0)}$ with uniform distribution and probability p from $X_l(k)$. Set $j := 0$. Set the set of barycenters $C := \emptyset$.
2. Find the next barycenter for each initial center $\mathbf{c}_i^{(0)} \in C^{(0)}$:

   (a) For each $\mathbf{c}_i^{(j)} \in C^{(j)}$ determine the set of k nearest neighbors kNN_i with an Euclidean distance measure. Then extend kNN_i: $\text{kNN}_i := \text{kNN}_i \cup \mathbf{c}_i^{(j)}$.
   (b) Compute the mean $\tilde{\mathbf{c}}_i^{(j+1)} := \frac{1}{k} \sum_{\mathbf{x} \in \text{kNN}_i} \mathbf{x}$.
   (c) For each $\tilde{\mathbf{c}}_i^{(j+1)}$ find the pattern $\tilde{\mathbf{x}}^{(j+1)} \in \text{kNN}_i$ with minimal Euclidean distance to $\tilde{\mathbf{c}}_i^{(j+1)}$.
   (d) Set $\mathbf{c}_i^{(j+1)} := \tilde{\mathbf{x}}^{(j+1)}$.
   (e) Add all centers $\mathbf{c}_i^{(j+1)}$ that did not change to C and the others to $C^{(j+1)}$.
   (f) If $C^{(j+1)} \neq \emptyset$, set $j := j + 1$ and continue with Step 2a.

3. Remove redundant barycenters (barycenters that are included several times) from C.
4. Run a modified c-means clustering algorithm (cf. the algorithm we introduced in [11]) starting with the barycenters in C.
5. Create rule prototypes with centers $\mathbf{w}_j^{(I,H)}$ (cluster means) and radii $\mathbf{r}_i$ (empirical standard deviations) resulting from that clustering. Ignore the prototypes that correspond to already existing active rules.

The chance to find the actual cluster centers with Step 2 is very high. Of course, this technique is prone to produce suboptimal results, particularly in sparse data areas which are not close to actual cluster centers. However, corresponding prototypes can easily be detected due to the sparseness of the assigned data.

Prototype & Rule Assessment: In a dynamically changing environment the assessment of rules should have some dynamic behavior as well. That is, it only makes sense to integrate an offered rule when currently data is observed that could be classified by the evaluated rule. To achieve such a functionality, it is necessary to have a certain memory ability. We introduce the following mechanism which is inspired by Markov chain theory (see, e.g., [13]).

An evaluated rule j is assigned a fitness value $f_j \in \mathbb{R}$. Measured data points cause a movement within this interval. Good evaluations increase the fitness, bad evaluations decrease it. Once the fitness value reaches one of the interval boundaries, the corresponding rule is either accepted ($f_j \geq 1$) or discarded ($f_j \leq 0$). By default, a rule is assumed to be discarded and, therefore, the fitness value generally must tend to sink. This is controlled by a parameter $1 > \lambda_{\text{penalty}} > 0$. Good evaluations of rule measures can compensate the penalizing effect and even increase the fitness. This is controlled by a parameter $1 > \lambda_{\text{reward}} > 0$ (typically, $\lambda_{\text{reward}} > \lambda_{\text{penalty}}$). The evaluation of a rule j can be described by the following algorithm:

1. Initially, set $f_j(k) := 0.5$.
2. For each observed pattern $\mathbf{x}(k+1)$:
   (a) $f_j(k+1) := \lambda_{\text{reward}} \cdot f_j(k) - \lambda_{\text{penalty}}$.
   (b) If $f_j(k+1) \geq 1$ accept the rule.
   (c) If $f_j(k+1) \leq 0$ discard the rule.

Currently, we apply two kinds of measures to evaluate rules: rule activation and premise dissimilarity. The measure for rule activation is simply defined as the activation $a_j^{(H)}$ of the hidden neuron in the corresponding classifier. The measure for premise dissimilarity is based on the following similarity measure for multivariate Gaussian functions (cf. [14,15]): The similarity of two univariate Gaussians $\varphi_{(i,j)}$ and $\varphi_{(i,k)}$ can be determined geometrically. Let A and B be the areas under $\varphi_{(i,j)}$ and $\varphi_{(i,k)}$, respectively. Then, the similarity is defined by:

$$sim(\varphi_{(i,j)}, \varphi_{(i,k)}) \stackrel{\text{def}}{=} \frac{A \cap B}{A \cup B}. \tag{5}$$

Approximating the quadrature of a Gaussian using the logistic function, this measure can be computed efficiently. To get a similarity measure for rule premises of two rules j and k, this measure has to be extended to multivariate Gaussians. Under the assumption of axes-oriented premises, this can simply be done by multiplying the similarities of the univariate Gaussians:

$$sim(j,k) \stackrel{\text{def}}{=} \prod_{i=1}^{|U_I|} sim(\varphi_{(i,j)}, \varphi_{(i,k)}). \tag{6}$$

The dissimilarity is then computed by

$$dissim(j,k) \stackrel{\text{def}}{=} 1 - sim(j,k). \tag{7}$$

Rule Integration & Broadcasting: The rule integration process is right now simply realized by inserting rules without further actions. In the future, further improvements will be done, e.g., in order to keep the number of linguistic terms as low as possible. In [3] we describe how this can be done in principle. Also, there are no special methods for communication up to now – we broadcast the rules to a pre-defined set of neighbored organic nodes. In the future, we will develop an environment-awareness component that selects agents that are known to be interested in certain functional knowledge or that are trusted, for instance (e.g., because they are known as being competent).

3   Experimental Results

In this section we present two scenarios that demonstrate the feasibility and and the advantages of the methods described in Section 2. The first is an artificial scenario, which allows us to have full knowledge about the scenario's dynamic (i.e., data with underlying distributions). The second scenario is taken from the field of intrusion detection in computer networks to show the applicability to a real-world problem.

3.1 Artificial Scenario

In this experiment we simulate three agents (organic nodes) that measure and classify data. Agent 1 will be confronted with a new situation (i.e., novel data), acquire new functional knowledge (rules), and broadcast this knowledge to agents 2 and 3. Agent 2 will accept this knowledge and profit from that decision. Agent 3 will discard this knowledge. In a nutshell, the experiment shows how agent 2 is enabled to handle a certain situation before being directly confronted with a similar situation.

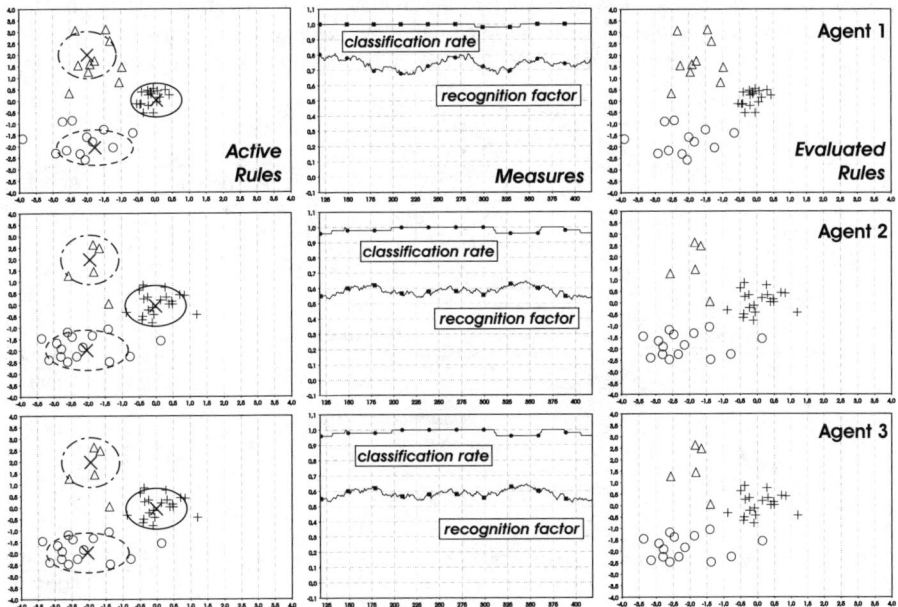

Fig. 4. Phase 1: The three agents were initialized with three rules each (*ellipses and diagonal crosses in the first column*). They measure data (*small circles, triangles, and crosses in first and third column*) that can be classified with a high classification rate (*upper curve in second column*). The recognition factors (*lower curve in second column*) are at an acceptable level, too.

The data are generated utilizing Gaussian mixture distributions and – depending on the underlying five Gaussians – assigned to one of three classes.

Agent 1 is equipped with a novelty detection component using $\vartheta_{\text{novelty}} = 0.4$ as decision threshold. The recognition factor is computed using a sliding window of 50 data points. In the prototype generator, the probability is set to $p = 0.1$. The prototype evaluation is parameterized with $\lambda_{\text{reward}} = 0.04$ and $\lambda_{\text{penalty}} = 0.01$ using the rule activation measure. Agent 2 and agent 3 both work on the same data starting with the same initial classifier and they do not have a novelty detection and a prototype generation component. The rule evaluation components are parameterized with $\lambda_{\text{reward}} = 0.02$ and $\lambda_{\text{penalty}} = 0.01$. Agent 2 utilizes the premise dissimilarity measure, and agent 3 uses the minimum of the premise dissimilarity measure and the rule activation measure.

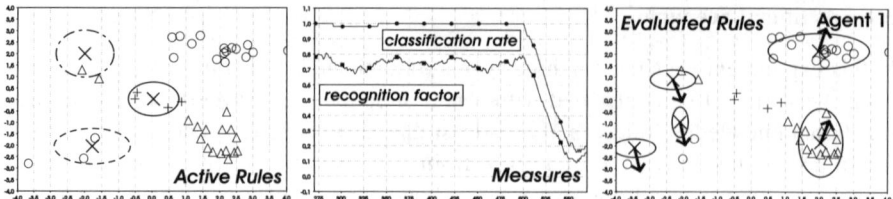

Fig. 5. Phase 2: Agent 1 measured data for which no appropriate rule exists (new phenomena) and, therefore, the recognition factor decreased. Agent 1 decided to create five new rule prototypes (*ellipses with center crosses in third column*) by means of an on-line clustering mechanism. Three of these prototypes (*on the left side*) are rated quite low (*arrows facing down*) as only a few data points activate these prototypes. The other two prototypes (*on the right side*) are rated high (*arrows facing up*).

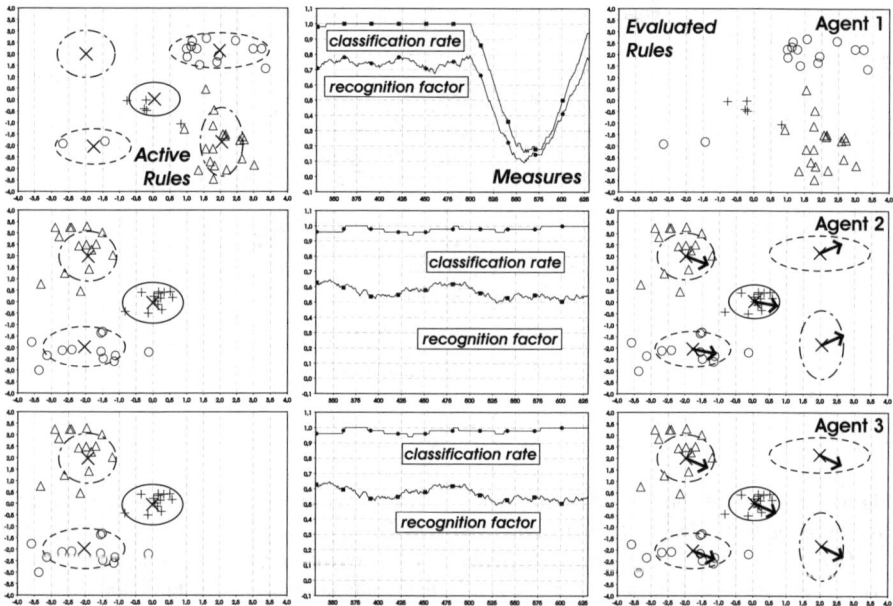

Fig. 6. Phase 3: Agent 1 accepted two of the rule prototypes (*new ellipses with center crosses in first column of agent 1*) and asked a human expert to label them (i.e., to assign them to classes). Then, agent 1 committed these new rules – i.e., they became active – and sent all rules to the other agents (broadcasting mechanism). Afterward, the recognition factor and the classification rate of agent 1 increased again. The other two agents received the new rules (*ellipses with center crosses in third column of agent 2 and agent 3*) and they evaluate them (*arrows in third column of agent 2 and agent 3*) using different measures: Agent 2 uses a premise dissimilarity measure, agent 3 uses a combination of premise dissimilarity and activation measures.

The experiment is shown in Figures 4 – 8. Each figure corresponds to a certain phase of the experiment (time step). The rows of the figures correspond to the different agents. The first column shows the two-dimensional input space of the active classifier of the

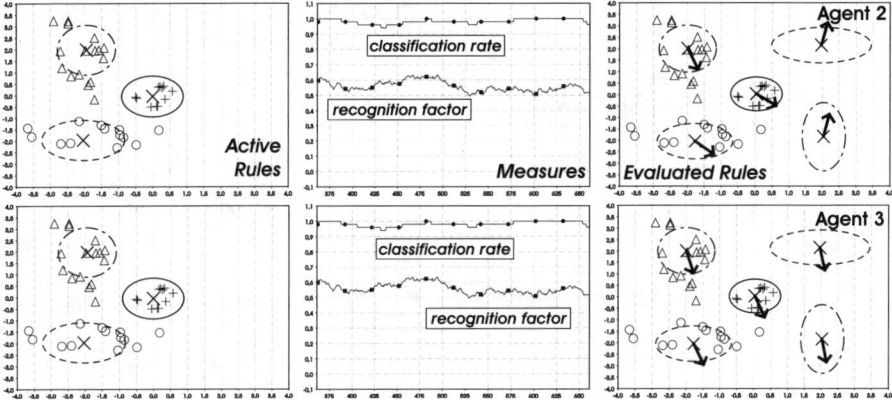

Fig. 7. Phase 4: Agents 2 and 3 show different behavior concerning the rating of observed rules as they use different measures

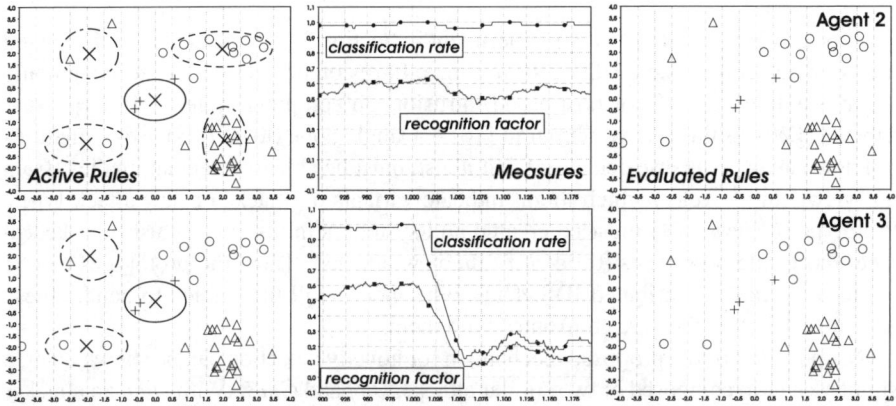

Fig. 8. Phase 5: Agent 2 and agent 3 now also measure data in an area of the input space for which appropriate rules were provided by agent 1. Agent 2 is able to classify the data correctly, whereas agent 3 is not.

respective agent with the 40 most recent data points and the active rules. Rules are symbolized by a level curve of the corresponding Gaussian (ellipse) and the position of the center (big **x**). The class assignments of rules and data points are indicated by different line and symbol types. The second column depicts the classification rate of the active classifier and the recognition factor for novelty detection. The third column shows again the two-dimensional input space and the data points. If applicable, currently observed rules are drawn here. The rating of these evaluated rules is indicated by the direction (between up and down) of thick arrows originating at the rule centers (e.g., ↑: very good, ↓: very bad, →: undecided). It must be kept in mind that the agents do not see the class labels of measured data or their own classification rate. This information is only provided in the figures for a better understanding of the agents' behavior.

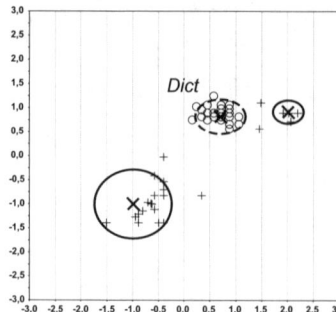 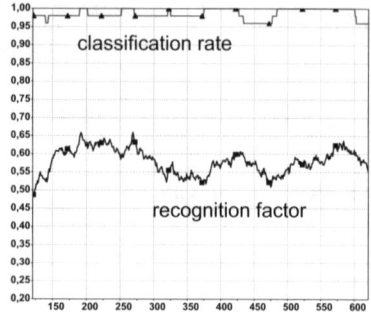

Fig. 9. Misuse detection: The left graph shows the rules (*centers shown as diagonal crosses, ellipses illustrate the shape of the rules*). The right graph displays the classification rate and the recognition factor. In this first phase of the scenario data arrives that can be well recognized (i.e., only *Dict* and non-attack). The classification rate is high and the recognition factor is about 0.6.

3.2   Intrusion Detection – Combination of Misuse and Anomaly Identification

Exactly the abilities set out in our experiment in Section 3.1 are needed in a distributed intrusion detection system (IDS), for instance. Intrusion detection aims at recognizing and preventing network- or host-based intrusions in computer systems. In large, distributed IDS – as the one we set out in [16] – a single node (local IDS) uses signatures of known attacks (misuse detection), but it also must be able to detect anomalies (from our viewpoint: novelty) which might indicate variants of attacks or new attack types. Prototypes of new rules must be created and system administrators must be asked to label those rule prototypes (to identify the new attacks). Then, the prototypes can be broadcasted in the distributed IDS. Thus, other nodes will be enabled to handle these new attacks before they are confronted with them.

The potential of the proposed mechanism for novelty detection can be shown using a simple scenario composed of two similar attacks, *Dict* and *Guest*. With *Dict*, an attacker aims at gaining access by guessing different username/password combinations. This process may use a large, pre-defined dictionary. *Guest* is a variant of *Dict* that tries to get access without passwords or with weak passwords. For the experiment we use data from a DARPA intrusion benchmark set which contains about 140 features that were extracted from TCP/IP packet header information [17]. Here, a feature selection was conducted and two appropriate features (uncorrelated and with high information gain) were selected. A data scenario has been constructed using 650 patterns from the *Dict* set randomly mixed with 650 non-attack patterns. After that, 360 patterns from the *Guest* set (containing duplicates) have been mixed with 360 non-attack patterns. These data are concatenated to build a sequential scenario to simulate a first phase where a *Dict* attack occurs (here regarded as misuse) and a second phase with a *Guest* attack (here regarded as anomaly). An RBFS is initially created using the first 500 patterns only, i.e., a misuse detector for the *Dict* attack is trained. After that, the remaining patterns are processed without adaptation.

Figure 9 shows the RBFS in the first phase (time step 600). The rules contained in this network are displayed on the left side. The centers of the RBF which describe

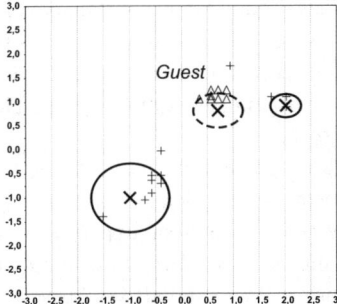 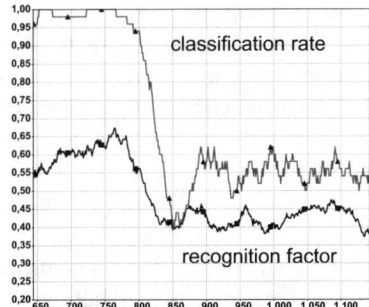

Fig. 10. Anomaly detection: In this second phase (after about 800 time steps) no more *Dict* data arrive, but patterns from the *Guest* set. The classification rate deteriorates as the new attack is misclassified (labeled as *Dict*). The recognition factor decreases, too (now about 0.4). The *Guest* patterns can be seen in the left figure (small triangles).

the premises of rules are displayed as diagonal crosses and their shapes are shown as ellipses again. Additionally, the most recent 100 observed patterns are displayed where small crosses denote non-attack patterns and circles represent *Dict* patterns. In the right graph, the classification rate and the recognition factor are displayed. In this phase, only patterns are observed that can be well recognized by the RBFS and, thus, the classification rate is high and the recognition factor is about 0.6. Figure 10 shows the scenario at later time step (about 1100). At time step 800 the observation of *Dict* patterns stopped and *Guest* patterns arrived. The classification rate deteriorates as the patterns are misclassified (i.e., labeled as *Dict*). Simultaneously, the recognition factor decreases to about 0.4. This shows how the recognition factor indicates the observation of a novel phenomenon (anomaly). As this factor is computed without knowledge of the correct class label, this experiment demonstrates the potential of the mechanism for real on-line scenarios. In the case of the detection of an anomaly, it is possible to react by informing a human expert or by generating a new rule prototype that must be labeled by an expert.

4 Conclusion

In this article we demonstrated the exchange of rules within an intelligent distributed system. By means of functional knowledge exchange, the overall system will be provided with extended self-optimization abilities. Up to now, we focused on various aspects of a future cognition layer of an organic node.

The techniques described here could and will be improved further. We will focus on the self-awareness and the environment-awareness components of the organic node. Additional measures for the assessment of classifiers and rules must be defined to detect concept drift, for instance. New training techniques are needed to enforce the comprehensibility of rules and classifiers. We must improve the techniques for rule integration: Similar rules (or linguistic terms) must be fused (cf. [3]) to avoid that the number of rules increases permanently. Obsolete rules must be detected and deleted. Also, mechanisms for an active measurement of other nodes' competence will be developed. Additionally, based on assessments of the other nodes we will improve the current broadcasting mechanism to reduce communication effort. Finally, we have to develop methods for an assessment of the overall system's emergent behavior (cf. [18]). That is, we want to quantify the impact of functional rule exchange.

Furthermore, we plan to investigate our methods in the field of robotics, specifically, soccer-playing robots. Compared to the intrusion detection application, it will be possible to create a self-optimizing system without any need of a human expert. Due to the fact that the system's success can be measured by means of available information (e.g., scoring is a success), automatically generated reinforcement signals could be applied for training (reinforcement learning; see, e.g., [19]).

Acknowledgements

This work is supported by the German Research Foundation (DFG), grant SI 674/3-1.

References

1. Allrutz, R., Cap, C., Eilers, S., Fey, D., Haase, H., Hochberger, C., Karl, W., Kolpatzik, B., Krebs, J., Langhammer, F., Lukowicz, P., Maehle, E., Maas, J., Müller-Schloer, C., Riedl, R., Schallenberger, B., Schanz, V., Schmeck, H., Schmid, D., Schröder-Preikschat, W., Ungerer, T., Veiser, H.O., Wolf, L.: Organic Computing – Computer- und Systemarchitektur im Jahr 2010. VDE/ITG/GI-Positionspapier (2003)
2. Zadeh, L.A.: What is soft computing? Soft Computing – A Fusion of Foundations, Methodologies and Applications 1(1) (1997) 1
3. Buchtala, O., Sick, B.: Techniques for the fusion of symbolic rules in distributed organic systems. In: Proceedings of the IEEE Mountain Workshop on Adaptive and Learning Systems (SMCals/06), Logan. (2006) 85 – 90
4. Duda, R.O., Hart, P.E., Stork, D.G.: Pattern Classification. John Wiley & Sons, Chichester, New York (2001)
5. Dunham, M.H.: Data Mining: Introductory and Advanced Topics. Pearson Education, Upper Saddle River (2003)
6. Fayyad, U.M., Piatetsky-Shapiro, G., Smyth, P.: Knowledge discovery and data mining: Towards a unifying framework. In: Proceedings of the Second International Conference on Knowledge Discovery and Data Mining (KDD 1996), Portland. (1996) 82 – 88
7. Mainzer, K.: Self-organization and controlled emergence. In Bellman, K., Hofmann, P., Müller-Schloer, C., Schmeck, H., Würtz, R.P., eds.: Organic Computing – Controlled Emergence. Number 06031 in Dagstuhl Seminar Proceedings, Internationales Begegnungs- und Forschungszentrum (IBFI), Schloss Dagstuhl, Germany (2006) (on-line: http://drops.dagstuhl.de/opus/volltexte/2006/577).
8. Kephart, J.O., Chess, D.M.: The vision of autonomic computing. Computer 36(1) (2003) 41 – 50
9. Branke, J., Mnif, M., Müller-Schloer, C., Prothmann, H., Richter, U., Rochner, F., Schmeck, H.: Organic computing – addressing complexity by controlled self-organization. In: Proceedings of the 2nd International Symposium on Leveraging Applications of Formal Methods, Verification and Validation (ISoLA '06), Paphos. (2006)

10. Poggio, T., Girosi, F.: A theory of networks for approximation and learning. A.I. Memo No. 1140, C.B.I.P. Paper No. 31, Massachusetts Institute of Technology – Artificial Intelligence Laboratory & Center for Biological Information Procesing – Whitaker College (1989)
11. Buchtala, O., Neumann, P., Sick, B.: A strategy for an efficient training of radial basis function networks for classification applications. In: Proceedings of the IEEE-INNS International Joint Conference on Neural Networks (IJCNN 2003), Portland. Volume 2. (2003) 1025 – 1030
12. Buchtala, O.: Transformation von Radialen-Basisfunktionen-Netzen in Fuzzy Systeme. Master's thesis, University of Passau, Department of Mathematics and Computer Science (2005)
13. Grinstead, C.M., Snell, J.L.: Introduction to Probability. 2 edn. American Mathematical Society, Providence (1997) (also available on-line: http://www.dartmouth.edu/~chance/teaching_aids/books_articles/probability_book/book.html).
14. Jin, Y., von Seelen, W., Sendhoff, B.: An approach to rule-based knowledge extraction. In: Proceedings of IEEE International Conference on Fuzzy Systems (FUZZ-IEEE 1998), Anchorage. Volume 2. (1998) 1188 – 1193
15. Jin, Y., von Seelen, W., Sendhoff, B.: Extracting interpretable fuzzy rules from RBF neural networks. Internal Report 2000-02, Institut für Neuroinformatik (INF), Ruhr-Universität Bochum (2000)
16. Buchtala, O., Grass, W., Hofmann, A., Sick, B.: A fusion-based intrusion detection architecture with organic behavior. In: The first CRIS International Workshop on Critical Information Infrastructures (CIIW'05), Linköping. (2005) 47 – 56
17. Hofmann, A., Horeis, T., Sick, B.: Feature selection for intrusion detection: An evolutionary approach. In: IJCNN 2004: Proceedings of the IEEE-INNS-ENNS International Joint Conference on Neural Networks, Budapest. Volume 2. (2004) 1563 – 1568
18. Müller-Schloer, C., Sick, B.: Emergence in organic computing systems: Discussion of a controversial concept. In Yang, L.T., Jin, H., Ma, J., Ungerer, T., eds.: Autonomic and Trusted Computing, Proceedings of the 3rd International Conference ATC-06, Wuhan. Number 4158 in LNCS, Springer Verlag, Berlin, Heidelberg, New York (2006) 1 – 16
19. Merke, A., Riedmiller, M.: Karlsruhe Brainstormers – A reinforcement learning approach to robotic soccer. In Birk, A., Coradeschi, S., Tadokoro, S., eds.: RoboCup 2001: Robot Soccer World Cup V. Number 2377 in LNCS. Springer-Verlag, Berlin, Heidelberg (2002) 435 – 440

Architecture for Collaborative Business Items

Till Riedel[1], Christian Decker[1], Phillip Scholl[1], Albert Krohn[1],
and Michael Beigl[2]

[1] TecO, Universität Karlsruhe (TH), Vincenz Prießnitz Str. 3, 76131 Karlsruhe, Germany
{riedel,cdecker,scholl,krohn}@teco.edu
[2] IBR, Universität Braunschweig, Mühlenpfordtstraße 23, 38106 Braunschweig, Germany
beigl@ibr.cs.tu-bs.de

Abstract. Sensor network technology is pushing towards integration into the business world. By using sensor node hardware to augment real life business items it is possible to capture the world and support processes where they actually happen. Many problems of the business logic running our world can be efficiently implemented "on the item". In order for these smart items to couple back to the virtualized world of business processes it necessary to design a uniform system abstraction for enterprise systems. Service oriented architectures are the tool to describe functionality apart from its concrete implementation. This paper describes a system and the experiences made integrating wirelessly networked smart business items into high-level business processes.

Keywords: Wireless sensor networks, service oriented architecture, business logic, distributed systems.

1 Introduction

The complex and interwoven business logic demands more and more information sources to enable reliable, flexible and efficient processes. Information has become one of the key values of today's business world. At the same time technologies emerge that can lead to a ubiquity of information sources. Especially sensor nodes are a promising platform for enabling the digitalization of our environment. Augmenting real life business items like goods or people with tiny wirelessly connected computing and sensing platforms creates a broad range of possibilities.

However, the enormous amount of unfiltered information arising from a steadily growing amount of sensors can soon become a problem of scalability. The need for continuous evaluation of sometimes-unreliable information often contradicts the goal of reliable, flexible and efficient business processes. We think that only *integrating sensor networks into business logic* (as in [2]) will thus fall short of a scalable solution. However, *integrating business logic into sensor networks* we believe can become a future direction of computing systems.

The information processing capabilities of sensor networks can implement business logic in a collaborative fashion without pushing unnecessary information to backend systems.

P. Lukowicz, L. Thiele, and G. Tröster (Eds.): ARCS 2007, LNCS 4415, pp. 142–156, 2007.

Fig. 1. Smart chemical drums equipped with sensor nodes

The following motivating scenario, for which we build a trial installation in a BP chemical plant in Hull, UK, shall outline sensor network capabilities in the context of business processes. In this example containers of chemical goods – the business items – are equipped with wireless sensor nodes (see Fig. 1) in order to enforce pre-defined storage conditions. The regulations are encoded in backend enterprises systems as part of the workflow with the business items. However, the supervision on the regulation is delegated to the business items themselves. In this scenario, the nodes on the items communicate with each other and collaboratively supervise the environment around the containers. Business logic on the nodes utilizes the sensor information on presence on number of other containers, chemical content, and environmental conditions, e.g. temperature, as input to collaboratively reason on eventual violation of storage regulations. If detected, the nodes in this scenario are able to warn locally by triggering a visual alert signal and report this incident back for purposes of logging to a backend system. In contrast to technologies like Radio Frequency Identification (RFID), the process on detection and enforcement is completely distributed among the participating business items. This enables a continuous, very accurate in-situ supervision with no additional information overhead for back-end systems.

Sensor networks already have the means for executing such business logic (e.g. in [1]). However, in order to ensure that backend end sensor network seamlessly work together a technical framework is needed that provides interconnectivity between sensor networks and server based enterprise systems. Both systems have optimized ways of processing and communicating data. All assumptions about increasing efficiency by combining both only hold if we don't lose their original efficiency along the path of integrating both systems. Based on the implementation of the business logic of the smart drum scenario we show how efficient an unwanted and potentially dangerous storage combination can be identified on item level and how this can increase the reliability of systems where logic is coupled with physical entities.

In this paper we describe a service-oriented architecture for seamless integrating business logic executed on sensor networks. The goal of the architecture is to delegate parts of the business logic from resource intensive backend systems to thin sensor node technology. In the next section we analyze the general structure of business logic and the requirements for mapping them to sensor networks. Furthermore we discuss the technical and structural challenges for an integrated architecture. We continue describing a pragmatic architecture for enabling sensor supported business logic, the CoBIs Gateway Architecture. It links services executing logic on top of the sensor network through UPnP (Universal Plug and Play) proxies to the logic running in backend systems. In the last section we present our experiences gathered while applying this technology within a real world trial.

2 Analysis

A business process describes the transformation of resources in order to achieve a measurable business relevant outcome. Business processes can be split into process tasks, which can be delivered by different service providers. The sequence of such tasks can e captured and modeled in so-called business logic.

The term *business logic* is used to distinguish the processing part from presentation and storage within classical 3-tier architectures. The term "functional process logic" might be a better term describing the same thing. Classical business logic is built on top storage layer most commonly called CRUD (Create, Read, Update and Delete). Because business logic components still share a lot of intimate knowledge about the data layout CRUD layer parts of the reusability of business logic is often limited. Services oriented architectures have recently been used to abstract from this rather tight coupling to the data layer by defining functional interfaces across all components.

In our view (see [3]) a service consists of a well-defined functionality and a well-defined interface for accessing this functionality from a client. To distinguish the service abstraction from others we demand that the client is independent of the concrete implementation of the service and that the each service is independent from the internals and the state of any other service. The discriminating factor of service-oriented architectures is the loose coupling between services. This enables exchanging functional entities and gives us the possibility to seamlessly integrate new technology.

2.1 Collaborative Business Items

Business logic acts on a virtual representation of physical world. Every business item that is part of a process is modeled to have an equivalent in the virtual world. Business rules and workflows describe how to act on those objects. Business logic often relies on user or machine interfaces to update their state according to the real world process in parallel. The state between the real world and the virtual world is thus only synchronized at certain checkpoints in space and time.

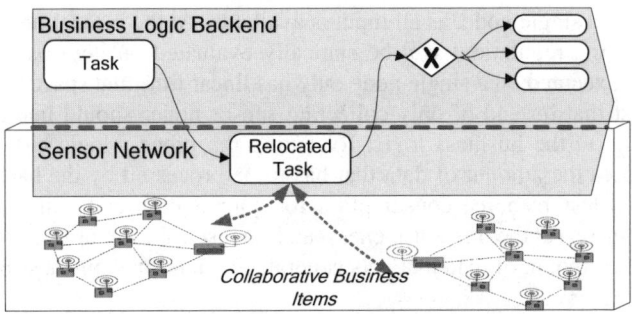

Fig. 2. Relocated business logic

Collaborative Business Items (CoBIs) push the interface between real world processes and business logic out onto the business item itself (see Fig. 2). This allows us close the gap between the virtual and physical world. Sensor nodes attached to business item as the drums in the introductory example can directly act as service providers. CoBIs add the following key capabilities normal business items (also compare [3]):

- Computing
- Data Storage
- Monitoring
- Controlling
- Communication

Wireless sensor nodes such as the Berkley Motes [6], Ambient uNodes [7] or Particle Computers [8] can be used as enabling platforms for embedding those features into physical business items.

2.2 Logic on the Item

Functionality implemented on a single sensor node is subject to many constraints. For instance, the micro-controller is often a resource-restricted 8-bit processing unit, providing typically up to 512KB Flash memory for programs and only a small amount (around 2KB) of RAM for volatile data.

However, in spite of the limited amount of data space and computation power on sensor nodes, they can provide complex services to business applications by in network collaboration. There are two key properties of business logic that can compete with and outplay server-based alternatives.

The first is the distributability of business logic. Because business logic is item related it can often be split on item level. The processing power then can scale linear with the number of collaborative business items. While backend systems often need complex strategies to scale with an always-increasing number and speed of inputs, scaling is an intrinsic feature of sensor networks.

The second key property of most business logic is that it exposes a high locality concerning their information working set. As an example matching storage regulations against environmental conditions such as temperature or humidity can be done locally on a single node, as all input is available. At the same time the number of possible matching regulations can be statically evaluated on item basis, so that the logic actually executed on a single node only has linear time and space requirements.

We suggest that instead of only collecting sensor nodes should interpret data and pass on results to the business logic. Executing the process logic close to the data source decreases the amount of data that has to be processed by the backend system. This results in less resource consumption for computation and communication and can in turn increase the responsiveness and the scalability of the whole system considering the amount of data that gets generated by a normal business process.

2.3   Architectures for Collaborative Business Items

Considering the growing number of sensor network platforms for services relevant to business logic the backend systems would have to interface a number of different sensor networks hosting different sensor types and using different data encoding.

Uniform service concepts for sensor networks have already been designed in the past however they either see sensor networks only as a data source like [2] or at least restrict the logic which is processed on the item to query statements [10]. Other systems like [9] also suggest new service architectures for backend systems.

In contrast to this related work we see the challenge in fitting sensor networks into existing business service architectures. Seeing a sensor networks as regular service providers frees business developers from the need to develop proprietary connectors and leaves sensor network technology the freedom of optimized implementations. The coupling point between the two worlds is a service interface that has to be provided by a collaborative item architecture.

3   Key Design Challenges

In this chapter we describe the key design challenges of a service-oriented architecture (SOA) integrating collaborative business items.

3.1   Interfaces

Supporting standard RPCs to the services running on the nodes could lead to a painless integration with backend systems. However, it can be shown that RPC is not the right abstraction to directly support on sensor nodes (see [4]). RPC is very restrictive about the calling semantics, which makes stub generation easy, but proves to be inefficient for the communication requirements of sensor networks.

Data packets in sensor networks are often built in a way to support cross-layer protocol optimizations. For this purpose the data is encapsulated in an efficient encoding that can easily be parsed by the system. Because of different protocol stacks and operating systems, this leads to very different presentations of structured information in a packet. As an example the Particle platform use a tuple-oriented data format enforcing strictly typed information. Motes use Active Messages allowing a direct mapping to the component interfaces of TinyOS. A client would have to support all different encoding in such a heterogeneous environment.

Even if we can understand the message encoding, we will still fail to extract sensible information from the data. If transport container and environment have both humidity sensor embedded, it is not possible to make a statement about neither absolute nor relative humidity, because both sensors will most likely have different sensitivity or different resolutions.

The goal has to be that interfacing sensor network services are not more complex than using backend services. All domain knowledge needed to understand interfaces to sensor network services needs to be made explicit in standardized service interfaces.

If we acknowledge the necessity of proprietary protocol layers, this means that sensor network messages need to convert to a uniform message encoding. We propose the use of "active" service descriptions. Additional to the interface descriptions they can contain information to generate transformation stubs implementing endpoints for both sensor network and backend service communication.

3.2 Addressing

Besides interfaces, addressing can be identified as one of the most essential parts about service interaction. It is closely coupled to topics like transport and routing, which are also key elements of wireless sensor network research [12,13]. For us it is important not to impose implications on concrete implementations of such protocols and algorithms via the addressing scheme. From the application's point of view, however, a common method for addressing is necessary for a truly service-oriented view on the network. Because the semantics of an address is determined by the concrete implementation, we need means for address translation in our system.

Arbitrary resolving scheme can be applied to describe the different needs for addressing. Like in a DNS system semantic hierarchies can be constructed across an address space. Once resolved, the client can use the address to communicate with the system hosting a service using the target address space.

In highly dynamical systems like item networks (in contrast to rather static sensor networks) this generates consistency problems. To illustrate this we may think of a temperature service in out smart drums scenario. Getting the temperature can easily be executed on any sensor node in the network. Typically higher-level logic is, however, not really interested in the temperature of some node A but in e.g. the temperature at a location L. If node A is in location L we can nonetheless execute the service there by resolving L to A. If node A physically moves, however, this leads to an obvious problem when calling the service again.

Another kind of service mobility leads to the same problem: migrating a service to a neighboring node. Once again the service cannot be reached using A's address. Those problems occur, because we once again replicated part of the service state (namely the current host) in the backend system. Exactly this we stated as a problem before, as now the problem is keeping virtual and physical world consistent. Constantly requesting new addresses leads to an immense overhead on the network traffic.

Once we also push the logic of matching a concrete address into the sensor network, we see this problem disappear. An efficient implementation could e.g. involve location-based routing algorithms. We can easily imagine other routing schemes. Therefore we propose to introduce service proxies in order to hide the addressing information behind an IP-based addressing scheme. Those proxies represent a "service running on an address". This way we can avoid pushing the routing functionality into the client.

3.3 Discovery

By saying that we map service addressing and interfacing to IP-based proxies we only shifted the problem of service binding to IP technologies. However, the problem of service discovery can be handled fairly efficient in those networks. Two main approaches can be identified here: infrastructure-based and infrastructure-less approaches.

An example of an infrastructure-based service binding and discovery approach is described the Web Service Interoperability Standard, namely the UDDI registry (www.uddi.org). Without going into detail about the specific up- and downsides of specific discovery implementations, it can be said that infrastructure based approaches are useful for rather static service landscapes and often have a broader scope than just announcing functionality. They also may represent a bottleneck in a distributed system or at least need special care to setup.

The service enabling of sensor networks should be a rather "plug-and-play" oriented approach that simplifies the use of a specific technology and should not generate additional infrastructure dependencies. For those purposes infrastructure-less discovery seems to be the better choice. This kind of discovery either uses multicast announcements and queries or distributed hash tables to announce services throughout the network. Because service mapped to real world items via wireless interfaces can easily disappear or appear, services may move with the nodes. Therefore announcement-based discovery interfaces also provide monitoring functionality for the liveliness of a service. Multicast announcements can provide a powerful means to program dynamical business logic, that acts on the availability of a service and also can take trigger necessary steps if a service is unavailable.

3.4 Lifecycle Management

The first thing to do if some logic fails to discover and bind necessary service functionality is to try to deploy this functionality to the sensor network. Because services can be mapped to multiple nodes within the network it makes sense to install one lifecycle management interface per network. The whole sensor network acts as a container for services.

Creating container interfaces for all service hosts would create the need to generate proxies for all addressable entities. Those are not known beforehand, as they can be arbitrarily defined. In order to solve this problem we use deployment descriptions that can have the power of selection statements of any query language. Those descriptions are used to instantiate the service on the target host generating a local proxy. Addressing the service instance is implemented by the sensor network routing services, thus the expressiveness of deployment description is also directly depended on their addressing modes.

We assume precompiled binaries called *service executables* as the input for the sensor network, which are then passed to the single nodes. As stated before we are totally agnostic about the implementation language and content of the service executables. The lifecycle management service only needs a counter part on each node in the network to forward the binary to. Once the deployment request was issued it is left to sensor nodes to execute it and initiates a *service instance*. The success of the deployment can be verified end-to-end using service discovery. Functions for removal and temporary disabling of a service are added to the service interface of each proxy.

4   CoBIs Gateway Architecture

In this chapter we describe our implementation of a service-oriented architecture for enabling sensor networks to run business logic. The Collaborative Business Item Gateway Architecture implements an UPnP to sensor network gateway. We chose the Universal Plug and Play (UPnP) standard because of its lightweight, infrastructure less and yet complete approach. The Standard includes the Simple Object Access Protocol (SOAP), the Simple Service Discovery Protocol (SSDP) and the General Event Notification Architecture (GENA). Because UPnP uses Internet technology it can easily be included in most business applications. The implementation, however, is not UPnP specific but can be easily ported to e.g. Web Services.

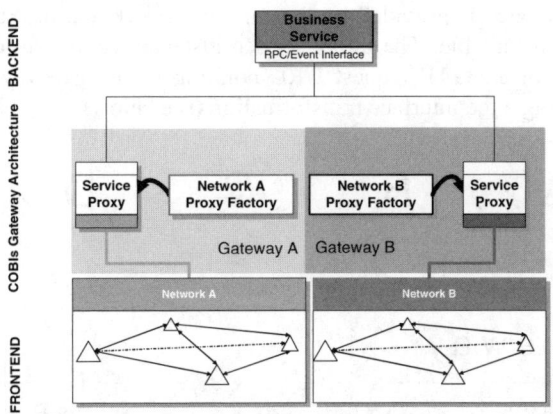

Fig. 3. The CoBIs Gateway Architecture

Key element of this architecture is the dynamic instantiation of service proxies. Proxies can be accessed like native UPnP devices, providing detailed service descriptions for the implemented functionality. The proxy itself, however, only exists as a virtual representation of the service interface. Request issued to the service proxy are transformed by the gateway to sensor network messages and vice-versa. From the backend the gateway itself is only visible for deploying new services to the sensor network. Multiple gateways can be instantiated simultaneously to include several sensor networks into the architecture. The general architecture is depicted in Fig. 3.

4.1 Gateway Devices

The platform gateway is a application level bridge that handles all aspect of communication with the backend system. This includes all protocol levels beginning with the physical layer bridging from Ethernet to the wireless network ending with the application level bridging for service interaction from proprietary interfaces to RPC-style service interfaces. The idea of platform gateway is that it capsules all domain knowledge needed for communication with a sensor network platform. Because the interfaces to the backend system is always of the same type this allows exchanging the platform as well as using multiple sensor network platforms in parallel.

Because we handle the platform gateway as a monolithic software component in our architecture this does not mean that it is a single machine. The system has been designed to be able to both different levels of bridging as well as proxies to different services distributed on multiple machines. The only interface to the gateway from the backend system is the lifecycle management service that enables pushing new service binaries to the network.

This and all interfaces dynamically created service proxy are announced in the network. UPnP handles Service discovery natively once a proxy is initiated. All proxy instances provide a pointer to their description via GENA when requested by some client. For the gateway this means that proxy services have to be instantiated whenever a new service is provided by the sensor network and destructed when the service becomes unavailable. The proxy service instance itself is a only dispatching path associated with a SOAP request URL pointing to the gateway and a service description including a the interface transformation (see below).

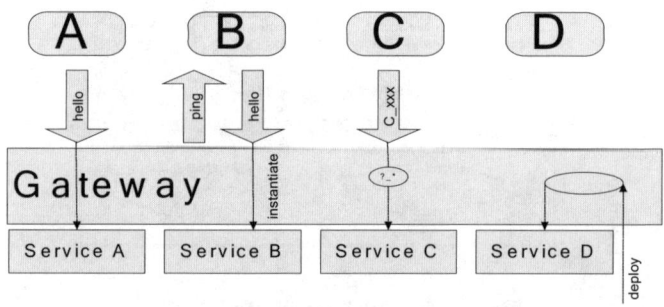

Fig. 4. Discovery methods supported by the gateway

Fig. 4 shows four means to initiate a service instantiation. They are provided by Discovery Services running on the gateway and can be exchanged. In the first case the gateway parses predefined "hello" and "bye-bye" packages (Fig. 4A). Querying actively instructs all services to issue hello packets by sending broadcast pings (Fig. 4B). Passive discovery omits special hello packets completely by adding a service identifier to all packets (Fig. 4C). Proxies can also be permanently installed on deployment, requiring manual removal (Fig. 4D). This however disables support for liveliness monitoring.

4.2 Interface Transformations

We integrate the platform specific semantic transformation for the RPCs into the XML-based UPnP service description. Because the UPnP demands flexible XML parsing we can use the same description for providing the client description to the control point. Leaving the transformations inside the descriptions allows easy debugging of the transformations and allows the management services to analyze the running system.

The descriptions are automatically parsed by UPnP stack implementation, which already provides the UPnP RPC dispatching and eventing facilities. We integrate our transformation logic into the device instantiation that is guided by the XML description. This allows a direct coupling between transformation and interface.

We suggest a simple template-based transformation, we have successfully used for creating interfaces for Particle Computers and Ambient uNodes. Templates work as bi-directional transformation. Fig. 5 shows the model of the templates organized as an ingress filter tree that is loosely connected via a listener pattern to the UPnP protocol stack. When a RPC call is received each outgoing argument is serialized via the template the UPnP state variable and incoming arguments are parsed by inverting the process: the template is matched against the packet.

A template entry can be of three types of data, wildcard or don't care. If a matching template is found the information under the wildcards is extracted/serialized together with any optional static data. UPnP typing information contained in the interface description can be used for de- and encoding. This suffices for most standard integer data, but the scheme can be extended to support more explicit interfaces without the increasing message size. In cases more complex parsing is needed one may specify platform specific extensions for message decoding.

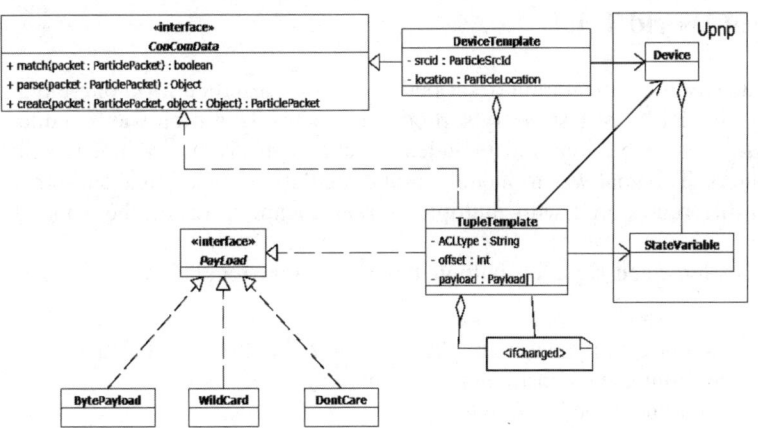

Fig. 5. Mapping message templates of RPC arguments

4.3 Message Primitives

We support seven message types for communication between sensor node services and backend-based services that can be accessed via the RPC interface:

- Non-blocking Send, Receive and Call
- Blocking Send, Receive and Call
- Callback

The classification of blocking and blocking messages concerns the behavior on the gateway not on the issuer as RPCs are always blocking. The non-blocking send can be used if no return types are given. It sends out a message to the sensor network and does not wait for an answer.

The non-blocking receive uses replicated services states on the proxy to answer the RPC. Here no function argument can be provided to RPC. An example would be a *getTemperature* action providing the most current temperature data. Caching data in the gateway allows the services on the sensor node to run in a duty cycle without loosing accessibility. The non-blocking call is a combination of the message types described before. It does not wait for data to arrive, but returns cached data. It can be used to signal the service that its data was received and consumed by the backend. The blocking call can be used to implement real RPC for a service. It sends out data and awaits an answer before returning. It can also be used to provide support for acknowledged that means blocking sending support as well as blocking receive by omitting either in or out parameters. For blocking operation we wait for a message matching the associated return argument. Timeout values may be specified individually for each function.

The callback provides support for a node service to asynchronously trigger the Business Logic in the backend. An example would be an alerting service that needs to result in rapid action within the backend. UPnP's General Event Notification Architecture (GENA) handles the callback subscription. Subscriptions are thus handled by the gateway.

5 Real World Trial

The gateway implementation described above was installed for a one-month trial at one of the BP's largest acetyls production site. The trial was conducted in a declassified storage area and included 21 chemical drums equipped with Particle Computers. The goal was to model storage regulation of chemical substances stored in two different stored with multiple storage locations inside the same store (see Fig. 6).

We implemented logic for multiple types of storage regulation:

- per chemical storage limit
- incompatibility classes of chemicals stored in the same location
- environmental constraints (maximum/minimum temperature)
- maximum time in storage

The Business logic for the storage regulations was modeled in SAP's EH&S (Environment, Health and Safety) system. The system was used to parameterize a so-called hazardous goods service on the nodes. This service needed input from a location service, which was implemented using a simple infrastructure-based infrared location system. A temperature monitoring service gave input to check environmental storage regulations. Additional management functionality (voltage, duty cycle) was provided to manage the networks functionality from Smart Items Management Console developed by SAP Research. The gateway software was run on two 200 MHz embedded Linux MIPS systems in combination with a with 2GHz Intel Windows XP server also running the monitoring logic.

Fig. 6. Storage location enabled with Collaborative Business Items

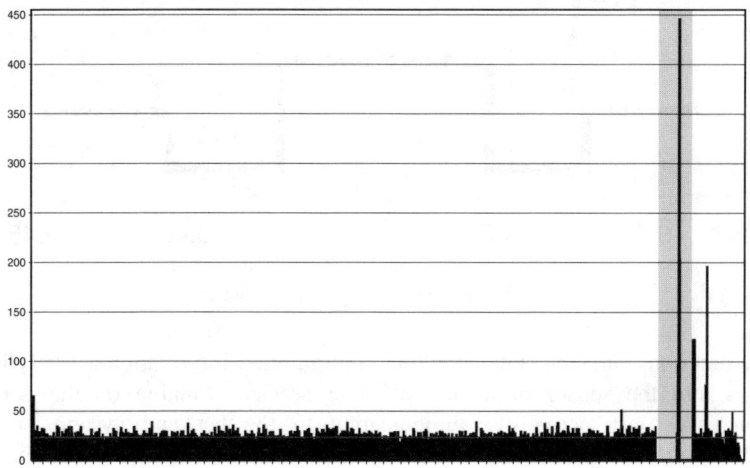

Fig. 7. Message load per minute to backend system

5.1 Trial Evaluation

Surprisingly most of the problems encountered on site were related to technologies not specific to our architecture. Using Internet technologies the hope beforehand was to seamlessly integrate into any existing network using those technologies. As BP provided us with a wireless 802.11 based infrastructure we were able to connect our gateways without the need of complicated wiring on site. Probably because of the rather humid weather conditions close to the cooling towers the 802.11 network showed a high packet loss at times. This packet loss did not very much affect TCP traffic but unacknowledged traffic like UDP. UPnP uses UDP multicast as part of its discovery scenes and also the DHCP based dynamic addressing for the gateways used unacknowledged transport for discovery.

The logic pushed to the sensor network worked reliable. Additionally to the month long trial we performed an extensive test set at the end of the trial to confirm the correct behavior of the system. Both the business software running on an SAP application server as well as the sensor network performed their services by specification. The average message load to the business logic was only about 23 messages per minute (see Fig. 7), mostly resulting from voltage monitoring need for the management application. We were however able to put our system into an overload situation (right part of Fig. 7), when simultaneously generating alerts from all business items. This was due to the GENA implementation of the used UPnP stack, which set up new HTTP connections for each event and each subscription.

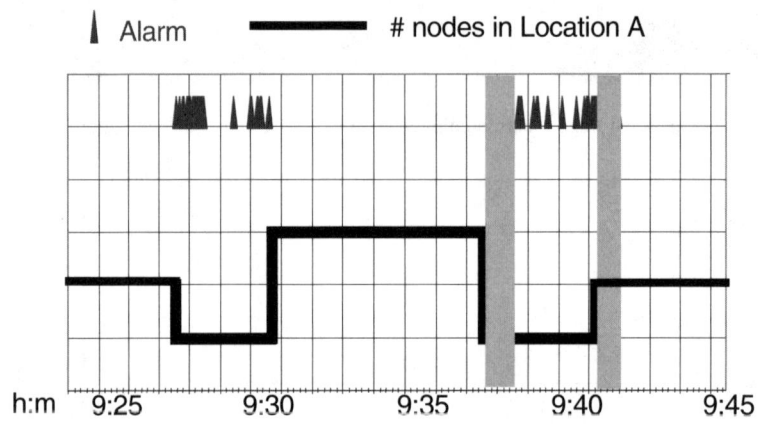

Fig. 8. Delayed issuing of alarm action

Only functions carrying state changing configuration information actively injected messages into the sensor network. All other services running on the nodes ran autonomously, only actively communicating to the backend system on events subscribed by the business logic. This scheme made the reactions of the sensor network especially in critical situations very robust as the message load was in some sense predictable within the system and message prioritization could be decided within the node network. This at times (due certain amount of message loss in unacknowledged traffic) leads to an inconsistent view of the system between sensor network frontend and backend (business logic/management) systems. Most events were retransmitted repeatedly so that the error was temporal and the correct logic was executed with a delay (compare Fig. 8).

6 Conclusion

In this paper we have demonstrated how a system can be build to easily integrate arbitrary sensor networks into business logic. For this purpose we did not stop at the point where we use this technology as a data source, but rather tried to unleash the computation power that results from deploying networked sensor nodes into real life environments. We showed that service oriented architectures are way to abstract from classical system designs using only backend driven business logic. The proposed system enables sensor networks to actively take part in distributed processes.

The implementation of our UPnP sensor network gateway shows that this technology can easily be used to adapt existing sensor node platforms to a service oriented business architecture. While trying to make the least possible assumptions about functionality of the services executed by sensor network during the design of our architecture, we showed by trial a specific use case that this technology can be successfully applied in real life settings. In spite of the still prototypical nature this system we hope that our experiences can help the deployment and integration of sensor node technology into business applications in near future.

Acknowledgements

The work presented in this paper was fully funded by the European Community through the project CoBIs (Collaborative Business Items) under contract no. 4270. We further like to thank Paul McCune and the people at BP Chemicals Saltend for their support during the application trial.

References

[1] M.Strohbach, H.-W.Gellersen, G.Kortuem, C.Kray. *Cooperative Artefacts: Assessing Real World Situations with Embedded Technology.* Ubicomp 2004
[2] C. Bornhövd, T. Lin, S. Haller, J. Schaper, Integrating Automatic Data Acquisition with Business Processes *Experiences with SAP's Auto-ID Infrastructure,* Proceedings of 30th VLDB Conference, Toronto, Canada
[3] Z. Nochta, N. Oertel, P. Spiess. *Relocatable Services and Service Classification Scheme,* CoBIs Deliverable Report, http://www.cobis-online.de/files/Deliverable_D101.pdf, 2005
[4] U. Saif, D. J. Greaves, *Communication Primitives for Ubiquitous Computing or RPC Considered Harmful,* 21st ICDCSW, p. 0240, 2001.
[5] C. Decker, P. Spiess, L. Moreira sa de Souza, M. Beigl, Z. Nochta.: *Coupling Enterprise Systems with Wireless Sensor Nodes: Analysis, Implementation, Experiences and Guidelines,* Pervasive Technology Applied @ PERVASIVE, May 7, 2006, Dublin, Ireland
[6] J. Hill, R. Szewczyk, A. Woo, S. Hollar, D. Culler, K. Pister, *System Architecture Directions for Networked Sensors* ASPLOS-IX, 2000
[7] P. Havinga, *The Quest for Low Cost, Ultra Low Power Wireless Networks for smart environments,* Ambient Systems white paper, 2006

[8] C. Decker, A. Krohn, M. Beigl, T. Zimmer, *The Particle Computer System* Proceedings of the ACM/IEEE Fourth International Conference on Information Processing in Sensor Networks, Los Angeles, 2005

[9] J. Shneidman, P. Pietzuch, J. Ledlie, M. Roussopoulos, M. Seltzer, and M. Welsh, *Hourglass: An Infrastructure for Connecting Sensor Networks and Applications,* Harvard Technical Report TR-21-04, 2004

[10] S. Madden, M.Franklin, J. Hellerstein, W. Hong. TinyDB: An Acqusitional Query Processing System for Sensor Networks. ACM TODS, 2005

[11] *Universal* Plug and Play Device Architecture, Microsoft Corporation, 1999.

[12] C. Intanagonwiwat, R. Govindan and D. Estrin, *Directed Diffusion: A Scalable and Robust Communication Paradigm for Sensor Networks.* In Proceedings of the Sixth Annual International Conference on Mobile Computing and Networks (MobiCOM 2000), August 2000, Boston, Massachusetts.

[13] J. Kulik , W. Heinzelman , and H. Balakrishnan, *Negotiation-based protocols for disseminating information in wireless sensor networks.* Wireless Networks, March 2002.

Autonomic Management Architecture for Flexible Grid Services Deployment Based on Policies

Edgar Magaña[1,3], Laurent Lefevre[2], and Joan Serrat[3]

[1] Cisco Systems, Inc.
10 West Tasman, San Jose, CA 95134, USA
emagana@cisco.com
[2] INRIA RESO / LIP Laboratory
UMR 5668 (CNRS, ENS Lyon, INRIA, UCB), France
laurent.lefevre@inria.fr
[3] Universitat Politècnica de Catalunya
Jordi Girona 1-3, Barcelona, Spain
serrat@tsc.upc.edu

Abstract. This paper describes a dynamic, scalable and flexible Policy-based Management Architecture (PbMA), which is characterized by a reliable and autonomous deployment, activation and management of Grid Services. This architecture follows the implied conditions by the Open Grid Services Architecture (OGSA) standard. Although applicable to any user profiles, our system is essentially intended for non-massive resource owners accessing large amounts of computing, software, memory and storage resources. Unlike similar architectures, it is able to manage service requirements demanded by users, providers and services themselves. This architecture is also able to manage computational resources in order to fulfill Quality of Service (QoS) requirements, based on a balanced scheduling of resources exploitation. Our approach is scalable and flexible by extending itself the management components and policies interpreters needed to control multiple infrastructures regardless network technology, operative platform or administrative domain. The management architecture shows its reliability through a Grid Service deployment example.

Keywords: Network Computing, Policy-based Management Architecture, Quality of Service, Grid Services.

1 Introduction

The impact of the Grid in the academic and industrial sectors is growing every day, although it was originally understood as an emerging technology that provides an abstraction for resources sharing and collaboration across multiple administrative domains [1]. The current generation of Grid Architectures heavily relies on the program designer or users to express their requirements in terms of resource usage. Such requirements are usually hard-coded in a program using low-level primitives, but Grid needs mechanisms to handle computational resources as much efficiently

P. Lukowicz, L. Thiele, and G. Tröster (Eds.): ARCS 2007, LNCS 4415, pp. 157–170, 2007.
© Springer-Verlag Berlin Heidelberg 2007

and dynamically as possible. In other words, Grid Services require important innovations to guarantee network Quality of Service (QoS) levels, management autonomy, scalability, and fault-tolerance to support Grids consisting of thousands of nodes [19]. It is worthy to say that the term "resources" covers a wide range of concepts including physical entities (computation, communication, storage, etc), information storage (databases, archives, instruments, etc), individuals (people and their expertise), capabilities (software packages, brokering and scheduling services) and frameworks for access and control [3]. Therefore, Grid Management frameworks have a very heterogeneous environment to handle, which should be compatible with the emerging Grid Services specifications [6]. Also it is necessary to take into account the innovations in Web Services related to maintaining the state of the resources behavior (stateful), in order to fulfill the implicit requirements by new generation of Grid Services. Thereby, relevant research effort is needed to innovative Grid Systems and also to make the current Grid models suitable for those emerging usage scenarios.

Policy-based Technology has attracted significant industry and academic interest for management purposes. Presently, it is promoted by the Distributed Management Task Force (DMTF) and is standardized within the Internet Engineering Task Force (IETF) Policy Working Group [5]. Policies are defined as rules that specify the expected behavior of the managed system under certain conditions and they are enough abstract to apply across a variety of different devices so there is no need to create separate rules for each policy client. This paper covers the presentation of a management architecture to benefit common users with access to Grid Infrastructures wherever they are and whenever they need adding the "stateful" resources parameters of the new generation of Grid Services offering to reduce procurement, deployment, maintenance, and operational costs through multiple infrastructures regardless network technology, operative platform or administrative domain. The ability to link a Grid Infrastructure (GI) to users anytime and anywhere addresses new challenges, the straighter of them is to be able for handling resources and services, which might appear and disappear in a completely dynamic way. Policy-based management systems offer a more autonomous and flexible management approach by allowing the possibility of customizing network behavior to user requirements.

The paper is structured as follows. Section 2 presents related work. In Section 3, we detail the components of the Policy-based Management Architecture. In this Section, we also explain the implemented mechanisms to extend its management facilities to show its scalability and flexibility properties. Section 4 explains the policy schema that is used to deploy and to manage Grid Services. In Section 5, we present the functionality of the overall architecture and its ability to handle Grid Services' requirements implied by OGSA standard. A set of early experiments are provided in Section 6. Finally, conclusions and future work are described in Section 7.

2 Related Work

Only a few systems for monitoring, scheduling and managing resources of Grid environments are very well known and exploited. A good example of QoS provisioning in Grid is the Globus Architecture for Reservation and Allocation (GARA) [7], which is presented as an extension of Globus Resource Management

Architecture (GRMA). It introduces a generic resource object, which encompasses network flows, memory blocks, disk blocks as well as processes and adding reservation functionality as a first class entity in the resource manager architecture. Although GARA fulfils important resource management necessities, it has limitations supporting Service Level Agreement (SLA) [15] protocols and it lacks of OGSA compatibility [2], therefore it is not yet compatible with recent versions of Globus Toolkit [4]. On the other hand, the Open Grid Forum (OGF – www.ogf.org) is working on better solutions for QoS through its Grid Resource Allocation Agreement Protocol Working Group (GRAAP-WG), which has produced a "state of the art" document, laying down properties for advanced reservation in Grids [16]. Basically our approach makes use of their most relevant recommendations.

Moreover, the Grid community claims for a major entity in charge of providing high-level management to allow quick and autonomous deployment, activation and reservation of Grid Services as well as to handle QoS parameters. In this field, some projects appeared with proposals to improve the management of the Grid such Condor-G [8], Control Architecture for Service Grids [11], Data Grid [9] and Nimrod-G [10]. Policy-based middleware systems for Grid Services were presented in [12] and [13], involving technologies as active networks and the GARA architecture but without any constrains regarding OGSA compatibility. Some of the above developments are just functional improvements within the context of their respective projects, whilst Condor-G and G-QoSM [17] show their drawbacks of not being completely autonomous. Although G-QoSM is coping similar features which are presented in this paper, it is not very flexible as well. It bases its reliability into a central component, the middleware Resource Manager (RM), which may present overload problems for large amount of Grid Service's requests.

3 Policy-Based Management Architecture

Policy-based Management (PbM) [5] is a very suitable technology to manage complex heterogeneous environments such as Grid Computing. The Autonomic Management Architecture for flexible Grid Services deployment is an implementation of this technology.

This architecture deals with three different sources of resource requirements. The users QoS necessities, resource provider's availability (i.e. amount of resources free to execute new services) and services specifications according to Open Grid Services Architecture (OGSA). It is designed as a hierarchically distributed architecture, consisting of two levels; the Network Management System (NMS) and the Element Management System (EMS). The proposed hierarchical levels combine the benefits of management automation with reduction of management traffic and distribution of management monitoring activities. In this way we assure a high level of scalability in this approach. This hierarchical approach is shown in Figure 1.

The NMS is the entry point of the management architecture. It is the recipient of policies, which may have been the result of network operator management decisions or Service Level Agreements (SLAs) between Grid Infrastructure Providers (GIPs) and Grid Services Consumers (GSCs). The SLA requires reservation of resources per service as well as configuration of the network topology, which is automated by means of policies sent to the NMS. Network-level policies are processed by the NMS

Policy Decision Points (PDPs), which decide when policies can be enforced. When enforced, they are delivered to the NMS Policy Enforcement Points (PEPs) that map them to element level policies, which are, in turn, sent to the EMSs. EMS PDPs perform similar processes at the element level. Finally, the Grid-Node (GN) PEPs execute the enforcement actions at the Grid Infrastructure.

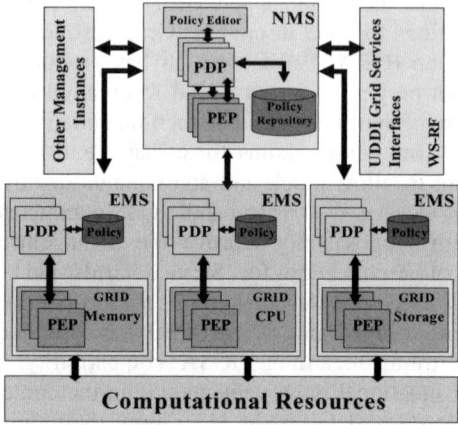

Fig. 1. The Hierarchical Policy-based Management Architecture

3.1 Components of the Network Management System

The components of the proposed PbMA for Grid Services Management are illustrated on Figure 2. They have been developed in order to support service deployment, decision-making with regards to resources control, communication interfaces with WS-Resource Framework [18] and Inter-Domain Communication respectively. We proceed now to present the details of all of them. As the NMS and EMS have similar functionality and components, we focus on the NMS and wherever applicable we note the differences between them.

Policy Editor: It exists only at the network level. It offers a GUI and a tool-set in the form of templates and wizards for the composition of policies. These are generic enough to accommodate different types of policies, thus exploiting the extension capabilities of the architecture.

Policy Manager: The role of this component is to provide a higher layer abstraction (or adaptation) that offers a global view of Grid resources to upper layer Grid services or applications in a more QoS deterministic way. The Policy Manager receives the SLA that has been agreed between GIPs and GSCs. This input, together with Grid Service resources and topological requirements received from Service Descriptor Component, is used to start the Grid Services management process.

Service Descriptor: This component is designed to retrieve the Grid service resource requirements from the WS-Resource Properties Document. In fact, it integrates a set of processes usually entrusted to the user of a Web Service, namely, to invoke a service name registry, to request a service instantiation, to check whether into the

WSDL definition of the Web Service Interface there is a declaration for a WS-Resource properties document and, finally, to retrieve all the properties elements from this one.

Domain Manager: The PDP Manager receives policies to be dispatched to the appropriate Policy Decision Point (PDP). If the corresponding PDP is not installed, it requests its download and installation. In this way, the management functionality of the system can be dynamically extended at run-time when is needed. The PDP Manager also controls the lifecycle of these PDPs.

Policy Conflict: When policies arrive at specific PDP should be checked for possible conflicts against other policies previously processed within that PDP. However, the PBMA has multiple PDPs each covering a particular functional domain (e.g. QoS, Grid Service, etc.).

Policy Decision Point (PDP): The PDP component checks for possible syntactic and semantic conflicts in policies, solves detected conflicts, makes decisions about when a policy should be enforced, forwards the policies that need to be enforced to PEP components, answers requests for decisions about configuration actions coming from the managed device and controls the policy validity period in order to uninstall expired policies.

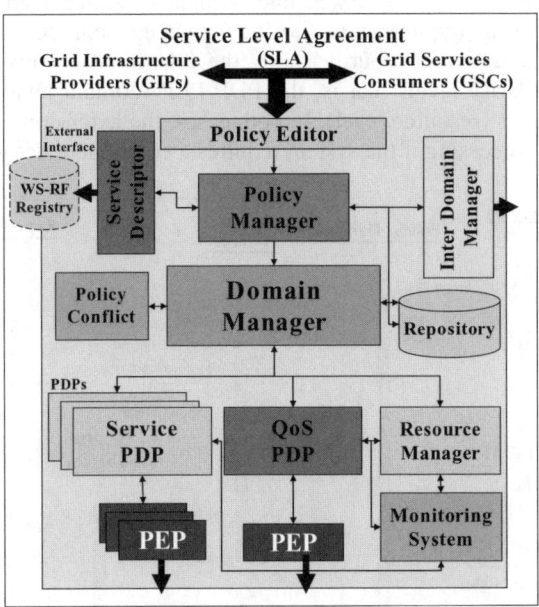

Fig. 2. The Network Level Management Architecture

Resource Manager: The role of the resource manager mainly focuses on assessing resource utilization. This component maintains information about the nodes and links of the system and can compute possible end-to-end routes for a given service, based on the network topology and resource information obtained by the monitoring system.

Monitoring System: The monitoring system is logically and physically distributed in the overall management infrastructure as well as on the Grid nodes. The monitoring system effectively collects analyses and provides the necessary information needed by the PDPs to make appropriate decisions. It is worthy to mention that the monitoring system will use the Globus Metacomputing Directory Service (MDS) [4].

Inter-Domain Manager (IDM): The IDM is in charge of implementing end-to-end negotiation of service deployment into separate Grid nodes that belong to different administrative domains, managed by different organizations.

Policy Enforcement Point (PEP): It receives policies and translate them into the appropriate commands offered by the API of the managed device, for instance in commands of the Globus Toolkit [4]. In this way the management framework is able to support heterogeneous managed devices just installing the appropriate PEP component for a particular type of managed device.

Repository: The primary role of the database is to meet the management framework software components needs. Typical information stored in the database could be policies, domain components, profiles, access rights, topological information, etc.

3.2 Architectural Flexibility and Extensibility

The architecture is dynamically extensible with new management functionality to handle new appearing requirements (i.e. new management domains in charge of different functional activities). Furthermore, the functional extension is available at two distinct granularity levels that is, the PDP-PEP (domain level) and the action-condition interpreters (resource level). In both cases, the extension might be triggered during the policy processing if the system requires a component not yet installed.

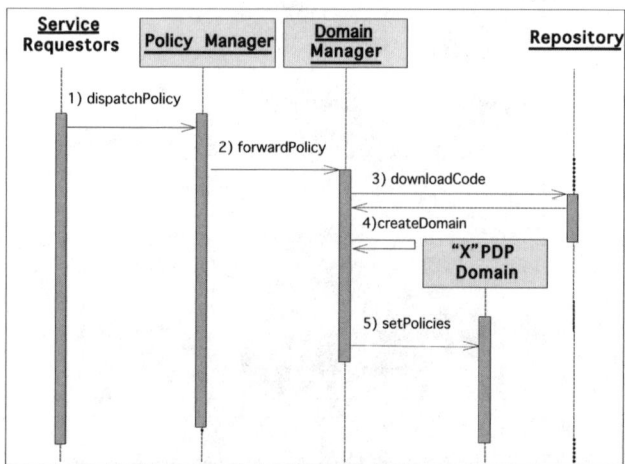

Fig. 3. Extending Management Domains

Extending Management Domains: It is the first level of extensibility in this approach. When a new management activity is required and it is not already installed on the system, a new functional management domain deployment process is triggered automatically by the system. This action is one of the main activities of the Domain Manager Component. The first instance is the Domain Manager, which is responsible for forwarding received policies to the appropriate Domain Policy Decision Point (PDP). If the corresponding PDP is not installed, the Domain Manager requests the Components Repository to download and install it, thereby extending the management functionality of the system as required.

The sequence diagram in Figure 3 illustrates how the aforementioned extension is achieved. Then, a request is raised to the Repository that initiates the installation of that component. Once the PDP is ready, the Domain Manager forwards the policy to it normally. Often, the needed functional extension would not imply the introduction of a complete management service but just the extension and/or modification of one already available.

Extending Management Action – Condition Interpreters: The extension of a PDP by dynamically installing new Action and Condition Interpreters is another option for extending the management functionality in addition to the dynamic installation of a PDP previously described. Two key classes within the PDP component are the action and condition interpreters. They provide action and condition processing logic for some policy types of those handled by the PDP. Each PDP has at least one action and condition interpreter although it might have more. Within each PDP, although drawn separately, there is a generic Action Interpreter class and a class loader class. The generic Action Interpreter class receives all requests to Action Interpreters and demultiplexes them. When a requested Action Interpreter object is not found it interacts with the Repository to download the needed code. Figure 4 shows the interactions occurring when this happens.

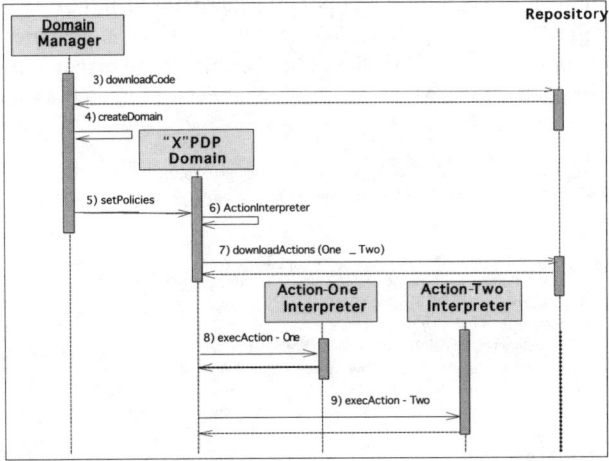

Fig. 4. Extending Management Actions – Conditions Interpreters

Removing Management Domains and Interpreters: Any new deployed instance has a limited time of live. It is a counter that is decremented once the component has finished its corresponding management activities. Once the deadline is reached the component is removed from the system. Therefore, this architecture keeps certain level of autonomy when a service instances are no longer needed

4 Management Policies Structure

Policies specify the actions that should be applied when particular conditions are met. Notifications are used to communicate to higher-level management instances or even user applications reports containing information such as the enforcement result of a policy or group of policies, the resource consumption, a performance or fault management event, etc. Policies are expressed in XML (eXtensible Markup Language) [21] and transmitted using SOAP (Simple Object Access Protocol). SOAP permits the transmission of policies as plain XML, ensuring interoperability.

Data-types are supported and there is the ability to specify relationships and constraints between different elements of a document. The architecture presented takes advantage of the properties of the XML Schema to reflect the access rights of users in relation with management functionality. That is, the framework is capable of dynamically creating, as result of management policy enforcement, restricted XML Schemas for particular users against which, XML user policies are validated. When a user obtains certain management responsibilities from the network operator, by means of delegation, it effectively is assigned one or more restricted XML Schemas, which delimit the types of policies and the policy action and condition values allowed to that user. The policy structure used in our approach is based on the IETF Policy Core Information Model [5] though simplified by defining as mandatory only those features essential for policy processing. Hence, the size of policies is considerably reduced (around five times smaller than following the PCIM model) and their processing is simpler. A section of the policy schema used in our evaluation tests is depicted in Figure 5. The policy rule consists of seven elements. First, the *PolicyRuleName* uniquely identifies the policy within the management infrastructure. Hence, in addition to the policy type identifier a sequence number is included.

```
<xsd:schema targetNamespace=  http://nmg.upc.edu/~emagana/pbma_Schema
xmlns:xsd="http://www.w3.org/2001/XMLSchema" xmlns="http://nmg.        upc.edu/~emagana/pb_Schema ">
<xsd:element name="PolicyRule">
    <xsd:complexType>
        <xsd:sequence>
            <xsd:element name="PolicyRuleName" type=   "format_Type "/>
            <xsd:element name="PolicyRoles" type=    "... "/>
            <xsd:element name="UserInfo" type=   "... "/>
            <xsd:element name="PolicyRuleValidityPeriod" type=      "... "/>
            <xsd:element name="PolicyDomain" type=    "... " minOccurs ="1"/>
            <xsd:element name="Conditions" type=    "... " minOccurs ="1" maxOccurs ="unbounded"/>
            <xsd:element name="Actions" type=    "... " minOccurs=" 1" maxOccurs="unbounded"/>
        </xsd:sequence>
    </xsd:complexType>
</xsd:element>
```

Fig. 5. Policy Schema Example

The *PolicyRoles* element identifies the Roles to which the policy applies. That is, all network elements developing a role listed in the policy is expected to respond to it.

The *UserInfo* contains the identifier of the user that is introducing the policy in the framework. This identifier is used to select the restricted XML Schema against which the user policy should be validated with. The policy expiration date is contained within the *PolicyRuleValidityPeriod* element. Usually the expiration date is just given with the day and hour the policy starts and finishes being valid. Nevertheless, filters specifying concrete months, days and hours during which the policy is not valid can be also introduced. The *PolicyDomain* element is used for the correct processing of policy sets. A policy set is a group of policies that should be processed in a particular way, i.e. atomically, sequentially, etc.

The *Conditions* element includes all policy conditions. Conditions can be either compound or simple and refer to an hour of the day, an IP flow, a concrete notification or a managed device status. The modules needed to monitor these conditions, if any, are also extracted from the Conditions element information. This element is optional; when not included, the framework interprets that the policy action should be enforced directly. The *Actions* element contains the action type and parameters as well as information about the module responsible of enforcing this action. At least one Actions element is mandatory in all policies but there can be more that one. The defined policies have been categorized according to the domain of management operations, policies that belong to a specific domain are processed by dedicated Policy Decision Points PDPs and PEPs.

5   Management and Deployment of Grid Services

The Grid service management process starts when an authorized user requests a Grid Service to the PBMA through its policy editor interface. At this time the client will specify the name of the requested Grid service and the QoS requirements for its deployment. Once the PBMA receives the client request, it exchanges information related to the service in order to process the resources service requirements.

5.1   Service Level Agreement (SLA)

Previously to any Grid Service request, every GSC has concluded a SLA with at least one GIP. In this agreement a specific QoS level will be satisfied for every service running on their providers' infrastructure. We assume that QoS is quantified in levels like "diamond" (efficiency guaranteed), "gold" (completely distributed), "silver" (searching alternatives) and "bronze" (best effort).

5.2   Grid Service Requirements – Network Level Policy Creation

The PBMA has to configure its Grid Infrastructure (GI) with the constraint of matching the requested client needs with the available resources at the Grid target nodes where a Grid Service instance will be running. Our architecture merges the requirements from the client, the providers' resources availability as well as the service specification requirements and creates the Network Level (NL) Policy for its corresponding Grid Service. We would like to highlight that, this features is one of the most important novelties in this approach. In the context of this work, the Policy Manager has to contact with the Service Descriptor in order to receive the Grid service specifications, which will be extracted from WS-Resource properties document. This document has been associated with the WSDL porType defined by Grid Service Instance requested via UDDI Registry. In Figure 6 we show fragments of the above mentioned documents that were used during the evaluation phase described on Section 6.

OGSA Compatibility: As previously stated, the proposed architecture is compatible with OGSA. The Service Descriptor Component uses XML requests to tag service data, SOAP to transfer it through the network, WSDL for describing the services available and finally UDDI is used for listing what services are available. SOAP, as a data communication format, offers different advantages in order to extract the information necessary for our architecture. Essentially, we parse the SOAP files into the OGSA Policy Descriptor and just extract the Grid service specifications mentioned above. The Service Descriptor sends back the parsed information in Java format to be processed by the Policy Manager and thus to create the NL policy.

Inter-Domain Communication: Just in the case that the requested service needs resources that belong to different administrative domains, the Policy Manager contacts the Inter Domain Manager to start the resources negotiation with another domain. Due to the fact that communications are always under XML format, there should not be fallen problems regarding communication amongst management entities. At this moment, the second domain is starting its own resource management and scheduling processes.

5.3 Network Level Policy Analysis Sequence

The Policy Manager (PM) sends the just created NL Policy to the PDP Manager. It later evaluates the conditions of the policy and tries to match them with the availability of resources on the Grid Infrastructure. In order to complete these functions, it firstly has to analyze the QoS of the requested service. To do so it contacts the Resource Manager (RM), whose task is to analyze the network topology and resources information received from the monitoring system. The checking of availability resources is realized for every resource involved in the establishment of a service into the Grid topology (i.e. memory, bandwidth, storage, etc).

5.4 Selection of Grid Target Nodes

The RM tries to find suitable set of Grid nodes that satisfies the requirements given by the NL-Policy. If the search is not limited by other constraints, a set of different nodes will result. All these Grid nodes are candidates for the allocation of the service because they fulfill the resource and topological requirements expressed in the NL-Policy. However, a service has additional requirements specified in the Grid Service Data. The RM does not have access to such information because this lies within the domain of the Service PDP. Therefore, the PDP Manager decides the set of final nodes to allocate the service.

5.5 Grid Nodes Configuration – Element Level Policy Creation

At this time, the PDP Manager has to decide which resources will be part of the final set of Grid nodes that will be configured to execute the Grid service with the specified

QoS level. In this component, an appropriate algorithm carries out the selection of the best nodes and forwards the policy to the QoS PDP. Next, the QoS PDP sends the decision to its corresponding PEP. The QoS PEP will transform the request into a set of appropriate element-level QoS policies (one policy for each of the Grid nodes selected) and it will send the policy into the EMS of the established nodes. Once the QoS policy is enforced, the EMS calls an activation method of Globus interfaces for each node, thus ending the configuration process.

5.6 Grid Service Resources Activation

Since the enforcement of the QoS policies has successfully terminated in all nodes, the PDP Manager starts processing the activation policies by forwarding them to the corresponding PDP to be evaluated. If there are no conditions or actions to be processed at the network level, it forwards the policy to the PEP of the involved Grid nodes. The activation PEP enforces the policy that assigns the resources using the interfaces offered by the Globus ToolKit. The result of the Grid resources activation is forwarded back to the network-level through the element-level for control and fault management proposes. At this moment the service requested is running with the agreed QoS level. The Monitoring System Component updates the QoS PDP with resource utilization. Therefore, in case that one or more of the Grid nodes can not offer anymore their resources, the PBMA will restart the Grid Service Management Process to find new Grid nodes which offer similar resources.

```
<!-- ========= WSDL Interface for Newton's Method Application ============ -->
<?xml version="1.0" encoding="UTF-8"?>
<wsdl:definitions targetNamespace="http://nmg.upc.es/Newton'sMethodExample" ...>
 <wsdl:types>
   <xsd:schema
     <xsd: import
   targetNamespace="http://nmg.upc.es/NewtonsMethodExample_Properties"
     <xsd:attribute name="ResourceProperties" type="xsd:Newton's Method"/>
     ...
 <!-- == WS-Resource Properties Document for Newton's Method Application == -->
 <wsdl:portType name="Newton's Method"
     wsrp:ResourceProperties= "intf:GenericMethodProperties">
     <xsd:sequence>
       <xsd:element maxDistribution="5" minDistribution="1" name=" " ...
       <xsd:element amountMinMemory="20" amountMaxMemory="250" name=" "...
       ... "wsa:EndpointReferenceType"/>
 </wsdl:portType>
```

Fig. 6. WSDL and WS-RF Interfaces

6 Early Experiments

For the evaluation of the architecture, we have used a set of heterogeneous nodes (i.e. Intel[®] and AMD[®]) with different operating platforms (Windows 2000 and Linux Fedora 4) as well as different amount of resources to share, such as our Grid Infrastructure. The only homogeneous feature of these nodes is that all of them have Globus Toolkit installed and they have been signed by the same Certificate Authority (CA), our own CA and not the standard Globus-CA for security reasons. A random process generator was used to dispatch processes to the network nodes so that we emulated normal "working day" conditions for all the nodes involved to assure results according to real Grid environments. The entire architecture was programmed in Java platform. The components of the PBMA were CORBA objects. Policies were expressed in XML and the interfaces were implemented with the standard Interface Definition Language.

The Grid service application selected to be distributed along the Grid environment was Newton's Method [14], a generalized process to find an accurate root of the equation $f(x) = 0$. This method has many physical and astronomical application and basically was selected because, the number of the algorithm's iterations is considerable, therefore the amount of process resources needed is really significant, and also because this method does not imply a complex code to implement.

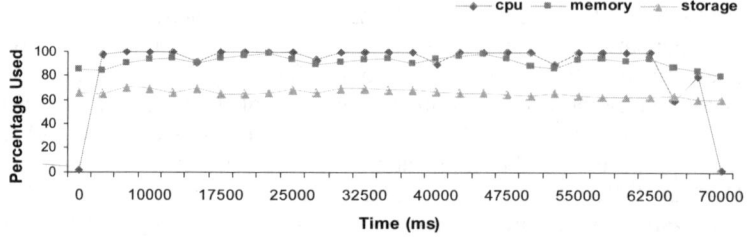

Fig. 7. Resources' Performance with Best Effort Management Policy

The first results are shown in Figure 7. In this experiment we have plotted the percentage of resources (processor, memory and storage) used by Newton's application during five sequences with around of one thousand iterations for each one of them (i.e. the same application will be executed five times in order to obtain a more precise result). It is clear that during the application processing time the resources are at their maxim used capability and any other process will be queued and executed after completing the equation. In this example, the total time is around seventy thousand milliseconds (70,000ms), whilst in some astronomic applications it could reach days in simulation.

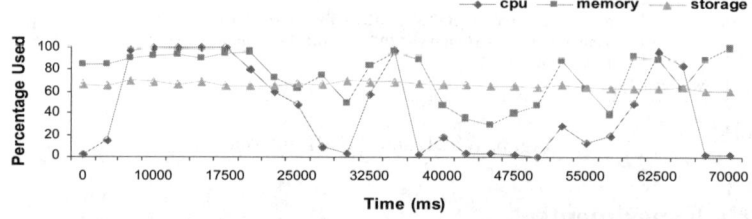

Fig. 8. Resources' Performance with Golden Level Management Policy

The following test in our evaluation process is the insertion of a QoS Policy in the application above described. In this test, the policy demands the maximization of resources exploitation in a minimal amount of time (diamond QoS level policy). To obtain this level of efficiency, the PBMA used different nodes for each application sequence with a maxim of five nodes. These service requirements were obtained from the WS-Resource Properties document, presented in Figure 6.

The Grid nodes continue their normal activity until the system detects that, the used percentages of resources are at low levels in some of them. At this moment, their resources are able to be shared with any client within the Grid Infrastructure. The management system distributes the application to the set of selected nodes and remotely executes the applications. Figure 8 plots the resources monitoring activity in one of the selected nodes. In order to compare the new processing time versus the previous graph we have selected the same node. The analysis of this graph illustrates the benefits of our approach because the system offers substantial savings on time and used resources. Moreover, this graph shows the time needed by the server to gather the results provided by the other selected servers and to obtain the final result.

Finally, Table 1 shows the elapsed times measured during policy Grid Service resources merging, resources selection, setting nodes up and deployment of the service application. The driving force to improve this architecture will be the possibility to reduce these times, which in turn are directly proportional to the efficiency of the Autonomic Management Architecture for Grid Services Management.

Table 1. Consuming Times by Grid Service Deployment Process

Actions	Timing
Network Level Policy Creation (SLA - OGSA)	750 ms
Resource Monitoring	2200 ms
Resource Selection and Reservation	3525 ms
Policy Execution	30125 ms

Our current work also includes merging the presented architecture with autonomic gateways [20]. This design will help Grid's designers to evaluate and monitor more precisely the usage of their network resources. This architecture will be deployed on a large scale basis. It is currently evaluated on Grid5000 [19] platform.

7 Conclusions and Future Work

This paper describes the features, implementation details and advantages of the Autonomic Management Architecture for flexible Grid Services deployment based on Policies. This approach is able to deploy and to manage Grid Services instances into heterogeneous networks configured such as Grid Infrastructure. The PbMA extracts computational resources' requirements from WSDL documents. It merges these requirements with QoS necessities of Grid Services' consumers. Then, our approach collects information about resources availability in the Grid Infrastructure Providers. And finally, it schedules Grid Services on best available computational resources, getting a load balancing through all over the nodes.

Although our approach is focused on Grid environments where the wide range of nodes will offer small amounts of resources, the solution is not only limited to this domain of users/clients. The results presented in this paper show the advantages of our architecture. In fact, the reduction in process time, regarding a Grid Service did not manage by our approach, is around the 57% and the percentage of resources used per second in all Grid Infrastructure is much more bigger. Therefore our environment in terms of used resources is more efficient. We are offering to individual, small business and industrial users, an automatic and self-management effective access to massive amounts of computing, network and storage resources, reducing procurement, deployment, maintenance, and operational cost. The ongoing work involves the development of components necessary for supporting system fault tolerance. The last phase will consist of testing the performance of the architecture on more complex networks like the French initiative, Grid5000 [19].

References

1. Foster, C. Kesselman and S. Tuecke, "The Anatomy of the Grid: Enabling Scalable Virtual Organization". Int. Journal of Supercomputer Applications, vol. 15 no. 3, USA 2001.
2. Foster, C. Kesselman, J. Nick, and S. Tuecke, "The Physiology of the Grid: An Open Services Architecture for Distributed Systems Integration", GGF June 2002.
3. J. Nabrzyski, J. M. Schopf and J. Weglarz, "Grid Resource Management State of the Art and Future Trends" Kluwer Academic Publishers. Boston, USA October 2003.
4. The Globus Project Site: http://www.globus.org/
5. IETF Policy Site: http://www.ietf.org/
6. K. Czajkowski, D. Ferguson, I. Foster, J. Frey, S. Graham, I. Sedukhin, D. Snelling, S. Tuecke, W. Vambenepe, "The WS-Resource Framework" Globus Alliance Group http://www.globus.org/wsrf/ May 2004.
7. Foster, C. Kesselman, C. Lee, R. Lindell, K. Nahrstedt, and A. Roy. "A Distributed Resource Management Architecture that Supports Advance Reservation and Co-Allocation". In the International Workshop on Quality of Service, June 1999.
8. J. Frey, T. Tannenbaum, et al., "Condor-G: A Computation Management Agent for Multi-Institutional Grids", Proceedings of HPDC10, San Francisco, USA, August 2001.
9. DataGrid Site: http://www.eu-dataGrid.org
10. R. Buyya, "Nimrod/G: An Architecture for a Resource Management and Scheduling in a Global Grids" 4th International Conference on High Performance Computing in Pacific Region. Los Alamitos, USA, 2000.
11. S. Graupner, V. Kotov, A. Andrzejak and H. Trinks, "Control Architecture for Service Grids in a Federation of Utility Data Centers", HP Labs., Palo Alto, USA. August 2002.
12. K. Yang, A. Galis, C. Todd. "Policy-Based Active Grid Management Architecture". 10th IEEE International Conference on Networks ICON 2002.
13. E. Magaña, E. Salamanca and J. Scrrat, "Proposal of a Policy-Based System for Grid Services Management", APGAC'04/ICCS'04 Krakow, Poland June 2004
14. Steven C. Chapra and R. Canale, "Numerical Methods for Engineers: With Software and Programming Applications". McGraw-Hill ISBN: 0072431938.
15. Sahai, S. Graupner, V. Machiraju and A. Moorsel. "Specifying and Monitoring Guarantees in Commercial Grids through SLA". In 3rd IEEE/ACM CCGrid2003, Tokyo, Japan 2003.
16. J. MacLaren. Advance Reservations: State of the Art. GGF GRAAP: http://www.ggf.org/ Meetings/ggf7/sched-graap2.0
17. R. J. Al-Ali, K. Amin, G. Laszewski, O. Rana, et al., "Analysis and Provision for Distributed Grid Applications". Journal of Grid Computing, Kluwer, 2004.
18. WS-Resource Framework Site: http://www.globus.org/wsrf/
19. Grid 5000 Project. Web Site: http://www.Grid5000.org/
20. M. Chaudier, et al, "Towards the design of an autonomic network node", Int. Working Conference on Active and Programmable Networks (IWAN), Nice, France, Nov. 23, 2005.
21. A. Vedamuthu, et al., "Web Services Policy 1.5 – Framework". W3C Working Draft.

Variations and Evaluations of an Adaptive Accrual Failure Detector to Enable Self-healing Properties in Distributed Systems

Benjamin Satzger, Andreas Pietzowski, Wolfgang Trumler, and Theo Ungerer

Institute of Computer Science
University of Augsburg
D-86135 Augsburg, Germany
{satzger, pietzowski, trumler, ungerer}@informatik.uni-augsburg.de
http://www.informatik.uni-augsburg.de/en/chairs/sik/

Abstract. The initiatives *Organic Computing* and *Autonomic Computing* introduced challenging visions for future computer systems. They address the growing complexity of these systems that demands for new ways to control them. Future systems should be able to adapt dynamically to the current conditions of their environment. They should be characterised by so-called self-x properties like self-configuring, self-healing, self-optimising, self-protecting, and context-aware. For the incorporation of self-healing capabilities into distributed systems the detection of failures is a crucial part. Recently we proposed a new failure detector that can be described as an adaptive accrual algorithm. It has been designed for flexible generic usability as a basis to realise self-healing of distributed systems. This paper introduces variations of the proposed basic algorithm to improve its performance and provides an evaluation of all algorithms using message delay and loss models of the internet.

1 Introduction

Organic Computing (OC) [22,17,23] and Autonomic Computing (AC) [11,14] both identify the exploding complexity as a major threat for future computer systems and postulate so-called self-x properties for these systems. To achieve these goals both the OC [19] and the AC community [14] regard monitoring information as a basis for organic or autonomic systems.

The Autonomic Middleware for Ubiquitous eNvironments (AMUN) [25], also called Organic Computing Middleware for Ubiquitous Environments (OCμ), is a middleware for distributed systems. The OCμ architecture allows to plug in features as services and monitors to enrich the whole system with certain self-x properties, e.g. self-configuring [26], self-optimising [27], self-protecting [18], and self-healing [21].

A failure detection service is one fundamental part of the self-healing capabilities of OCμ. Failure detectors generally provide information on failures of components of distributed systems. Typically distributed systems consisting of a finite set of processes or nodes are considered with a local failure detector

P. Lukowicz, L. Thiele, and G. Tröster (Eds.): ARCS 2007, LNCS 4415, pp. 171–184, 2007.

attached to each process, see for example [4]. Failure detectors return a list of processes they are suspecting to have crashed.

This paper proposes variations of the adaptive accrual failure detector published in [21]. Accrual failure detectors decouple monitoring and interpretation. That makes them applicable to a wider area of scenarios and more adequate to build generic failure detection services. In this paper we evaluate the quality of the failure detector presented in [21], which we will call in the following *basic failure detector* or *basic algorithm*, in a more comprehensive way than this has been done before. Furthermore we present variations of the basic algorithm together with evaluations of these variations.

The paper is organised in five sections. Section 2 gives a short overview of the state of the art of failure detectors and related work. Section 3 presents the basic failure detection algorithm that serves as a basis of the proposed variations of Section 4. Then, Section 5 describes the simulation results. Finally, section 6 concludes the paper and gives an overview of future work.

2   State of the Art and Related Work

Completeness and accuracy. Several impossibility studies [3,15,8] show that perfect failure detectors cannot exist in asynchronous distributed systems. The major reason is the impossibility to distinct with certainty whether a process has failed or the communication network is just slow.

Chandra et al. [4] introduced the idea of failure detectors as an unreliable distributed oracle at which it is possible that (1) a process has failed but is not suspected as well as (2) a process is suspected but has not failed. Moreover a failure detector can change its mind for example stopping to suspect a process it previously suspected. In consequence the authors of [4] characterise failure detectors by specifying their properties regarding *completeness* and *accuracy*. Completeness refers to failure detectors eventually suspecting crashed processes, while accuracy restricts the mistakes that a failure detector can make.

Monitoring strategies. There exist two main monitoring approaches for failure detectors: *push* and *pull*. Assuming process p has a failure detector monitoring q. Using a push failure detector q has to send heartbeat messages to p. This information is used by p to draw conclusions about q's status. A simple failure detection algorithm using the push approach [5] works as follows: q sends heartbeat messages at regular time intervals Δ_i to p. When p receives a heartbeat messages it trusts q for a certain period of time Δ_{to}. If this period elapses without receiving a newer heartbeat p starts to suspect q.

In systems with a pull failure detection (e.g. [12]) the monitored node adopts a passive role. p monitors q by sending "are you still alive"-messages every Δ_i. If p doesn't receive an answer from q within a certain period of time Δ_{to}, p is suspecting q. Failure detectors using the push paradigm have some benefits compared to pull failure detectors. They need only half the messages for an equivalent failure detection quality. Furthermore it is rather hard to determine the timeout Δ_{to} as you have to take two messages into account which are both sent over the network and subject to network delays.

Adaptive failure detection. Adaptive failure detectors [7,5,10] are able to adjust to changing network conditions. The behavior of a network can be significantly different during high traffic times as during low traffic times regarding the probability of message loss, the expected delay for message arrivals, and the variance of this delay. Thus adaptive failure detectors are highly desirable.

Chen et al. [5] propose a well-known adaptive failure detection approach based on a probabilistic analysis of network traffic. The protocol uses sampled arrival times to compute an estimation of the arrival time of the next heartbeat. The timeout is set according to this estimation plus a constant safety margin, and is recomputed after each arrival of a new heartbeat.

Bertier et al. [1] combine Chen's estimation with another estimation developed by Jacobson [13] for a different context. Their approach is similar to Chen's, however, they don't use a constant safety margin but compute it with Jacobson's algorithm.

Accrual failure detection. The principle of an accrual failure detector, introduced by Hayashibara et al. [10], is not to output whether a process is suspected to have crashed or not. Rather they give a suspicion information on a continuous scale whereas higher values indicate a higher probability that the monitored process has failed.

Hayashibara et al. propose a so-called φ failure detector that is based on an estimation of inter-arrival times assuming that inter-arrivals follow a normal distribution. They also motivate the benefits of accrual failure detectors over conventional boolean failure detectors. As principal merit they indicate that accural failure detectors favour a nearly complete decoupling between application requirements and the monitoring environment.

Lazy failure detection. Lazy failure detection protocols [7] use application messages to monitor other processes whenever this is possible.

Our basic algorithm [21] can be classified as an adaptive accrual failure detector. It uses an approach to compute suspicion information based on a histogram density estimation. As this failure detection algorithm is further evaluated in the following and also the basis of the proposed variations in Section 4 it is now briefly revisited.

3   Basic Failure Detection Algorithm

We are considering two processes p and q where p is monitoring q. The only task of q is to send heartbeat messages to p every Δ_i seconds. Process p manages a list S where the inter-arrival times of the received heartbeats are stored. This list is called *sampling window* and has the maximal size η i.e. it always contains the last η calculated inter-arrival times. Furthermore p stores the time of the last received heartbeat called *freshness point*.

Process p computes the failure probability of q by counting the number of elements in S that are smaller or equal than the time that has been passed since the last freshness point. This time is denoted with t_Δ, the respectively elements of S with S^{t_Δ} ($S^{t_\Delta} = \{x \in S \mid x \leq t_\Delta\}$). The actual failure probability is the normalised number of elements in S^{t_Δ}: $\frac{|S^{t_\Delta}|}{|S|}$, where $|S|$ is the current size of S. The computation of the suspicion value is based on the estimation of the cumulative distribution function of the inter-arrival times using their cumulative frequencies.

For a more detailed explanation of this algorithm we refer to [21].

4   Variations of the Basic Algorithm

In the following we present two variations of the basic algorithm explained above.

4.1   A Different Freshness Point Strategy

The first variation of the basic failure detection algorithm that is presented in the following is inspired by the failure detector of Chen et al. [5]. One problem with the heartbeat sampling and freshness point strategy as used in the algorithm described above is the dependence of the failure probability on the previous heartbeat. Assuming again p is monitoring q and p is waiting for the i-th heartbeat from q, then the failure probability of q not only depends on the arrival time of the i-th heartbeat m_i, but it is also depending on the past receipt time of the $i - 1$-th heartbeat m_{i-1}. In fact this time has a big influence on the current failure probability. If m_{i-1} had arrived "fast" then p has to wait a longer time for m_i since the last freshness point which is the receipt time of m_{i-1} has been set early. Thus the failure probability will become higher as if m_{i-1} had arrived "late". In the latter case the freshness point had been set later and therefore the failure probability was lower.

To circumvent this dependency here it is proposed to use the sending time plus the average network delay instead of the receipt time for the freshness point. Being able to do this q has to piggyback the sending time of each heartbeat according to its local clock. p manages a variable $\Delta_\varnothing$ that represents the average sending delay of messages. With every receipt of a new heartbeat message p updates the variable $\Delta_\varnothing$. Let t_s be the time q sent the i-th heartbeat to p according to q's local clock. Let t_r be the time p received the i-th heartbeat according to p's local clock. Furthermore let n be the size of the sampling window. Then, $\Delta_\varnothing$ is calculated as the mean of $t_r - t_s$ of the last n received heartbeats. Please notice that the method introduced here to abolish the dependence on the last heartbeat is not based on synchronised clocks. The freshness point f is now calculated as $f = t_s + \Delta_\varnothing$ instead of $f = t_r$ assuming the i-th heartbeat has been received. Thus the freshness point and the failure probability isn't influenced by one previous heartbeat arriving early or late. The values of the sampling window S also change slightly as each sample is calculated as the time that has elapsed since the last freshness point to the receipt time of the actual heartbeat and the freshness points are now set differently.

Figure 1 shows the basic failure detection algorithm modified according to the concepts presented here.

Process q:
send heartbeat meassage to p every Δ_i and append sending time t_s to the message

Process p:

$f = -1$	//freshness point
$S = nil$	//S is initialised as an empty list
η	//max size of S (e.g. 1000)
$\Delta_\varnothing$	//the average sending delay of the last η sent messages
t_s	//the sending time according to q's clock

upon receive heartbeat message m_j at time t_r

$\Delta_\varnothing = \frac{\sum_{i=1}^{\eta} t_r - t_s}{\eta}$
   if $f == -1$ then $f = t_s + \Delta_\varnothing$
   else
      $t_\Delta = t_r - f$
      $f = t_s + \Delta_\varnothing$
      append t_Δ to S
      if size of $S > \eta$ then remove head of S endif
   endif

on call of get_failure_probability of q at time t

   $t_\Delta = t - f$
   $|S^{t_\Delta}|$ = number elements in S that are lower or equal t_Δ
   $|S|$ = number of elements in S

   return $\frac{|S^{t_\Delta}|}{|S|}$

Fig. 1. A failure detection algorithm with a different freshness point strategy

The influence of this variation on the performance of our failure detector will be analysed in Section 5. In the following a second variation of the basic algorithm is presented that is based on histogram smoothing.

4.2 Histogram Smoothing

In the basic algorithm as described in Section 3 the sampling window S is a list containing the last η sampled inter-arrival times. To compute a failure probability the cumulative frequencies of the entries in the sampling window are used. The resolution level of the cumulative frequencies is at the resolution of the data - no certain binwidth is used to cluster the data.

But there are some advantages that come along with the division of the data into bins. Then, the sampling window doesn't consist of the values of the sampled data, but only the information how many values a bin contains. For instance $S = [1.083s, 0.968s, 1.062s, 0.993s, 0.942s, 2.037s, 0.872s]$ could become to

$S = [0s, 1s) : 4, [1s, 2s) : 2, [2s, 3s) : 1$, while $[0s, 1s) : 4$ denotes four arrivals in the time interval $[0s, 1s)$. It is obvious that the use of such a bin-based representation of the data allows for a faster generation of a failure probability. That is valid due to the fact that the counting of elements that are lower or equal to a certain elapsed time only depends on the number of bins which is typically clearly smaller than the number of samples.

Another advantage of the representation of the sampling window as a histogram is that it allows for a simple smoothing process. The heartbeat data that are sampled over time are subject to random variations due to for instance the unpredictable behaviour of the network. There exist methods for reducing the effects of random variation of sampled data. The purpose for this is to reveal more clearly the underlying basic distribution of the data. A technique that can be used to achieve this is called *smoothing* that can be used to smooth histograms. A smoother can be seen as a kind of a weighted averaging process. The aiming value is transformed by an averaging of the values in its neighbourhood. The size of the neighbourhood that is taken into account has to be set in an appropriate way. The parameter that characterises this amout of neighbouring values is called *smoothing parameter*. Generally, the larger the smoothing parameter is, the smoother the result will be.

A very simple yet fast smoother is described in the following. This simple technique is suitable to use within the failure detector as the smoothing has to be repeated basically every time a new hearbeat is arriving. Each band of the histogram is smoothed by averaging over a moving window. The smoothing parameter k determines the size of the moving window which is set to $2k + 1$. If the window runs off the end of the histogram bands of size 0 are considered. For further readings about smoothings techniques we refer to [9,24].

The choice of the smoothing parameter is crucial. The larger the value k is, the smoother the resulting histogram. However, if k is chosen too large oversmoothing occurs with loss of essential histogram features. With the usage of histogram smoothing the basic failure detection algorithm only changes in calculating the failure probability based on the cumulative frequencies of the smoothed histogram instead of the unsmoothed cumulative frequencies. Using a smoothing technique can cause the failure detector to be more robust to random variations.

In the following the presented algorithms are experimentally evaluated.

5   Evaluation

This section presents results of four performance measurements of the basic failure detection algorithm in comparison with the failure detectors of Chen et al. [5], Bertier et. al [1], and the accrual φ failure detector of Hayashibara et al. [10] as well as the results of four measurements of the basic failure detector in comparison with algorithms implementing the variations presented in 4.

5.1   Experiment Setup

There exists an infinite set of environments regarding the computing devices and their interconnection in which we could test our failure detector. As we didn't want to pick one test environment we took the decision to generate the data for the evaluation. This has the benefit that the experiments are reproducible and independent of any special unique properties.

The data needed for evaluation consists of the arrival times of the heartbeat messages. We generated the arrival times based on studies of end-to-end internet packet delay and loss behaviour [2,20,16,6]. Bolot [2] and Mukherjee [16] reason that the internet end-to-end delay distribution they experienced in their experiments is best modeled by a shifted gamma distribution. Sanghi et al. [20] encountered packet loss rates between 2.1% and 10.1% in their measurements. Dam et al. [6] selected a site in the US, sent ping packets at regular intervals and noted the RTT for each ping packet. The closest gamma distribution fit for the packet delay of this experiment turned out to be a shifted gamma distribution with shape parameter 2.0 and scale parameter 2.8.

In the following experiments we distinguish between the *unconditional* loss probability and the *conditional* loss probability [2]. The unconditional loss probability ULP represents the mean rate at which heartbeats are lost. The conditional loss probability CLP determines the probability with which a heartbeat is lost given that the previous heartbeat has been lost. This can be used to model bursty loss behaviour.

We conducted four experiments to compare our basic failure detection algorithm against other state of the art failure detectors. Per experiment we generated one million heartbeat messages using a shifted gamma distribution with shape parameter 2.0 and scale parameter 2.8 to model the message delay. The heartbeat interval Δ_i has been set to 10 seconds. The experiments differ in the modeling of the message loss.

Experiment 1.1: ULP and CLP: 2% (non-bursty message loss behaviour).
Experiment 1.2: ULP and CLP: 10% (non-bursty message loss behaviour).
Experiment 1.3: ULP: 2%, CLP: 10% (bursty message loss behaviour).
Experiment 1.4: ULP: 10%, CLP: 50% (bursty message loss behaviour).

Within these settings we compare our basic failure detection algorithm with the well known failure detectors of Chen et al. [5] and Bertier [1] and the accrual φ failure detector of Hayashibara et al. [10]. We measure the algorithms' performance according to two metrics [5]:

mistakes λ_M: This measures the numbers of wrong suspicions per second.
detection time T_D: This is the average time that elapses since the crash of q until p starts to suspect q permanently.

Being able to compare the accrual and non-accrual failure detectors we have to transform the accrual failure detectors into conventional failure detectors. Therefore you just have to choose a threshold T. If the level of suspicion for q is lower than this threshold then q is not suspected to have failed. If the level of suspicion crosses T then q is assumed to have crashed.

The next barrier to compare the algorithms are their different tuning parameters. These influence the time when a failure detector starts/ends to suspect a process. For the accrual failure detectors the tuning parameter is the threshold T. For Chen's failure detector the tuning parameter is the safety margin α. This is a constant period of time that is added to the estimated heartbeat arrival time. The failure detector of Bertier has no tuning parameters. Being able to compare the different failure detection algorithms we measure the behaviour of each of the failure detectors using several values of their respective tuning parameters.

To compute the detection time of the failure detectors we assume that a crash would occur exactly after successfully sending a heartbeat message. Then we measure the time it takes until the failure detector reports a suspicion. This corresponds to the worst case situation. This method to compute the worst-case detection time has also been used in [10].

Finally we set the window size for all algorithms and experiments to 1000 samples. This means that the computations of the failure detectors rely only on the last 1000 heartbeat message samples. Furthermore we start our measurements not until 1000 heartbeats have been received to grant a warmup phase.

In the same manner as the comparison of the basic failure detection algorithm with Chen's, Bertier's, and Hayashibara's failure detectors we made four experiments to compare the basic failure detection algorithm with the proposed modifications of section 4. This means we used exactly the same parameters to generate the arrival times of the heartbeat messages. These experiments are named Experiment 2.1 - Experiment 2.4 respectively. The following three algorithms have been used within these experiments:

basic: Our basic failure detection algorithm (see Section 3).
fp: A variation of our basic failure detection algorithm using the different freshness point strategy (see Section 4).
smooth: A variation of our basic failure detection algorithm using the histogram smoothing technique with 100 bins and the smoothing parameter k set to 2 (see Section 4).

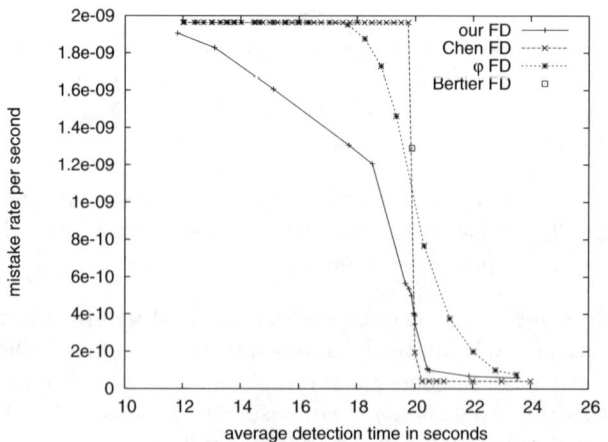

Fig. 2. Results of Experiment 1.1

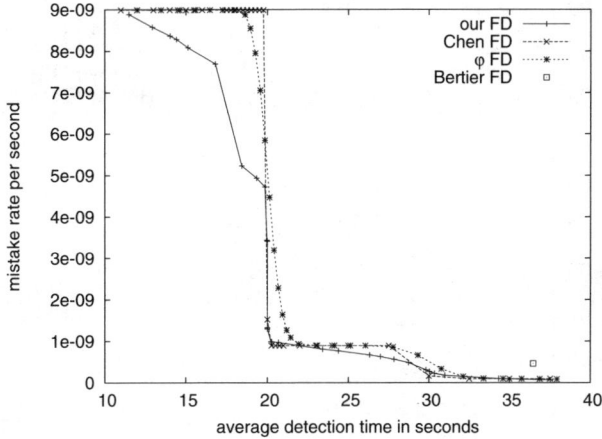

Fig. 3. Results of Experiment 1.2

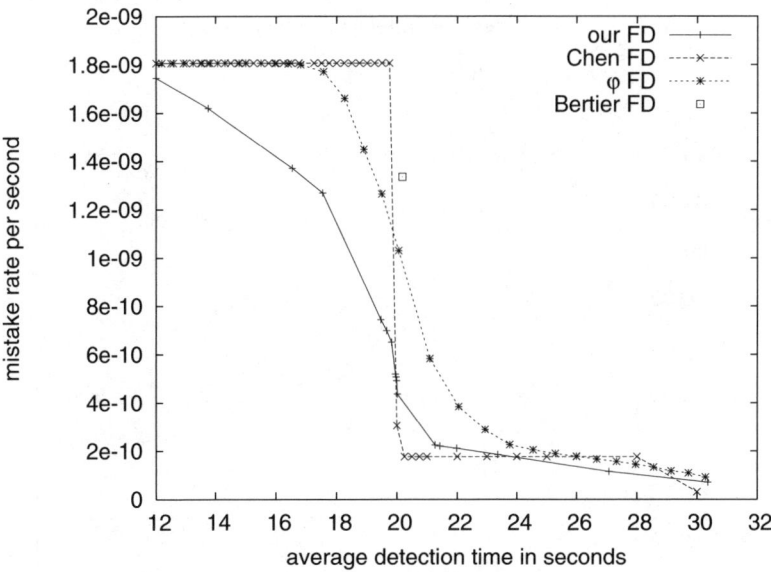

Fig. 4. Results of Experiment 1.3

5.2   Results

The results of the performance measurements of the eight conducted experiments are depicted in the figures 2 to 9. All figures show the detection time on the horizontal axis and the mistake rate at the vertical axis. Values near the lower left corner represent a short detection time with few mistakes. Every variation of the tuning parameters of the failure detectors represents a tradeoff between the failure detection speed and the mistake rate and produces one datapoint in the figures of the results of the experiments.

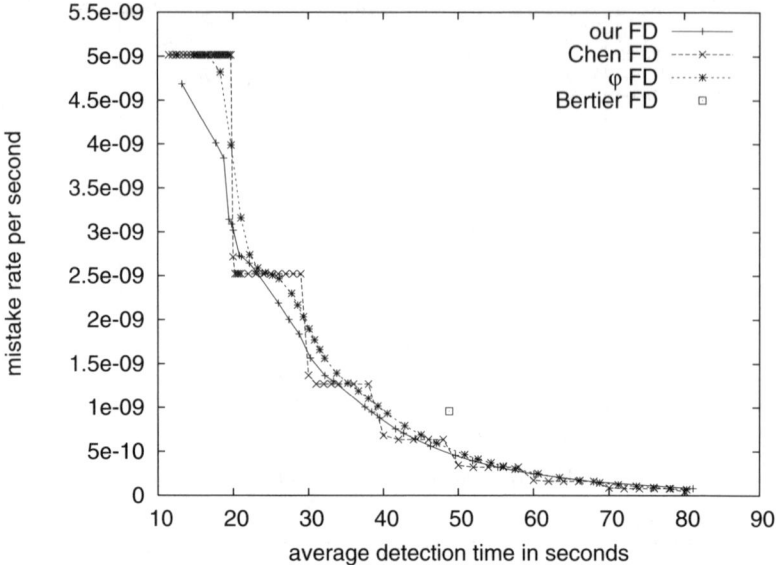

Fig. 5. Results of Experiment 1.4

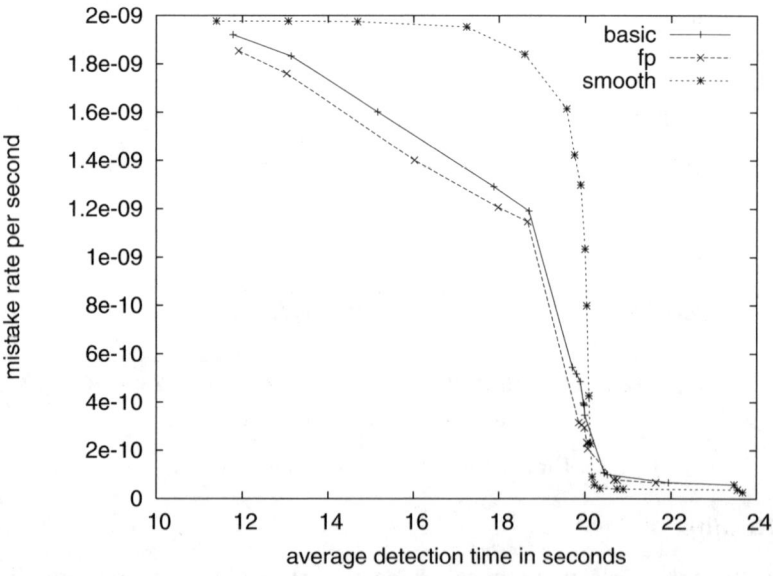

Fig. 6. Results of Experiment 2.1

The results of the experiments 1.1 to 1.4 show an excellent behaviour of our basic failure detection algorithm compared to the other three algorithms. Apart from this fact our failure detector is more flexible than Chen's, Bertier's, and other non-accrual failure detectors. In comparison to the φ failure detector of

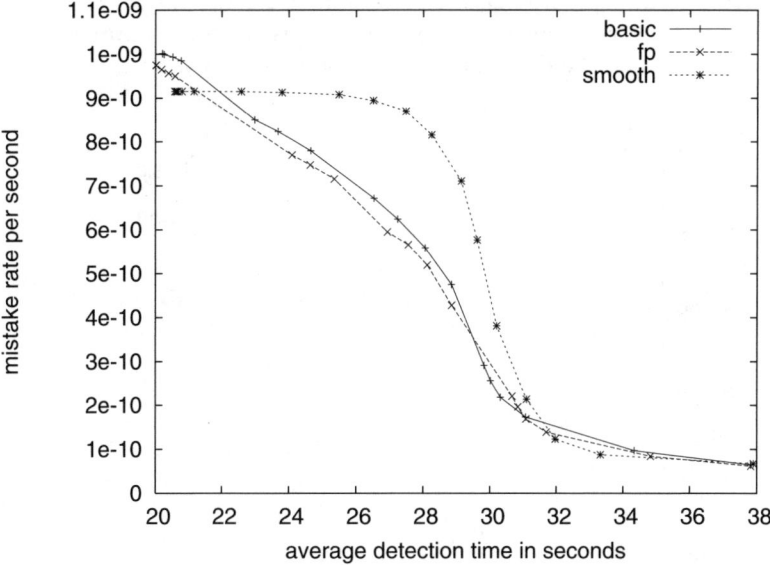

Fig. 7. Results of Experiment 2.2

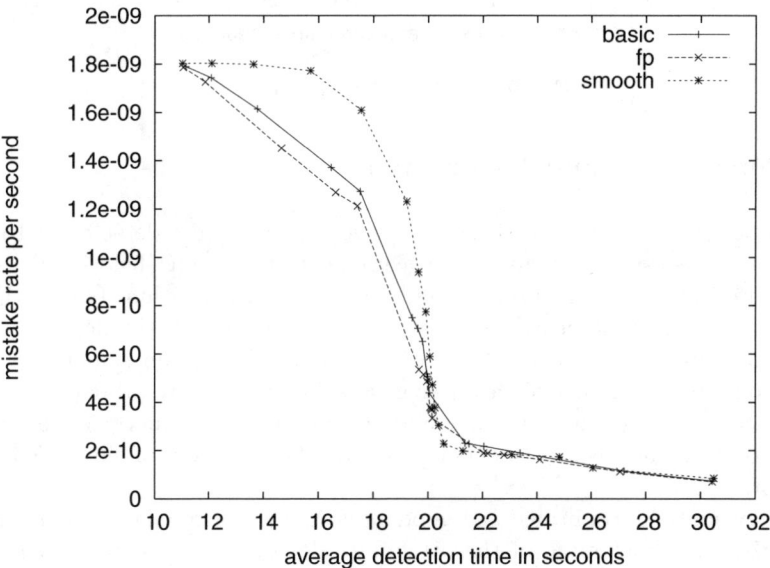

Fig. 8. Results of Experiment 2.3

Hayashibara that is also a flexible accrual failure detector our failure detector provides better evaluation results and is computationally much less expensive.

The results of the experiments 2.1 to 2.4 show that the variation of our basic failure detection algorithm that uses a different freshness point strategy provides the best results in these settings. The variant using the histogram smoothing performs mostly worse than its combatants. Thus we could improve our basic failure detector using the different freshness point strategy.

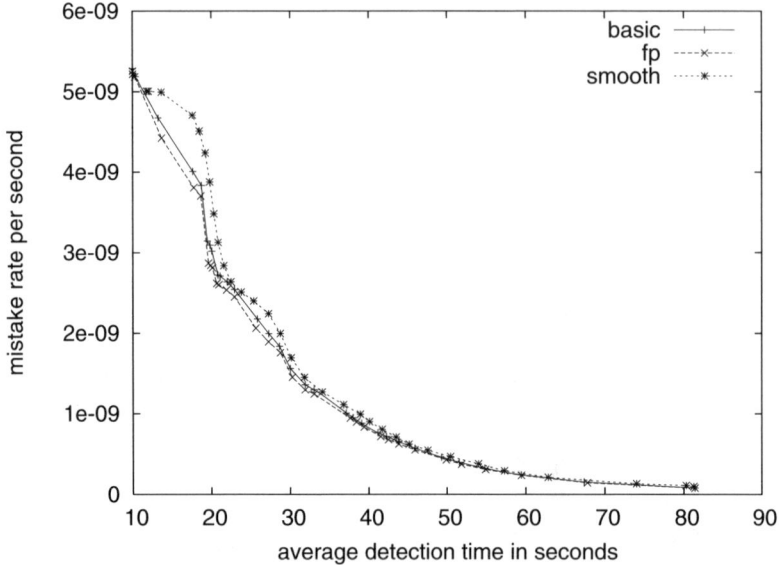

Fig. 9. Results of Experiment 2.4

6   Conclusions and Future Work

In this paper, we revised our basic adaptive accrual failure detector and made more comprehensive performance measurements than done before. We introduced two variations of our basic failure detection algorithm. One variation which uses a different heartbeat strategy serves to abolish the dependence of the failure probability on the last heartbeat. The histogram smoothing technique contains the ability for a faster generation of a failure probability, being more robust to random variations, and consuming less memory. We also conducted performance measurements to evaluate the variations of our basic failure detector.

The evaluations confirmed the good performance of our failure detector in comparison to other state of the art failure detectors. Furthermore our basic algorithm using a different freshness point strategy turned out to outperform our basic algorithm in the experimental settings.

In future work we plan to adress amongst others the reduction of overhead of our failure detector and to integrate the failure detector into a self-healing facility for the OCμ middleware.

References

1. M. Bertier, O. Marin, and P. Sens. Implementation and performance evaluation of an adaptable failure detector. In *DSN '02: Proceedings of the 2002 International Conference on Dependable Systems and Networks*, pages 354–363, Washington, DC, USA, 2002. IEEE Computer Society.

2. J.-C. Bolot. End-to-end packet delay and loss behavior in the internet. In *SIG-COMM*, pages 289–298, 1993.

3. T. D. Chandra, V. Hadzilacos, and S. Toueg. The weakest failure detector for solving consensus. *J. ACM*, 43(4):685–722, 1996.

4. T. D. Chandra and S. Toueg. Unreliable failure detectors for reliable distributed systems. *J. ACM*, 43(2):225–267, 1996.

5. W. Chen, S. Toueg, and M. K. Aguilera. On the quality of service of failure detectors. In *Proceedings of the International Conference on Dependable Systems and Networks (DSN 2000)*, New York, 2000. IEEE Computer Society Press.

6. K. K. Dam and L. M. Ni. Design and implementation of a network emulator. Technical Report MSU-CPS-ACS-98-16, Department of Computer Science and Engineering, Michigan State University, May, 1998.

7. C. Fetzer, M. Raynal, and F. Tronel. An adaptive failure detection protocol. In *PRDC '01: Proceedings of the 2001 Pacific Rim International Symposium on Dependable Computing*, page 146, Washington, DC, USA, 2001. IEEE Computer Society.

8. M. J. Fischer, N. A. Lynch, and M. S. Paterson. Impossibility of distributed consensus with one faulty process. *J. ACM*, 32(2):374–382, 1985.

9. W. Härdle. *Smoothing Techniques with Implementation in S*. Springer Verlag, Berlin, 1991.

10. N. Hayashibara, X. Défago, R. Yared, and T. Katayama. The f accrual failure detector. In *SRDS*, pages 66–78. IEEE Computer Society, 2004.

11. P. Horn. Autonomic computing: Ibms perspective on the state of information technology. *http://www.research.ibm.com/autonomic/*, 2001.

12. M. Horstmann and M. Kirtland. Dcom architecture. Technical report, http:// msdn.microsoft.com/library/backgrnd/html/ msdn_dcomarch.htm, July 1997.

13. V. Jacobson. Congestion avoidance and control. In *SIGCOMM '88: Symposium proceedings on Communications architectures and protocols*, pages 314–329, New York, NY, USA, 1988. ACM Press.

14. J. O. Kephart. Research challenges of autonomic computing. In *ICSE '05: Proceedings of the 27th international conference on Software engineering*, pages 15–22, 2005.

15. N. Lynch. A hundred impossibility proofs for distributed computing. In *PODC '89: Proceedings of the eighth annual ACM Symposium on Principles of distributed computing*, pages 1–28, New York, NY, USA, 1989. ACM Press.

16. A. Mukherjee. On the dynamics and significance of low frequency components of internet load. Technical Report MIS-CIS-92-83, University of Pennsylvania, December, 1992.

17. C. Müller-Schloer, C. von der Malsburg, and R. P. Würtz. Organic computing. *Informatik Spektrum*, 27(4):332–336, Aug. 2004.

18. A. Pietzowski, W. Trumler, and T. Ungerer. An artificial immune system and its integration into an organic middleware for self-protection. In *GECCO '06: Proceedings of the 8th annual conference on Genetic and evolutionary computation*, pages 129–130, New York, NY, USA, 2006. ACM Press.

19. U. Richter, M. Mnif, J. Branke, C. Müller-Schloer, and H. Schmeck. Towards a generic observer/controller architecture for organic computing. In C. Hochberger and R. Liskowsky, editors, *INFORMATIK 2006 – Informatik für Menschen*, volume P-93 of *GI-Edition – Lecture Notes in Informatics*, pages 112–119, Bonn, Germany, Sept. 2006. Köllen Verlag.

20. D. Sanghi, A. K. Agrawala, O. Gudmundsson, and B. N. Jain. Experimental assessment of end-to-end behavior on internet. In *INFOCOM*, pages 867–874, 1993.

21. B. Satzger, A. Pietzowski, W. Trumler, and T. Ungerer. A new adaptive accrual failure detector for dependable distributed systems. In *SAC '07: Proceedings of the 2006 ACM symposium on Applied computing*, New York, NY, USA, 2007. ACM Press.

22. H. Schmeck. Organic computing-vision and challenge for system design. In *Proceedings of the Parallel Computing in Electrical Engineering, International Conference on (PARELEC 2004)*, pages 3–3, Washington, DC, USA, 2004. IEEE Computer Society.

23. H. Schmeck. Organic computing. *Künstliche Intelligenz*, 05(3):68–69, July 2005.

24. B. W. Silverman. *Density Estimation for Statistics and Data Analysis*. Chapman & Hall/CRC, April 1986.

25. W. Trumler, F. Bagci, J. Petzold, and T. Ungerer. Amun - autonomic middleware for ubiquitous environments applied to the smart doorplate. In *Advanced Engineering Informatics*, volume 19, pages 243–252, Washington, DC, USA, 2005. ELSEVIER.

26. W. Trumler, R. Klaus, and T. Ungerer. Self-configuration via cooperative social behavior. In *Autonomic and Trusted Computing, Third International Conference (ATC 2006)*, 2006.

27. W. Trumler, T. Thiemann, and T. Ungerer. An artificial hormone system for self-organization of networked nodes. In *IFIP Conference on Biologically Inspired Cooperative Computing*, pages 85–94, Santiago de Chile, August 2006. Springer-Verlag.

Self-organizing Software Components
in Distributed Systems

Ichiro Satoh

National Institute of Informatics
2-1-2 Hitotsubashi, Chiyoda-ku, Tokyo 101-8430, Japan
ichiro@nii.ac.jp

Abstract. This paper presents a framework for deploying software components over a distributed system by using the notion of dynamics between components. It enables an application to be composed of one or more mobile components that can be deployed to different computers when the application is being executed. The key idea behind the framework is to provide components with deployment policies corresponding to gravitational and repulsive forces. The polices control the relocation relation between two components. As a result, a federation of distributed components can be moved and changed over a distributed system in a self-organizing manner. This paper also presents a prototype implementation of the approach and its applications.

1 Introduction

Distributed computing systems are composed of a number of software components running on different computers and interacting with one another via a network. The scale and complexity of modern distributed systems impair our ability to deploy components to appropriate computers using traditional approaches, such as those that are centralized and top-down. The structure of a distributed system may also be frequently changed by adding or removing components and changing the network topology. Applications, which consist of components running on different computers, must adapt to such changes. When computers are about to shut down, for example, the components running on them must be deployed elsewhere. Moreover, the requirements of the applications tend to vary and changed dynamically. For example, users in a ubiquitous computing setting may also want to constantly interact with their applications running on nearby stationary computers. When they move from location to location, the components that the application consists of should be dynamically deployed at computers that are near their current position and can offer the computational resources required by the components.

To solve these problems, we have developed a framework for dynamically dispersing software components over a distributed system. It provides components with their own relocation policies without the need of any global policies. As a result, it enables individual components or a group of components to migrate over a network in a self-organizing manner without losing their previous coordination. We have presented earlier versions of the framework in this paper in our previous papers [13,15]. These previous versions supported the attachment of components to other components, but not

P. Lukowicz, L. Thiele, and G. Tröster (Eds.): ARCS 2007, LNCS 4415, pp. 185–198, 2007.

the detachment of components, unlike this version. This problem is serious in implementing load-balancing and in fault-tolerant systems. The framework supports a mechanism for distributing components in addition to that for organizing a moving mass of components.

This paper describes our design goals (Section 2), the design of the framework, and a prototype implementation (Section 3). We also describe our experience with it (Section 4). We then briefly review related work (Section 5), provide a summary, and discuss some future issues (Section 6).

2 Basic Approach

Most modern large-scale systems consist of software components, which may run on different computers over a distributed system. The deployment of software components which a system consists of, seriously affects what a system can achieve and how efficiently it can achieve this. These components also need to be dynamically deployed and replaced at computers without them losing any previous coordination according to changes in the structure of the distributed system and the requirements of the system's applications. However, it is almost impossible for any centralized management systems to deploy components at appropriate computers because the systems have no global view of the distributed system as a whole. To solve this problem, our framework introduces two metaphors, i.e., *gravitational* and *repulsive* forces between components (Fig. 1). The former deploys components that coordinate with one another at the same computers or those nearby even when they move to other locations. The latter prevents specified components from being at the same or nearby computers. These component-deployment approaches are specified and managed as a relocation relationship between two components. That is, the framework enables each component to explicitly specify a deployment policy for its own migration as relocation between its current location and another component's location. An aggregation of components, each with its own deployment policies, can change its structure and move over a distributed system in response to changes in the underlying system and the requirements of the system's applications. All the deployment policies presented in this paper are managed in a non-centralized manner to maintain scalability and reliability.

Most interactions between components in object-oriented systems within a computer can be covered by three primitives: event passing, method invocation, and stream communication. Our framework enables these primitives to be available in partitioned systems on different computers. Achieving syntactic and (partial) semantic transparency for remote interactions requires the use of proxy objects that have the same interfaces as the remote components. The framework introduces such objects, called *references*, to track possibly moving targets and to interact with the these through the three primitives.

Remark

This framework was inspired by our earlier versions presented in previous papers [13,15]. The previous papers aimed at presenting the middleware for building and operating a large-scale system as a federation of one or more mobile components like the framework presented here, but they addressed ubiquitous computing environments

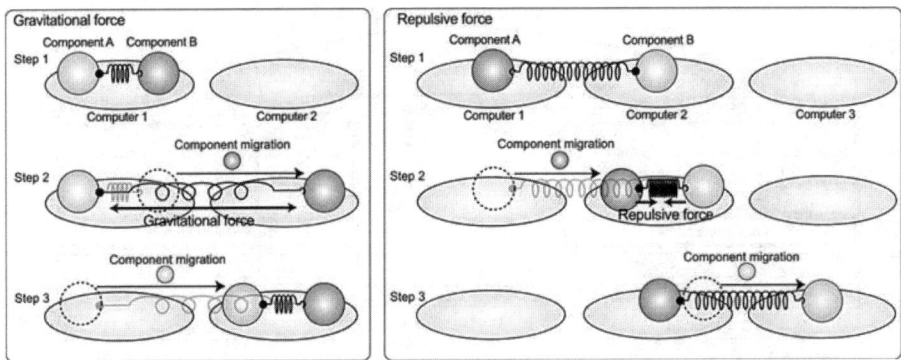

Fig. 1. Gravitational and repulsive policies

whose computers are heterogenous rather than large-scale distributed systems. The previous versions offered some of the gravitational relocation policies supported by this framework, but lacked any of the repulsive policies, which are essential in supporting load-balancing and fault-tolerant mechanisms. In fact, when many components are organized and deployed over a distributed system by only using the gravitational relocation policies, they tend to gather at several computers.

3   Design and Implementation

This framework consists of two parts: runtime systems and components. Each component in the current implementation is a collection of Java objects.

3.1   Component Runtime System

Each runtime system is running on a computer and is responsible for executing and migrating components to other computers. It establishes at most one TCP connection with each of its neighboring computers and exchanges control messages, components, and inter-component communications with these through the connection. Fig. 2 outlines the basic structure of a runtime system. Each component in the current implementation is a collection of Java objects in the standard JAR file format and can migrate from computer to computer and duplicate itself by using mobile agent technology [9].[1] When a component is transferred over the network, the component runtime system on the sending side marshals the code of the component and its state, e.g., instance variables in Java objects, into a bit-stream and then transfers them to the destination. The component runtime system on the receiving side receives and unmarshals the bit-stream so that the component can continue to be executed at the destination.

3.2   Component Programming Model

Each component runtime system governs all the components inside it and maintains their life-cycle states. When the life-cycle state of a component changes, e.g., when it is

---

[1] JavaBeans can easily be translated into components in the framework.

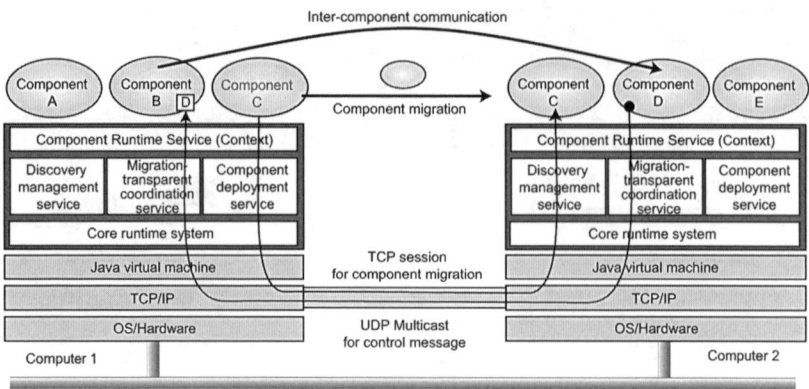

Fig. 2. Component runtime system

created, terminates, or migrates to another computer, the runtime system issues specific events to the component. This is because the component may have to acquire various resources, e.g., files, windows, or sockets, or release ones it had previously acquired. The current implementation uses Java's object serialization package for marshaling components. This package can save the content of instance variables in a component program but does not enable the stack frames of threads to be captured. Consequently, runtime systems cannot serialize the execution states of any thread objects. Instead, when a component is marshaled or unmarshaled, the runtime system propagates certain events to its components instructing them to stop their active threads and it then automatically stops and marshals them after a given period of time. Each component must be an instance of a subclass of the MComponent class. Here, we will explain the programming interface characterizing the framework.

```
class MComponent extends MobileAgent implements Serializable {
    void go(URL url) throws NoSuchHostException { ... }
    void duplicate() throws IllegalAccessException { ... }
    setPolicy(ComponnetProfile cref, MigrationPolicy mpolicy) { ... }
    setTTL(int lifespan) { ... }
    void setComponentProfile(ComponentProfile cpf) { ... }
    boolean isConformableHost(HostProfile hfs) { ... }
    void send(URL url, ComponentID id, Message msg) throws
        NoSuchHostException, NoSuchComponentException, ... { .... }
    Object call(URL url, ComponentID id, Message msg) throws
        NoSuchHostException, NoSuchComponentException, ... { .... }
    ....
}
```

A component executes go(URL url) to move to the destination host specified as a url by its runtime system, and duplicate() creates a copy of the component, including its code and instance variables. The setTTL() specifies the life span, called time-to-live (TTL), of the component. The life span decrements TTL over time. When the TTL of a component reaches zero, the component automatically removes itself.

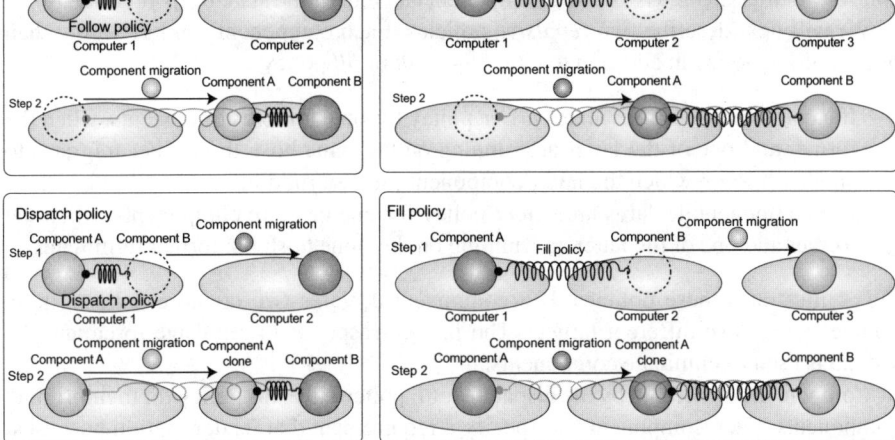

Fig. 3. Gravitational policies

Each component can have more than one listener object that implements a specific listener interface to hook certain events issued before or after changes are made in its life-cycle state. That is, each component host invokes the specified callback methods of its components when the components are created, destroyed, or migrate to another host.

3.3 Component Deployment Policy

A component can declare its own deployment policy by invoking the `setPolicy` method of the `MComponent` class while a component is running .

Let us now explain four *gravitational* policies (Fig. 3).

- If one component declares a *follow* policy for another, when the latter exists or migrates to a host, the former migrates to the latter's current or destination host.
- If a component declares a *dispatch* policy for another, when the latter migrates to another host, a copy of the former is created and deployed at the latter's destination host.
- If a component declares a *shift* policy for another, when the latter migrates to another host, the former migrates to the latter's source host.
- If a component declares a *fill* policy for another, when the latter migrates to another host, a copy of the former is created and deployed at the latter's source host.

The framework allows each component to have at most one gravitational policy for at most one component to reduce conflicts in individual or multiple policies. The *follow* policy is useful when relationships between components comprising an application need to be retained, and the *fill* policy is useful when components are distributed to hosts along the tracks of moving components. The deployment of one component depends on the location of another but the deployment of the latter does not need to depend on the location of the former. Instead, two components can explicitly declare policies for each other. When a component is created, the dispatch and fill policies can explicitly control whether the newly created component can inherit the state of its original.

We will next describe two *repulsive* policies. Each component can have more than one repulsive policy in addition to either the *shift* or *fill* policy.

- If a component declares an *exclusive* policy for one or more components, when the former and one of the latter are running on the same host, the former migrates to another host on which the latter components are not running.
- If a component declares an *extinct* policy for one or more components, when the former and one of the latter are running on the same host, the former terminates.

Fig. 4 illustrates these policies. If a component declares two or more polices, these policies must have different targets. The first corresponds to repulsive force and the second is used to eliminate components.

Components duplicated by the dispatch or fill policy have this policy for their original components. Each component can specify a requirement that its destination host must satisfy by invoking setComponentProfile(), with the requirement specified as cpf, where it is defined in CC/PP (composite capability/preference profiles) form [17], which describes the capabilities of the component host and the components' requirements. The class has a service method called isConformableHost(), which the component uses to determine whether the capabilities of the component host specified as an instance of the HostProfile class satisfy it requirements. Runtime systems transform the profiles into their corresponding LISP-like expressions and then evaluate them by using a LISP-based interpreter. When a component migrates to the destination according to its policy, if the destination cannot satisfy the requirements of the component, the runtime system recommends candidates that are hosts in the same network domain to the component. If a component declares repulsive policies in addition to a gravitational policy, the runtime system detects the candidates using the latter's policy and then recommends final candidates to the component using the former policy, assuming that the component is in each of the detected candidates.

3.4 Component Deployment Management

The policy-based deployment of components is managed by each component host without a centralized management server. Each component host periodically advertises its address to the others through UDP multicasting, and these hosts then return their addresses and capabilities to the host through a TCP channel.[2] (1) When a component migrates to another component host, each component automatically registers its deployment policy with the destination host. (2) The destination host sends a query message to the source host of the visiting component. There are two possible scenarios: the visiting component has a policy for another component or it is specified in another component's policies. (3-a) Since the source host in the first scenario knows the host running the target component specified in the visiting component's policy, it asks the

---

[2] We assumed that the components comprising an application would initially be deployed at hosts within a localized space smaller than the domain of a sub-network.

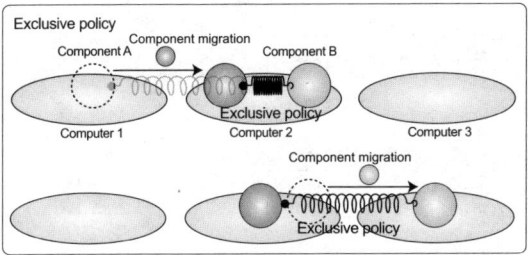 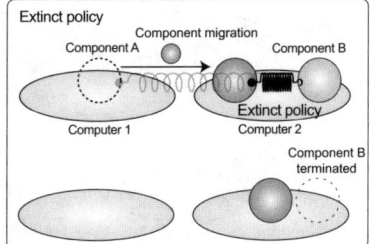

Fig. 4. Repulsive policies

host to send the destination host information about itself and about neighboring hosts that it knows, e.g., network addresses and capabilities. If the target host has retained the proxy of a target component that has migrated to another location, it forwards the message to the destination of the component via the proxy. (3-b) In the second scenario, the source host multicasts a query message within current or neighboring sub-networks. If a host has a component whose policy specifies the visiting component, it sends the destination host information about itself and its neighboring hosts. (4) The destination host next instructs the visiting component or its clone to migrate to one of the candidate destinations recommended by the target host, because this framework treats every component as an autonomous entity. Moreover, when the capabilities of a candidate destination do not satisfy all the requirements of the component, the component itself decides, on the basis of its own configuration policy, whether it will migrate itself to the destination and adapt itself to the destination's capabilities. The destination of the component may go into divergence or vibration mode due to conflicts between some of a component's policies, when it has multiple deployment policies. However, the current implementation does not exclude such divergence or vibration.[3]

3.5   Intercomponent Communication

The current implementation offers two communication policies for intercomponent interactions as follows:

- If a component declares a *forward* policy for another, when specified messages are sent to other components, the messages are forwarded to the latter as well as the former.
- If a component declares a *delegate* policy for another, when specified messages are sent to the former, the messages are forwarded to the latter but not to the former.

The former policy is useful when two components share the same information and the latter policy provides a master-slave relation between them. The framework provides three interactions: publish/subscribe for asynchronous event passing, remote method invocation, and stream-based communication as well as message *forward* and *delegate*

---

[3] From our experience with several applications, most components in a system have at most a gravitational or a repulsive policy. Therefore, we do not always feel the needs to resolve such conflicts.

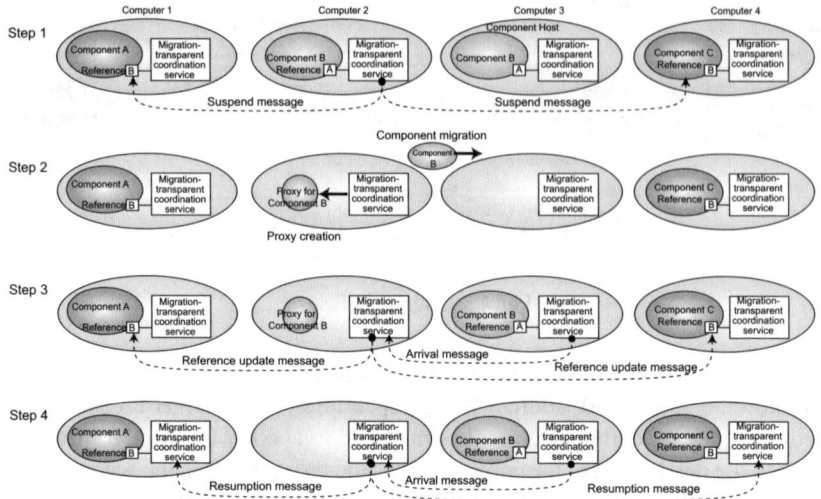

Fig. 5. Forwarding messages to migrated component

policies. Each runtime system offers a remote method invocation (RMI) mechanism through a TCP connection. It is implemented independent of Java's RMI because this has no mechanisms for updating references for migrating components. Each runtime system can maintain a database that stores pairs of identifiers of its connected components and the network addresses of their current runtime systems. It also provides components with references to the other components of the application federation to which it belongs. Each reference enables the component to interact with the component that it specifies, even if the components are on different hosts or move to other hosts.

Fig. 5 shows an approach enabling communication between a component moving from computer 2 to 3 and two components at computers 1 and 3. When a component, i.e., component B, requests the current runtime system to migrate to another computer, the system searches its database for the network addresses of runtime systems with components, i.e., computer 1 and 4. 1) It sends *suspend* messages to these systems to block any new uplinks from them to the migrating component with the destination's address. If the moving component contains references, the current runtime system sends the destination's address to the runtime systems that are running the components specified in the references so that they can update their databases. 2) It creates its own proxy at its current location and It migrates to its destination. 3) After the component arrives at its destination, it sends an *arrival* message with the network address of the destination to the departure runtime system and then *update* messages to the systems. 4) When the departure system receives the arrival message, it sends *resumption* messages with the address of the destination to runtime systems that may hold references to the moved component and then remove the proxy.

When a component begins to interact with another that is moving, the former can send messages to the source of the one that is moving before the basic algorithm above is completed. To solve this, a migrating component creates and leaves a proxy at the departure runtime system for the duration it takes the algorithm to finish. The proxy component receives uplinks from other runtime systems and forwards them to the moved component. Since not all components have to be tracked for other components to communicate with them, components can leave proxy components along their trail under their own control. Proxy components are also programmable entities, like components, so they can be modified based on application requirements.

3.6 Security

The current implementation is a prototype system to dynamically deploy the components presented in this paper. Nevertheless, it has several security mechanisms. For example, it can encrypt components before migrating them over the network and it can then decrypt them after they arrive at their destinations. Moreover, since each component is simply a programmable entity, it can explicitly encrypt its individual fields and migrate itself with these and its own cryptographic procedure. The Java virtual machine could explicitly restrict components so that they could only access specified resources to protect computers from malicious components. Although the current implementation cannot protect components from malicious computers, the runtime system supports authentication mechanisms to migrate components so that all runtime systems can only send components to, and only receive from, trusted runtime systems.

3.7 Current Status

A prototype implementation of this framework was constructed with Sun's Java Developer Kit, version 1.4 or later version.[4] Although the current implementation was not constructed for performance, we evaluated the migration of two components based on deployment policies. When a component declares a follow, dispatch, shift, or fill policy for another, the cost of migrating the former or its clone to the destination or the source of the latter after the latter begins to migrate is 92 ms, 116 ms, 89 ms, 118 ms, or 136 ms if the policy is follow, dispatch, shift, fill, or exclusive, where the cost of component migration between two computers over a TCP connection is 35 ms and the cost of duplicating a component in a computer was less than 7 ms.[5] This experiment was done with three computers (Pentium M-1.8 GHz with Windows XP and JDK ver.5) connected through a Fast Ethernet network. Migrating components included the cost of opening a TCP-transmission, marshaling the components, migrating them from their source computers to their destination computers, unmarshaling them, and verifying security.

4 Experience

This section presents several example applications that illustrate how the framework works.

---

[4] The functionalities of the framework, except for subscribe/publish-based remote event passing, can be implemented on Java Developer Kit version 1.1 or later, including Personal Java.

[5] The size of each of the three components was about 8 KB in size.

4.1 Dynamic Deployment for Duplicated Servers

We can easily implement distributed systems. Here, we present a fault-tolerant HTTP-based server to illustrate the use of these policies by combining gravitational policies, repulsive policies, and communication policies. Each component supports an HTTP server. It is a clonable component, where it and its clone declare forward policies for each other and its clone declares an exclusive policy for it. When a component is duplicated at a host, a clone is created at the host, but its exclusive policy deploys the clone at another host to distribute the original and cloned components at different computers to ensure tolerance against faults. When one of these receives messages from external systems, their forward policies send the messages to another so that they can share the same states. After the component duplicates itself, the cost of deploying its clone at another host is about 280 ms in the distribution system presented in the previous section.[6] This does not include the cost of terminating and restarting the HTTP server. The cost of forwarding a message is about 28 ms, where this is measured as the round-trip time and the message has no value. If the components declare delegate policies, they can support a master-slave instead of a duplication model.

4.2 Ant-Based Routing Mechanisms

Ants are able to locate a path to a food source using the trails of chemical substances called pheromones that are deposited by other ants. Several researchers have attempted to use the notion of ant pheromones for network-routing mechanisms [1,2]. Our framework allows moving components to leave themselves on their trails and to become automatically volatile after their life-spans are over. A component corresponding to an ant, A, corresponding to a pheromone is attached to another component corresponding to an ant according to the *fill* policy. When the latter component randomly selects its destination and migrates to the selected destination, the former creates a clone and migrates to the source host of the latter. Since each of the cloned components defines its life-span by invoking setTTL(), they are active for a specified duration after being created. If there are other components corresponding to pheromones in the host, the visiting component adds their time spans to its own time span. When another component corresponding to another ant migrates over the network, it can select a host that has components corresponding to pheromones with the longest time-spans from neighboring hosts. We experimented with ant-based routing for components using this prototype implementation with more than eight hosts. However, we knew that it would be difficult to quickly converge a short-path to the destination in real distributed systems, because routing mechanisms tend to diverge.

4.3 Component Diffusion in Sensor Networks

The third example is the speculative deployment of components like cell-lamellipodia. This provides a mechanism that dynamically and speculatively deploys components at sensor nodes when there are environmental changes. This mechanism was inspired by

---

[6] This experiment assumes that the destination of the clone has been statically derived.

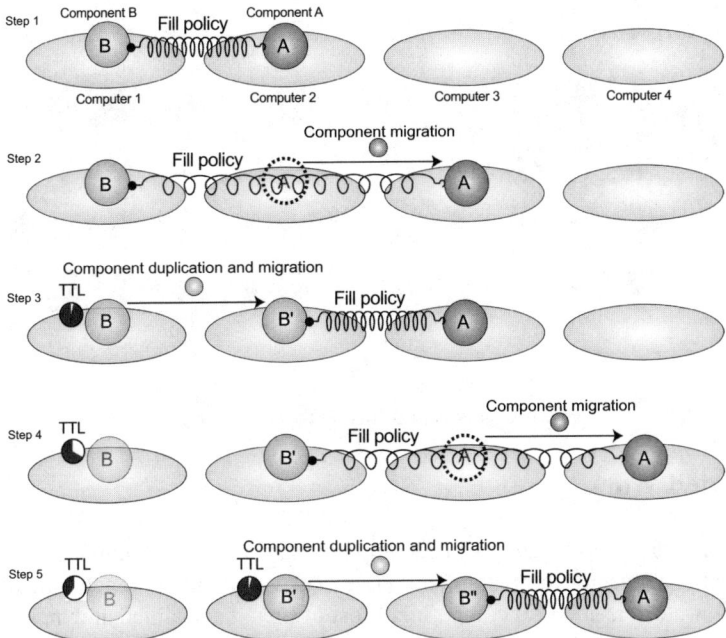

Fig. 6. Implementation of ant-based routing mechanism

lamellipodia in cells. It assumes that the sensor field is a two-dimensional surface composed of sensor nodes and it monitors environmental changes, such as motion in objects and variations in temperature. It is a well known fact that after a sensor node detects environmental changes in its area of coverage, some of its geographically neighboring nodes tend to detect similar changes after a period of time. It deploys monitoring components at sensor nodes, where each monitoring component can control and monitor its current sensor node and has its own TTL. Diffusion occurs as follows. When a component detects the presence of its target, it creates a specified number of its clones, e.g., two clones, where this number depends on the number of neighboring sensor nodes (Fig. 7). Each of the clones declares an exclusive policy for other monitoring components. It must migrate to a neighboring node according to the policy, because its original monitoring component is running on its current node. As a result, these clones are deployed at neighboring nodes around the target. When the target moves to another location, the monitoring components located at the nodes near the target detect the presence of the target and create their clones in the same way. We can provide application-specific components that declares a follow policy to monitor components. These components can be automatically deployed at nodes near the entity to annotate and assist the target. Each clone is associated with a resource limit that functions as a generalized TTL field. Although a node can monitor changes in interesting environments, it sets the TTLs of its components to their own initial values. It otherwise decrements TTLs as the passage of time. When the TTL of a component becomes zero, the component automatically removes itself.

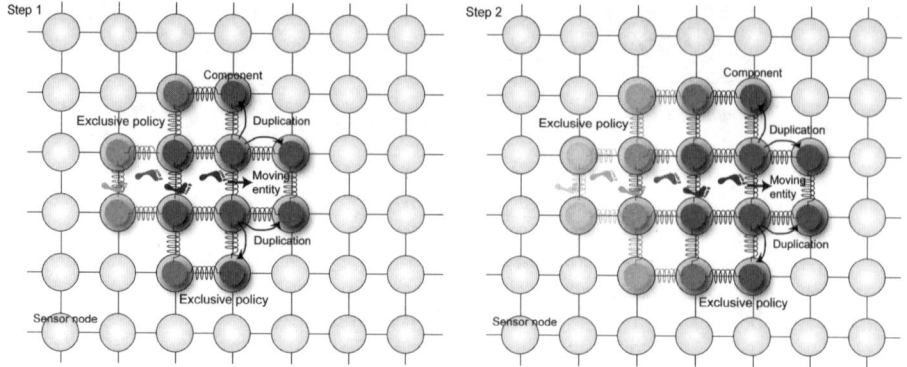

Fig. 7. Component diffusion for moving entity

5   Related Work

There have been several attempts to develop infrastructures for dynamically deploying components between computers in large-scale computing environments, e.g., workstation-clusters and grid computing. Most of them have aimed at dynamically deploying partitioned applications or systems to different computers in distributed systems to balance computational loads or network traffic. However, as they have explicitly or implicitly assumed centralized management approaches to deploying partitioned applications or systems to different computers, they have not allowed all partitioned applications or systems to have its own deployment approaches.

Of these, the FarGo system introduces a mechanism for distributed applications dynamically laid out in a decentralized manner [5]. This is similar to our relocation policy in the sense that it allows all components to have their own policies, but it is aimed at allowing one or more components to control a single component, whereas ours aims at allowing one component to describe its own migration. This is because our framework treats components as autonomous entities that travel from computer to computer under their own control. This difference is important, because FarGo's policies may conflict if two components can declare different relocation policies for one single component. Our framework is free of any conflict because each component can only declare a policy for its own relocation, and not for other components. Several researchers have introduced the dynamic deployment of partitioned applications as a technology that enables distributed computers to support various services, which they may not have initially been designed for, rather than to balance computational loads and traffic in a distributed system. For example, the Aura project [4] by CMU provides an infrastructure for binding tasks associated with users and migrating applications from computer to computer as users move about, like our framework does. Although Aura shares several common design goals with our framework, it focuses on providing contextual services to users rather than on integrating multiple computers to support functions and performance unattainable with a single computer. Like our framework, the Gaia project by the University of Illinois at Urbana-Champaign allows applications to be partitioned between different computers and move from computer to computer [8]. Gaia assumes that applications will be constructed based on a design pattern, called MPACC, which is is an extension of the MVC pattern [6], whereas our framework supports a variety of interactions between partitioned applications so that we do not have to assume any particular application model.

6 Conclusion

We described a framework for dynamically aggregating distributed applications in a distributed system based on physical dynamics. It was used to build an application from mobile software components, which can explicitly have policies for their own deployment. It enables a federation of components to be dynamically structured in a self-organized manner and to be deployed at computers as components that have gravitational and repulsive forces between them. We designed and implemented a prototype system for the framework and demonstrated its effectiveness in several practical applications. We believe that the framework provides a general and practical infrastructure for building distributed and mobile applications.

In concluding, we would like to identify further issues that need to be resolved. The current implementation relies on Java's security manager. Nevertheless, we are interested in security mechanisms for components that have their own deployment policies and plan on introducing various such policies to support adaptive applications over a distributed system. We also proposed a specification language for the itinerary of mobile software [14]. The language enables more flexible and varied policies for deploying the components to be defined.

References

1. O. Babaoglu and H. Meling and A. Montresor, Anthill: A Framework for the Development of Agent-Based Peer-to-Peer Systems, Proceeding of 22th IEEE International Conference on Distributed Computing Systems, July 2002.
2. G. Di Caro and M. Dorigo, AntNet: A Mobile Agents Approach to Adaptive Routing, Proceedings of Hawaii International Conference on Systems, pp.74-83, Computer Society Press, January, 1998.
3. E. Gamma, R. Helm, R. Johnson, and J. Vlissides, Design Patterns, Addison-Wesley, 1995.
4. D. Garlan, D. Siewiorek, A. Smailagic, and P. Steenkiste, Project Aura: Towards Distraction-Free Pervasive Computing, IEEE Pervasive Computing, vol. 1, pp. 22-31, 2002.
5. O. Holder, I. Ben-Shaul, and H. Gazit, System Support for Dynamic Layout of Distributed Applications, Proceedings of International Conference on Distributed Computing Systems (ICDCS'99), pp 403-411, IEEE Computer Soceity, 1999.
6. G. E. Krasner and S. T. Pope, A Cookbook for Using the Model-View-Controller User Interface Paradigma in Smalltalk-80, Journal of Object Oriented Programming, vol.1 No.3, pp. 26-49, 1988.
7. M. Román, C. K. Hess, R. Cerqueira, A. Ranganat R. H. Campbell, K. Nahrstedt K, Gaia: A Middleware Infrastructure to Enable Active Spaces, IEEE Pervasive Computing, vol. 1, pp.74-82, 2002.
8. M. Román, H. Ho, R. H. Campbell, Application Mobility in Active Spaces, Proceedings of International Conference on Mobile and Ubiquitous Multimedia, 2002.

9. I. Satoh, MobileSpaces: A Framework for Building Adaptive Distributed Applications Using a Hierarchical Mobile Agent System, Proceedings of IEEE International Conference on Distributed Computing Systems (ICDCS'2000), pp.161-168, April 2000.
10. I. Satoh, Building Reusable Mobile Agents for Network Management, IEEE Transactions on Systems, Man and Cybernetics, vol.33, no. 3, part-C, pp.350-357, August 2003.
11. I. Satoh, Configurable Network Processing for Mobile Agents on the Internet, Cluster Computing, vol. 7, no.1, pp.73-83, Kluwer, January 2004.
12. I. Satoh, Linking Phyical Worlds to Logical Worlds with Mobile Agents, Proceedings of IEEE International Conference on Mobile Data Management (MDM'2004), pp. 332-343, IEEE Computer Society, January 2004.
13. I. Satoh, Dynamic Federation of Partitioned Applications in Ubiquitous Computing Environments, Proceedings of 2nd International Conference on Pervasive Computing and Communications (PerCom'2004), pp.356-360, IEEE Computer Society, March 2004.
14. I. Satoh, Selection of Mobile Agents, Proceedings of IEEE International Conference on Distributed Computing Systems (ICDCS'2004), pp.484-493, IEEE Computer Society, March 2004.
15. I. Satoh, Organization and Mobility in Mobile Agent Computing, Programming Multi-Agent Systems (Postproceedings of 3rd Workshop on ProMAS'05), Lecture Notes in Computer Science, vol. 3862, pp.187-205, April 2006.
16. C. Szyperski, Component Software, Addison-Wesley, 1998.
17. World Wide Web Consortium (W3C), Composite Capability/Preference Profiles (CC/PP), http://www.w3.org/TR/NOTE-CCPP, 1999.

Toward Self-adaptive Embedded Systems: Multi-objective Hardware Evolution

Paul Kaufmann and Marco Platzner

University of Paderborn

Abstract. Evolutionary hardware design reveals the potential to provide autonomous systems with self-adaptation properties. We first outline an architectural concept for an intrinsically evolvable embedded system that adapts to slow changes in the environment by simulated evolution, and to rapid changes in available resources by switching to preevolved alternative circuits. In the main part of the paper, we treat evolutionary circuit design as a multi-objective optimization problem and compare two multi-objective optimizers with a reference genetic algorithm. In our experiments, the best results were achieved with TSPEA2, an optimizer that prefers a single objective while trying to maintain diversity.

1 Introduction

In the last decades, *natural computing* methods which take problem solving principles from nature have gained popularity. Among others, natural computing includes evolutionary computing. Evolutionary computing covers population-based, stochastic search algorithms inspired by principles from evolution theory. An evolutionary algorithm tries to solve a problem by keeping a set (population) of candidate solutions (individuals) in parallel and improving the quality (fitness) of the individuals over a number of iterations (generations). To form a new generation, genetically-inspired operators such as crossover and mutation are applied to the individuals. A fitness-based selection process steers the population towards better candidates.

Evolvable hardware denotes the combination of evolutionary algorithms with reconfigurable hardware technology to construct self-adaptive and self-optimizing hardware systems. The term evolvable hardware was coined by de Garis [1] and Higuchi [2] in 1993. In the last years, evolutionary techniques have generated astonishing circuits that are totally different from classically engineered circuits, and sometimes even superior, as presented by Thompson and Layzell [3]. Moreover, for specifications varying over time, evolutionary techniques achieved very promising results indicating their potential to construct self-adapting systems. Higuchi and Kajihara [4] presented case studies on evolved controllers for prosthetic hands and robot navigation. However, several problems

P. Lukowicz, L. Thiele, and G. Tröster (Eds.): ARCS 2007, LNCS 4415, pp. 199–208, 2007.

remain to be solved, the major ones being the scalability and the robustness of evolved hardware.

Our long-term goal is the development of autonomous embedded systems that implement hardware functions (circuits) characterized by their functional quality and resource demand. We plan to leverage on three concepts to achieve a flexible adaptation: First, an *intrinsic evolutionary search process* adapts the system to slow changes in the environment. Second, radical changes in available resources are compensated for by replacing the operational circuit with a *preevolved alternative* which meets the new resource constraints. To this end, we store at any time an approximated Pareto front of circuit implementations. Third, a *reconfigurable system on chip* platform is the technology allowing for the replacement of circuits during runtime and for the implementation of an intrinsically evolvable system.

The main contribution of this paper is the development and comparison of multi-objective evolutionary techniques for hardware design. In Section 2, we first outline our architecture concept and then we review the few approaches that treat evolutionary hardware design as a multi-objective optimization problem. Section 3 presents our basic hardware representation model and a baseline genetic algorithm as well as two multi-objective evolutionary optimizers for hardware design. Experiments and results are discussed in Section 4. Finally, Section 5 summarizes the paper and outlines further work.

2   Architecture Concept

2.1   Autonomous Subsystem and Fitness Evaluation

Fig. 1 shows the envisioned architecture concept for an intrinsically evolved subsystem. The currently instantiated circuit reads input signals from sensors, computes its function, and writes output signals to actuators. The instantiated solution has to meet area and speed constraints. The intrinsic evolutionary algorithm (EA) applies the genetic operators selection, crossover and mutation to the candidate solutions stored in the population data structure. The fitness evaluation is based on test vectors which are also stored in the subsystem. Depending on the application and system resources, the EA can run continuously or from time to time. At any time, however, the subsystem maintains a set of approximated Pareto points for the required circuit. Specifically, whenever a new solution is found with better quality than the currently instantiated one (while still meeting area and speed constraints), the subsystem's controller can replace the instantiated solution with the new one. In case of a rapid change in available resources, the controller selects one of the circuits from the Pareto set that meets the new constraints.

The fitness evaluation is highly application-dependent. For any reasonably sized circuit, we will not be able to store all possible input vectors as test vectors. A full test coverage is, however, only necessary for functions that reveal a binary

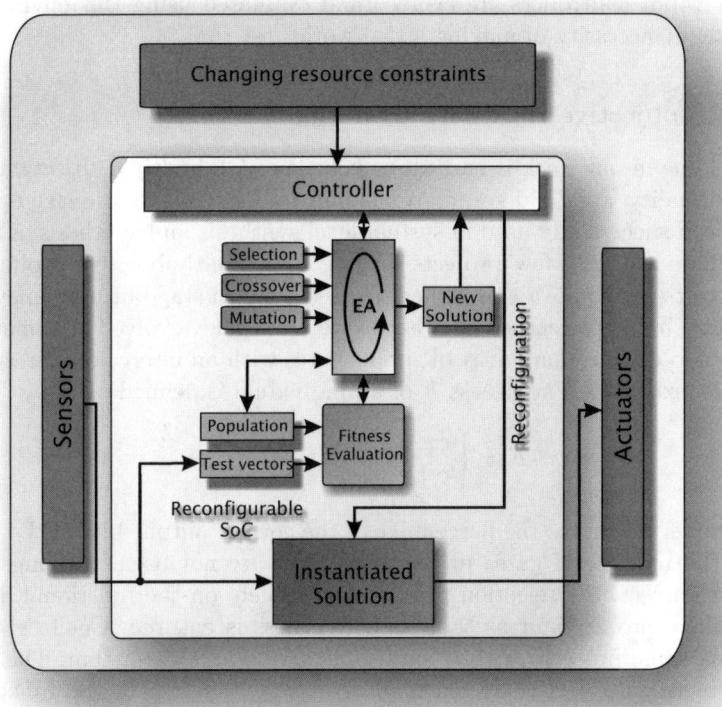

Fig. 1. Architecture concept for intrinsic evolution

correctness property. The prime example are arithmetic functions, where we typically accept nothing less than 100% correctness. Much of the recent work in evolvable hardware has been concerned with the design of arithmetic circuits. We use arithmetic functions as test functions for algorithm design and evaluation, but do not view them as main candidates for autonomous evolution.

The ideal candidates for autonomous evolution are functions that are rated by a quality metrics rather than a binary correctness. Damiani et al. [5] presented the example of a hashing function where quality is measured by the ability to distribute the input keys evenly. Other examples include image compression where the quality is expressed by the compression rate and prosthetic hand control of Higuchi and Kajihara [4], where the quality is given by the percentage of correct classifications. In all these applications, the optimal circuit depends on input data which varies with time. For these functions it suffices to store a certain amount of test vectors that can either be static or being sampled during runtime. For example, Keymeulen et al. [6] designed an adaptive robot control with the objectives to avoid obstacles and reduce the distance to a given target. The robot acquires spatial information about its environment, building a model of it. New robot controllers are evolved and evaluated using this environment model without necessity of making a real-world test run.

2.2 Multi-objective Hardware Evolution

A central issue in our work is hardware evolution with multiple objectives, e.g., functional quality, area and speed. While multi-objective evolutionary optimizers have been successfully used in system-level synthesis and synthesis of analog circuits, there are only few projects dealing with multi-objective evolution of digital circuits. Kalganova and Miller [7] used a multi-stage fitness function to optimize for circuit correctness and hardware area. They evolved arithmetic circuits on a two-dimensional array of simple gates with an interconnect restricted to feed-forward wires. The fitness F of an individual is defined as:

$$F = \begin{cases} c & \text{if } c < 100\%, \\ c + \gamma & \text{else} \end{cases} \tag{1}$$

The parameter c denotes the percentage of the correct output bits of the circuit and γ is the number of gates in the array that are not used. As long as the circuit is incorrect, the selection process bases solely on the functional quality. Area is taken into account as soon as correctness is ensured. Coello et al. [8] address the same problem with a multi-objective search algorithm. The initial single correctness objective is redefined in a way that treats the function of each circuit output as a separate objective. The evolutionary search algorithm has to first meet all these objectives, and then area is taken into account. This approach actually turns constraints into objectives but still uses a multi-stage fitness function to optimize for area.

In contrast to related work, we use a multi-objective EA to optimize for several objectives simultaneously. We are most interested in functions without correctness property. Hence, we do not have to turn (correctness) constraints into objectives. Resource constraints are satisfied by the system's controller that selects a proper circuit for instantiation. Research in multi-objective evolutionary algorithms has identified two key issues subsumed by Zitzler et al. [9]: minimizing the distance between the approximated and the real Pareto front, and maintaining a diverse population to avoid premature convergence to a single objective. The remaining part of this paper presents our work in multi-objective optimizers for evolving digital hardware. This is the central algorithmic challenge in building the autonomous system outlined in Fig. 1.

3 Evolutionary Hardware Design

We use the Cartesian Genetic Programming (CGP) introduced by Miller and Thomson [10] in our work. CGP is a structural hardware model where a circuit is formed by combinational logic blocks arranged in a two-dimensional

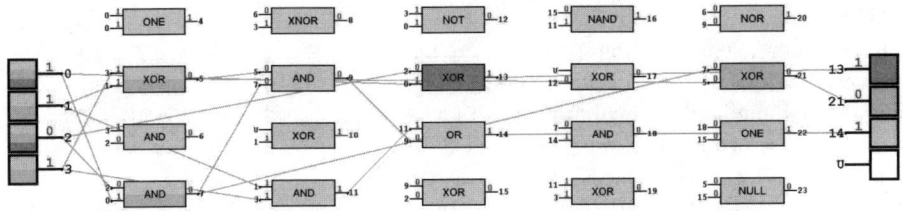

Fig. 2. 2×2 bit-adder evolved on the CGP model

array and an interconnect (wires) between the blocks. The array consists of $n_c \times n_r$ combinational blocks, n_i primary inputs, and n_o primary outputs. The primary inputs can be connected to the inputs of any logic block in the array. A logic block in column c has n_n inputs that can be connected to the columns $c - l, \ldots, c - 1$ of the array and to the primary inputs, respectively. This ensures that no combinational feedback loops are generated. A combinational block implements one out of n_f different logic functions of its inputs. An individual is defined by its chromosome (genotype) with a length of $n_c \cdot n_r (n_n + 1) + n_o$.

Fig. 2 presents an example of a successfully evolved 2×2 bit-adder on a GCP model instance with $n_i = 4, n_o = 4, n_c = 5, n_r = 4, n_n = 2, n_f = 9$, and $l = 4$. The nine possible logic block functions have been chosen as AND, ONE, XOR, NULL, NAND, NOT, NOR, OR, and XNOR.

In the following, we outline the three algorithms used for evolving circuits:

Reference Algorithm GA is a standard single-objective genetic algorithm. The parameters are set as follows: The top 5% of the individuals are selected and transferred without any modification to the next generation. The recombination probability is chosen to be 90%. The individuals are recombined uniformly. We choose the mutation rate such that only one combinational block or wire is mutated each time the mutation operator is applied. In our implementation each recombined child is mutated once. These parameter settings have been used for the experiments described in the following section.

SPEA2 is a recent multi-objective evolutionary optimizer introduced by Zitzler et al. [9] with a structure shown in Fig. 3. SPEA2 maintains two sets of individuals: an archive that contains non-dominated individuals and a breeding population. In each generation, the two sets are merged and the fitness of the individuals is evaluated. The non-dominated individuals are then copied to the new archive. If the archive exceeds a predefined maximum size, SPEA2 applies a nearest neighbor density estimation technique to thin out clusters on the Pareto front. The fitness assigned to an individual considers the number of individuals it dominates - the dominance count, the number of individuals that are dominators - the dominance rank, and a density estimate based on the k-th nearest neighbor method. All individuals undergo a binary tournament selection which selects parents for the recombination and mutation.

TSPEA2 is an algorithm we have devised to put an increased selection pressure on one objective while trying to keep diversity. This should be beneficial for evolving circuits with a correctness property. Compared to SPEA2, we expect degraded fitness values for the other objectives. Both SPEA2 and TSPEA2 use an archive and a breeding population and a selection scheme based on Pareto dominance ranking. TSPEA2, however, checks as a first selection rule in a binary tournament whether one of the two individuals dominates the other regarding the main objective. TSPEA2 has been motivated by an earlier algorithm MO-Turtle GA presented by Trefzer et al. [11], that preferred a main and several random objectives during the evolution of analog circuits.

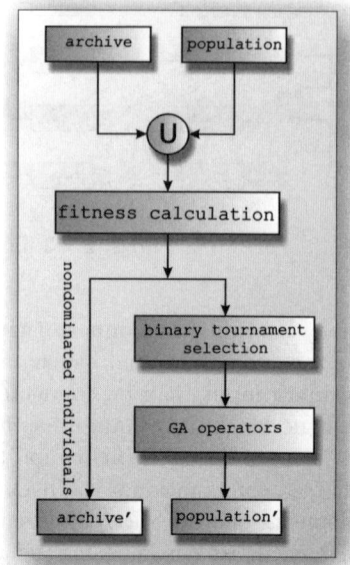

Fig. 3. Structure of the SPEA2 and TSPEA2 optimizers

4   Experiments and Results

We have evolved several test functions with GA, SPEA2, and TSPEA2. In this section, we report on typical results for a 6-parity function and a hashing function. While the 6-parity function is an example for a function with a correctness property, the hashing function is rated by a non-binary quality metrics. The functional set available for the logic blocks in the CGP model comprises the 9 functions shown in Fig. 2. The parameters for crossover and mutation used in SPEA2 and TSPEA2 are set as described in Section 3. The tournament selection operator is configured to execute two tournaments before selecting one of the competitors as a parent. For all evolutionary algorithms, we conducted 10 optimization runs with a maximum of 100.000 generations.

The delay of a circuit is in the range $\{0, \ldots, n_c + 1\}$. The fitness with respect to speed is determined as:

$$speed(c) = 1 - \frac{delay(c)}{n_c + 1} \qquad (2)$$

The speed equals 1 for the fastest possible circuit and 0 for a circuit that has no connection at all from primary inputs to primary outputs. The number of logic blocks used by a circuit, denoted as $used\_blocks(c)$, is in the range $\{0, \ldots, n_c \cdot n_r\}$. Based on this number, the fitness with respect to area is defined as:

$$area(c) = 1 - \frac{used\_blocks(c)}{n_c \cdot n_r} \qquad (3)$$

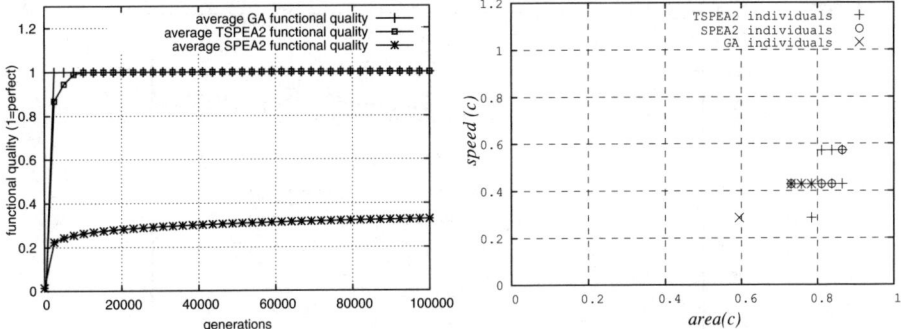

Fig. 4. Evolving the 6-parity function. Data from 10 experiments is shown.

A circuit with minimal area gets an area value of 1, a circuit that utilizes all available logic blocks has an area value of 0.

4.1 6-Parity

The used parameters for the CGP model are $n_c = n_r = n_i = 6$, $n_o = 1$, $n_n = 2$, $l = \frac{n_c}{2}$. For the parity function, a circuit's c fitness with respect to functional quality is defined as follows:

$$f(c) = \frac{1}{1 + \sum_{i \in \mathbb{B}^6}(\text{parity}(i) - c(i))^2}. \tag{4}$$

Thus, a correct parity function has a functional quality of 1. It is an easy task for a conventional GA to evolve a correct circuit for the 6-parity function. Using a population of size 100, only 69 generations were needed on average to evolve a fully functional circuit. In contrast to the GA, SPEA2 with an archive and population size of 100 evolved only four correct solutions overall and needed more than 30000 generations on average. With TSPEA2 preferring the functional quality, the search process converged faster with, on average, 903 generations to evolve a correct circuit. Fig. 4 shows the development of the average functional qualities for the three algorithms, and the speed and area parameters for correctly evolved circuits. Both SPEA2 and TSPEA2 found the same dominant solution. Moreover, TSPEA2 managed to discover a more diverse solution set compared to SPEA2. The conventional single-objective GA evolved correct circuits with inferior area and speed.

4.2 Hashing Function

The hashing function has been evolved previously by Tettamanzi et al. [5]. To be able to compare their experiments with ours, we used the same CGP-parameters: $n_c = 8$, $n_r = 8$, $l = 8$ and $n_n = 4$. The difference to our work is that Tettamanzi et al. restricted wires to connect only to logic blocks in the same row. We have relaxed this constraint which leads to an improvement using a conventional GA.

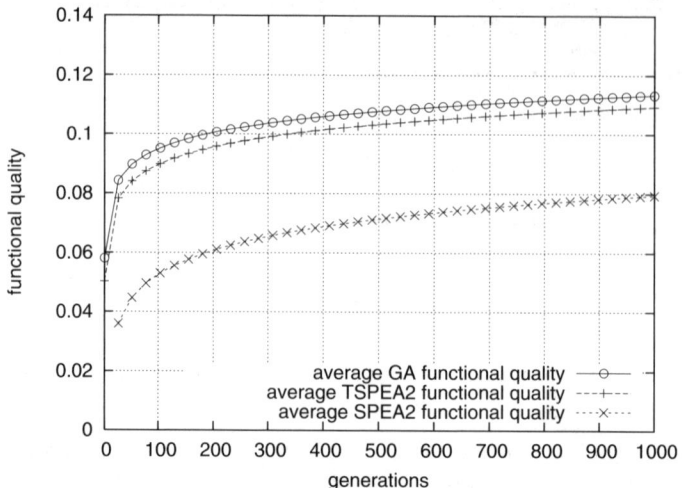

Fig. 5. Evolving the hashing function. Average fitness development over 10 experiments for GA, SPEA2, and TSPEA2.

The problem statement is as follows: Find a function $h : \mathbb{B}^{16} \rightarrow \mathbb{B}^8$ which maps a set M of 2^{12} keys to a set N of 2^8 indices in the most uniform way possible. The fitness function is defined as:

$$f(c) = \frac{1}{1 + \frac{1}{|N|} \sum_{i=1}^{|N|} (|\{j|j \in M, \ c(j) = i\}| - \frac{|M|}{|N|})^2} \tag{5}$$

Tettamanzi et al. [5] evolved the best individual with a fitness value of 0.097785 after 257 generations. On our less constrained CGP model, the single-objective GA reached easily an average fitness beyond 0.1, as is shown in Fig. 5. After 257 generations, the best individual showed a fitness of 0.116469. Fig. 5 shows the fitness development for GA, SPEA2, and TSPEA2. As expected, TSPEA2 performed close to GA while SPEA2 lagged behind. Fig. 6 displays the functional quality vs. area and speed. The figure shows two-dimensional projections of the Pareto front after 1000 generations. As expected, our experiments confirmed that a conventional GA optimizes the functional quality faster than SPEA2 and TSPEA2. Although TSPEA2 is close to GA measured in number of simulated generations, we have to note that simulating one generation in TSPEA2 takes about an order of magnitude longer than for GA. SPEA2 and TSPEA2 excel, however, in evolving solutions with improved area and speed. Table 1 lists the resulting functional qualities (best, worst and average case) after iterating for 1000 generations.

Comparing SPEA2 with TSPEA2, we note that SPEA2 did not evolve individuals with better area or speed. In fact, all individuals found by SPEA2 are dominated by individuals generated by TSPEA2. This is an interesting observation, as one would expect that TSPEA2, which prefers the functional quality over the

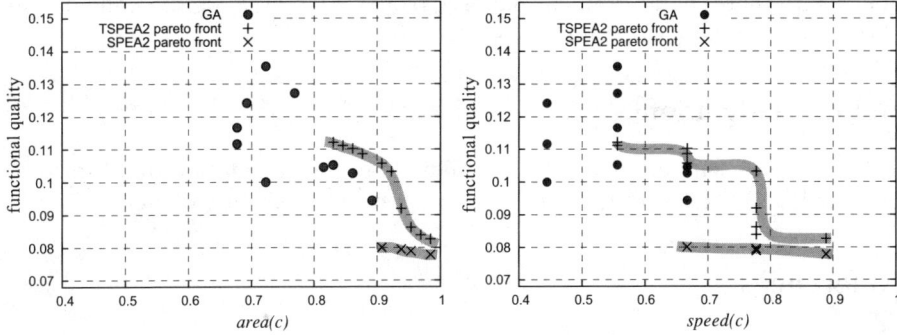

Fig. 6. Evolving the hashing function. 2D projections of the Pareto front for two typical populations. Also the objectives of the best individuals found by GA during the 10 experiments are plotted.

Table 1. Evolving the hashing function. Reached functional qualities after 1000 generations.

	GA	SPEA2	TSPEA2
best	0.135	0.084	0.125
worst	0.094	0.075	0.092
average	0.114	0.079	0.110

other objectives, leads to a somewhat deteriorated Pareto front. This result has been consistent over all simulation runs with the hashing function. A possible explanation is that in our experiments the objectives are not necessarily conflicting. Driving the evolution towards functional quality will then also improve area and/or speed. However, this may not be generalized as design experience shows that for many circuits the functional quality, speed and area are indeed conflicting.

5 Summary and Further Work

In this paper, we have outlined a novel architectural approach for self-adaptive autonomous embedded systems. Simulated evolution is used to adapt to slow changes in the environment; switching to preevolved alternatives is the proper reaction to drastic changes in the available resources. We have then focused on multi-objective evolutionary optimizers and compared the known algorithm SPEA2 with the newly devised technique TSPEA2 and a baseline GA. We have presented comparisons of these algorithms for two test functions.

An implementation of the overall system shown in Fig. 1 is ongoing. Further work will focus on the scalability problem and investigate variants of the CGP model with more coarse-granular building blocks. Moreover, we will validate our observations on larger test functions.

Acknowledgement

This work was supported by the German Research Foundation under project number PL 471/1-1 within the priority program *Organic Computing*.

References

1. de Garis, H.: Evolvable Hardware – Genetic Programming of a Darwin Machine. In: Proceedings International Conference on Artificial Neural Networks and Genetic Algorithms (ICANNGA), Springer (1993)
2. Higuchi, T., Niwa, T., Tanaka, T., Iba, H., de Garis, H., Furuya, T.: Evolving Hardware with Genetic Learning: A First Step Towards Building a Darwin Machine. In: Proceedings 2nd International Conference on Simulation of Adaptive Behavior (SAB), MIT Press (1993) 417–424
3. Thompson, A., Layzell, P.: Analysis of Unconventional Evolved Electronics. Communications of the ACM **42** (1999) 71–79 ACM Press.
4. Higuchi, T., Kajihara, N.: Evolvable Hardware Chips for Industrial Applications. Communications of the ACM **42** (1999) 60–66 ACM Press.
5. Damiani, E., Liberali, V., Tettamanzi, A.: Evolutionary Design of Hashing Function Circuits Using an FPGA. In: International Conference on Evolvable Systems (ICES), Springer (1998) 36–46
6. Keymeulen, D., Konaka, K., Iwata, M., Kuniyoshi, Y., Higuchi, T.: Robot Learning Using Gate-Level Evolvable Hardware. In: EWLR-6: Proceedings of the 6th European Workshop on Learning Robots, London, UK, Springer-Verlag (1998) 173
7. Kalganova, T., Miller, J.: Evolving More Efficient Digital Circuits by Allowing Circuit Layout Evolution and Multi-Objective Fitness. In: The First NASA/DoD Workshop on Evolvable Hardware, Pasadena, California, IEEE Computer Society (1999) 54–63
8. Coello Coello, C.A.: Treating Constraints as Objectives for Single-Objective Evolutionary Optimization. In: Engineering Optimization. Volume 32., Taylor and Francis (2000) 275–308
9. Zitzler, E., Laumanns, M., Thiele, L.: SPEA2: Improving the Strength Pareto Evolutionary Algorithm. Technical Report 103, Gloriastrasse 35, CH-8092 Zurich, Switzerland (2001)
10. Miller, J.F., Thomson, P.: Cartesian Genetic Programming. In: Proceedings of the European Conference on Genetic Programming, London, UK, Springer-Verlag (2000) 121–132
11. Trefzer, M., Langeheine, J., Meier, K., Schemmel, J.: Operational Amplifiers: An Example for Multi-objective Optimization on an Analog Evolvable Hardware Platform. In: International Conference on Evolvable Systems (ICES), Springer (2005) 86–97

Measurement and Control of Self-organised Behaviour in Robot Swarms

Moez Mnif[1], Urban Richter[2], Jürgen Branke[2], Hartmut Schmeck[2], and Christian Müller-Schloer[1]

[1] Leibniz Universität Hannover – Institute of Systems Engineering
Appelstr. 4, 30167 Hannover, Germany
{mnif,cms}@sra.uni-hannover.de
[2] Universität Karlsruhe (TH) – Institute AIFB
76128 Karlsruhe, Germany
{uri,jbr,schmeck}@aifb.uni-karlsruhe.de

Abstract. Today's technical systems are becoming increasingly complex. Future systems will consist of a multitude of complex soft- and hardware components, which interact with each other to satisfy global system functional requirements. This trend bears the risk of more and more breakdowns and other unexpected behaviour. Organic Computing (OC) has the vision of addressing the challenges of complex distributed systems by making them more life-like (organic), i.e. endowing them with abilities such as self-organisation, self-configuration, self-repair, or adaptation. This can only be achieved by giving the system elements adequate degrees of freedom. This may result in an emergent behaviour, which can be positive as well as negative. Therefore, we need an observer/controller architecture, which allows for self-organisation but at the same time enables adequate reactions to control the – sometimes completely unexpected – emerging global behaviour.

In this paper, we give an introduction to a generic observer/controller architecture, adapt this framework to a scenario of a self-organising robot swarm, and show how to control and prevent global, collective, unwanted behaviour based on observations of the local behaviour of the distributed agents.

Keywords: Organic Computing, emergence, observer/controller architecture, multi-agent systems.

1 Introduction

In our everyday life we are surrounded by computers, embedded computing devices, and other technical environments. The impressive progress in computing technology over the past decades has not only led to an increase of comfort, but also to an increase of the complexity of technical systems. The problem of the increasing system complexity will be one of the main challenges for computer science for future years. From an engineering point of view, it is yet unclear how to design such distributed and highly interconnected systems in a manner

P. Lukowicz, L. Thiele, and G. Tröster (Eds.): ARCS 2007, LNCS 4415, pp. 209–223, 2007.
© Springer-Verlag Berlin Heidelberg 2007

that makes them reliable and usable in dynamically changing environments. The designer will neither be able to predict all possible system configurations nor to prescribe proper behaviours for all cases. Also, we need to relieve the user of having to control all system parameters in detail, allowing him instead to influence the system at a higher level, e. g. by setting goals.

Organic Computing (OC) has the vision to address complexity of today's and future distributed technical systems by making them more life-like, and endowing them with properties such as self-organisation, self-configuration, self-repair, or adaptation [1], and calls therefore for new design principles. OC systems consist of a large population of interacting components, which organise themselves and adapt to new situations. To achieve the OC goals it is obvious we need to give the system components some degrees of freedom. As a consequence, the whole system can develop an unexpected emergent behaviour, which can be positive as well as negative. But there are additional drawbacks. In order to adapt to new situations and to learn new logic and behaviour, a technical system has to be allowed to make errors, and it may react more slowly. Also, system behaviour might become less predictable.

Emergent phenomena are often characterised as "the whole is more than the sum of its parts", where the study of the local behaviour of the individual components reveals little about the global system-wide behaviour. Emergence and self-organisation have been studied in the area of multi-agent systems [2,3], and we know many examples from nature, e. g. a flock of birds, the clustering behaviour of densely packed chickens in cages [4], or detecting the foraging behaviour of ants [5]. Humans decide intuitively on the occurrence of emergence, but in technical scenarios we need a more coherent definition of this phenomenon. Inspired by the examples from nature we can say that a precondition for emergence is a large population of interacting elements without central control, and, following the definition by Müller-Schloer and Sick [6], we define emergence as self-organised order.

In order to balance creative self-organised bottom-up processes and top-down control, an observer/controller architecture has been proposed in [7,8], similar to Autonomic Computing's MAPE (monitore, analyse, plan, and execute) cycle [9]. The observer monitors the overall system, uses metrics to quantify and detect emergent behaviour and aggregates its observations as a vector of situation parameters, which is sent to the controller. The controller evaluates the measured situation parameters with respect to the goal defined by the user, and attempts to influence the system, if an intervention is required. In technical applications, the controller has to use and to compare the results of different metrics to determine the most appropriate action. But in the scenario of this paper, we simplify the measurement method and the controller decides based on one metric only.

In this paper, we introduce in Sect. 2 the concept of a generic observer/controller architecture with the goal to observe emergent effects (positive or negative), and to control such emergent behaviour with respect to a global objective function. In Sect. 3 we show how to compute the emergent behaviour according to an entropy-based measurement method introduced by Mnif and

Müller-Schloer [4]. Our test scenario of swarm robots is introduced in Sect. 4. We explain the idea behind this simulation, describe the implementation using simple control mechanisms, and present first experimental results that validate the generic observer/controller architecture. The paper concludes with a summary and a discussion of future work.

2 Observer/Controller Architecture

To achieve the goals of OC and to endow technical systems with more life-like properties, they need some degrees of freedom. While this has apparent benefits and will relieve soft- and hardware designers from specifying every single detail, unexpected behaviour can occur. Therefore, the process of self-organisation has to be controlled. As described in [7] we propose a regulatory feedback loop of observation and control (see also Fig. 1 [8]).

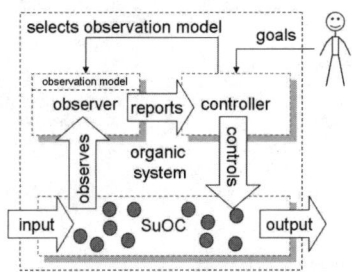

Fig. 1. Observer/controller architecture

The observer measures the collective behaviour (micro and macro level properties) of the system under observation and control (SuOC) through sensors and reports aggregated system parameters characterising the global state and the dynamics of the system to the controller. The controller evaluates the measured situation parameters with respect to the goal defined by the user and influences the system through actuators. Together with the layer of observer and controller the SuOC forms what we call an "organic system".

Figure 2 shows the generic observer/controller architecture in more detail, and we refer the interested reader to [8,10] for more information on this architecture. In the following we summarise just the main principles for a better understanding of the subsequent sections.

The observer consists of a monitor, a log file, a pre-processor, a data analyser, a predictor, and an aggregator. Using a prespecified sampling rate system data of the SuOC is sampled. This data specifies global system attributes, and individual data monitors attribute values on the level of single elements. For prediction and calculation of time-space-patterns in the data analyser all monitored data are stored in a log file. Some derived attributes can be computed from the raw data in the pre-processing unit (e. g. an attribute velocity can be derived from

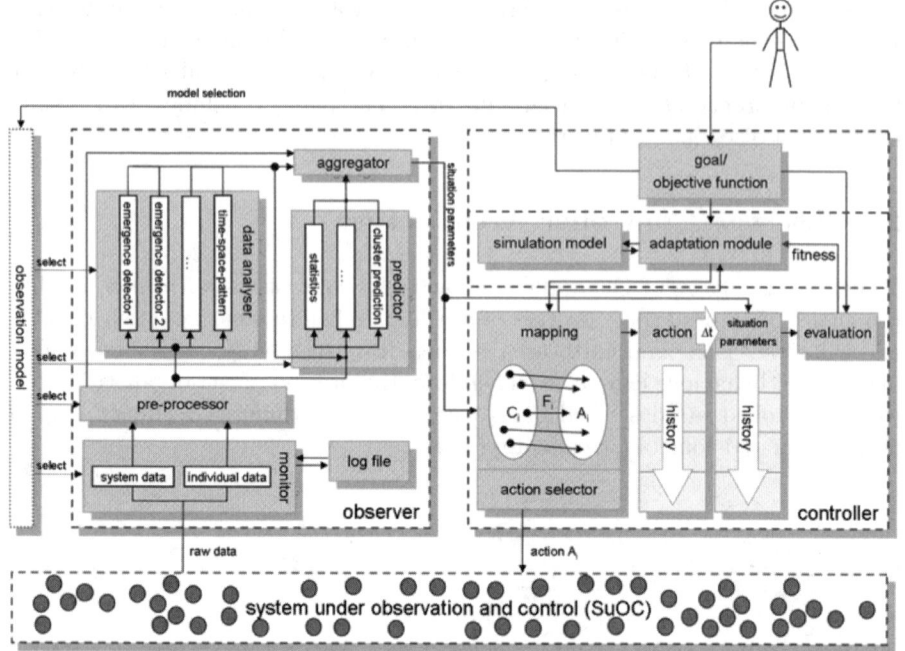

Fig. 2. Generic observer/controller architecture

attributes x-coordinate and y-coordinate collected in subsequent time steps). A set of detectors measures the pre-processed data and computes system detectors like emergence values with respect to the definition in [4]. In a more or less simultaneous step a future system state is predicted by the predictor. The results of data analyser, predictor, and possibly some unprocessed data coming from the pre-processor are forwarded to the aggregator and transmitted to the controller as a set of situation parameters.

The controller guides the self-organisation process between the elements of the SuOC and uses for its decisions the situation parameters from the observer. The controller interferes only when necessary and may affect the SuOC by influencing: (1) the local decision rules of the SuOC elements, (2) the system structure including e. g. the communication between the SuOC elements or the number of elements, or (3) the environment, which will indirectly influence the system by changing the data observed by the SuOC elements through their local sensors.

For selecting the Action A_i that is most appropriate for the measured situation, the controller has to decide quickly and to react in real-time. Every selected action together with its correspondent situation parameters is stored as a tuple in a kind of memory for an evaluation process in the following iteration of the observer/controller loop. The success of an action is determined with respect to the global objective function, and the adaptation module adapte the mapping from situations to actions correspondingly. Furthermore, the adaptation module may use a simulation model to anticipate consequences of specific actions before testing them on the real system, and to plan completely new actions.

The generic architecture has to be customised to different scenarios by adapting the various components of the observer (including the observation model) and the controller. In Sect. 4 we describe how we have implemented the architecture for a scenario of swarm robots to observe, predict, and control their emerging clustering behaviour.

3   Quantitative Emergence

Our objective is to quantify emergent behaviour in technical scenarios (e. g. a swarm of autonomous robots, a large ensemble of elements/agents), and to trigger certain actions in order to avoid (negative emergence) or to strengthen (positive emergence) certain system behaviour, whenever a SuOC does not fulfil the system goal and specified objective function that is given by the human system developer or designer.

But to do so, we need a definition of "emergence". Following Müller-Schloer and Sick [6] traditional and philosophical definitions of emergence are either too weak or too strong, do not characterise really our technical-oriented approach, and are not sufficient from the viewpoint of OC which bases its notions on engineering aspects.

Thus, we propose a definition which is based on measurements of local agent behaviour, and we define emergence as outlined in Sect. 1 as the formation of order from disorder based on self-organising processes. The proposed definition is based on Shannon's information theory, in particular on the information-theoretical entropy, and the quantitative definition of emergence is based on the assumption that unexpected behaviour can be observed in terms of patterns in space and/or over time. For more details the interested reader is referred to [4]. In the following, we briefly skech the idea for observation of emergent phenomena by computing a system fingerprint.

One type of detectors which can be used within the data analyser mentioned in Sect. 2 is the *emergence fingerprint* which reflects the order pattern of the common attributes of the system elements. A large population of agents that interact with each other on the base of local rules and without central control leads to a macroscopic behaviour that shows new properties which do not exist at the agent level. This macroscopic pattern is perceived as order and can be quantified by the emergence fingerprint.

The definition of this fingerprint relies on an entropy measure, a well-known metric for order. Low entropy is equivalent to a higher system order and vice versa. With other words: The more structure is present (unequal distribution), the more order is present. The fingerprint's change over time gives an indication of the dynamics of the system (e. g. transition from chaotic to an ordered state).

The main idea of computing the fingerprint of a system is first to identify the common attributes of its elements, second to build for each attribute a relative frequency of the occurrence of each value (which can be considered as a probability distribution), third to compute the entropy related to each attribute on basis of Shannon's information theoretical definition of the entropy:

$$H_A = \sum_{j=0}^{j=N-1} p_j \cdot ld\frac{1}{p_j} \tag{1}$$

in which A represents a given attribute and N the number of the different attribute values (the unit of measurement is $bit/attribute$). Finally, one has to compute the degree of emergent order of each attribute according to:

$$M_A = H_{A_{max}} - H_A \tag{2}$$

$H_{A_{max}}$ is the entropy of the attribute A in case of an equal distribution of attribute values (lowest level of order). The set of all M_k (k denoting an attribute) values of the SuOC constitutes a vector which is called system fingerprint (see also Fig. 3). Instead of the emergence M, we can also compute a relative emergence m according to:

$$m_A = \frac{H_{A_{max}} - H_A}{H_{A_{max}}} \quad (0 \le m_A \le 1) \tag{3}$$

In Sect. 4 we use such a fingerprint to quantify and to control the emergent behaviour of a swarm of autonomous robots.

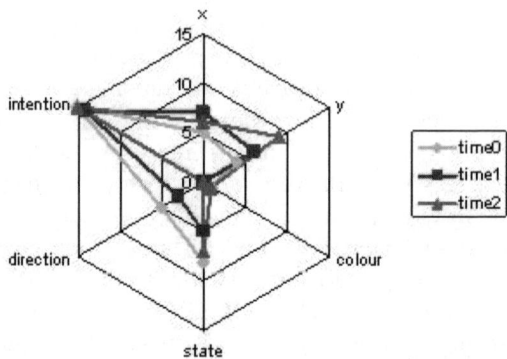

Fig. 3. Fingerprint with different attributes at three specific times, visualised as a *n*-dimensional Kiviat graph (one dimension for each attribute). The fingerprint shows e.g. the attributes x, y, *direction*, *colour*, etc., and the emergence fingerprint at the times *time0*, *time1* and *time2*.

4   Experimental Results

In this section, we describe how to implement the generic OC architecture in a technical domain: a swarm of robots. The application described below is inspired by nature, and shows clustering from a macroscopic point of view as an emergent behaviour of local interactions. The goal of our work is to get a better understanding of the global emergent behaviour of distributed agents. Our chosen application serves also as a test bed to validate the generic observer/controller architecture from Sect. 2.

The simulation reproduces the collective cannibalistic behaviour of densely packed chickens in cages (cooperation with the University of Veterinary Medicine Hannover), and tries to explain the unwanted behaviour of clustering, which is frequently observed when a chicken is wounded, and which leads to a major loss of animals (up to 50% of the animals). If chickens perceive a wounded chicken, they chase this chicken and pick on it until it dies.

Chasing and picking wounded chickens leads to the emergent building of chicken swarms (or clusters). A swarm disperses when the wounded chicken is killed. The emergent behaviour is spatial, but swarms move over time. This is a case of "negative", i. e. unwanted, emergence, because the global goal should be to reduce the chicken death rate.

While simulating this behaviour, order patterns emerge as expected in form of chicken swarms. In agriculture at the moment these patterns are interpreted by human experts. But our goal from the viewpoint of OC is to observe, classify, and control this behaviour automatically. To achieve the goal of maximising the lifetime of chickens, we use the observer/controller paradigm. The observer reports a quantified context of the underlying system to the controller. The controller evaluates the situation and reacts with adequate control actions to disperse chicken swarms or to prevent their formation.

Certainly, OC focuses on the problem of increasing complexity in technical scenarios, and we admit that the chicken simulation has no obvious technical relevance on its own. But, in the simulation, a chicken is directed by a predefined set of rules, and will be influenced by the behaviour of other chickens in its local neighbourhood or by changes in the environment, e. g. by noise that frightens them and feed that attracts them. Therefore, chicken are considered as autonomous robots or agents with simple rules und local goals, they aim for surviving as long as possible and they are attracted by wounded conspecifics.

There are many analoguous technical scenarios with a similar structure, as described e. g. in [11]. This analogy justifies from our point of view our biologically inspired simulation and shows a technical relevance.

4.1   Experimental Environment

We should mention that the notions of robot, agent, or chicken have the same meaning in our context. We use the item 'chicken' in analogy to the bio-inspired paradigm, but we abstract from the animal and presume that the chicken is an autonomous robot or a technical agent, which shows no life of its own and instead reacts as specified by its developer.

A chicken is characterised by the attributes position (x-, y-coordinates), heading and energy, and works according to the finite state machine (FSM), depicted in Fig. 4. Whether a chicken is wounded or not depends on its energy level. We have defined 5 different internal states of a chicken:

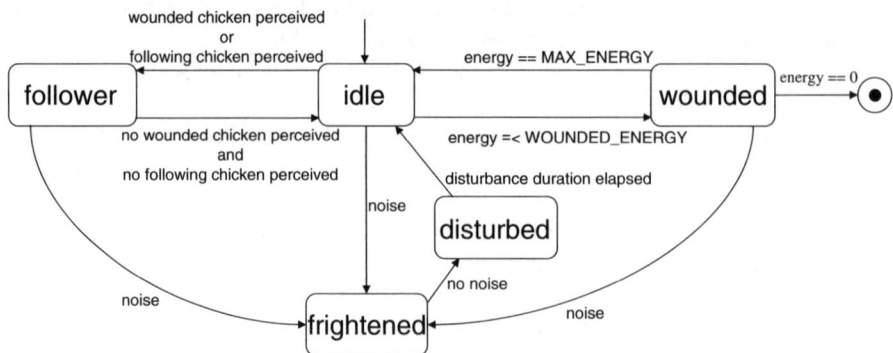

Fig. 4. Finite state machine of a chicken representing the local behaviour rules of a single chicken

Idle: In this state a chicken is not wounded. The chicken moves according to a simple mobility model, is initially placed at a random position in the cage/playground, and chooses randomly a new position to move in its direct neighbourhood. Arrived at its destination the chicken chooses another random one for the next simulation step. In one simulation step a chicken can move from one field to the fields in its direct neighbourhood. Typically, a chicken can choose its new position from a set of 8 possible positions, or it stays where it is.

Follower: A chicken moves to this state if a wounded chicken or a following chicken is perceived (the distance to this chicken is lower than a fixed perception horizon given as simulation parameter). Then the chicken tries to get as near as possible to the wounded (or the following) chicken, it minimises the distance between the injured chicken and itself. If it is immediately close to the wounded chicken it begins to pick on it (and the wounded chicken looses energy).

Wounded: A chicken gets to the wounded state if its energy level passes a threshold that is lower than a given energy level, which is set by a simulation parameter (see Tab. 1). In this case the wounded chicken tries to escape attacks, it maximises the distance between itself and the chickens in its direct neighbourhood, and it heals with each tick of the simulation (incrementation of energy), only if it is not picked.

Frightened: A chicken gets frightened, if it perceives noise. It moves as fast as possible (one field per simulation step) outside the fields that are affected by noise. If the chicken has been in the state of a follower, it tries to maximise the distance between the wounded chicken and itself.

Disturbed: After leaving the noise affected fields a chicken changes to the disturbed state for a fixed duration (see Tab. 1). A chicken doesn't react to wounded chickens and moves randomly as it does when idle.

Table 1. Simulation parameters

Agent number	20
Playground	30 × 30 fields
Simulation time	1000 ticks
Generation of wounded chickens	Every 60 ticks after a wounded chicken is killed or healed
Maximal number of concurrently wounded chickens	1
Maximal energy level of a wounded chicken	100 energy units
Energy lost per received pick	1 energy unit
Energy level of randomly generated wounded chicken	70-80 energy units (Uniformely distributed)
Healing rate	1 energy unit/tick
Perception horizon of a chicken	15
Number of directions	8
Duration of noise control	15 ticks
Intensity of noise control	Circular area with a radius of 7 fields
Critical emergence threshold	0.42
Disturbance duration of a chicken after a controller intervention	5 ticks
Waiting time after a controller intervention	2× duration of noise control

To achieve first experimental results we simplify our scenario and blind out the effects of feed. That is the reason why the FSM, as pointed out, is quite simple and has 6 different states only. The behaviour with feed will be integrated into future simulations.

Our simulation environment is set up with the parameters as listed in Tab. 1. We observe a scenario of 20 chickens that move randomly in the playground, which has the dimension of 30 × 30 fields (see Fig. 5). Typically, every chicken can move to eight different directions (e. g. see the possible next positions of chicken 9 in Fig. 5). If a chicken is set to a border or it reaches an edge of the playground it is obvious that five (see chicken 17) or three possible movements are left for these special situations. All chickens move with the same speed.

4.2 Observer/Controller Architecture Applied to Scenario

The generic observer/controller architecture is applied to the scenario as depicted in Fig. 6. The monitor collects data from the system at a fixed sampling rate (one data set each simulation tick). This data set consists of a system-wide attribute (the number of chicken killed in the last sampling period) and of individual attributes of each chicken (x-coordinate, y-coordinate and heading). The functionality of the pre-processor is reduced to passing the individual parameters to the data analyser and the number of killed chicken to the aggregator. The data analyser determines the emergence fingerprint of the system by computing

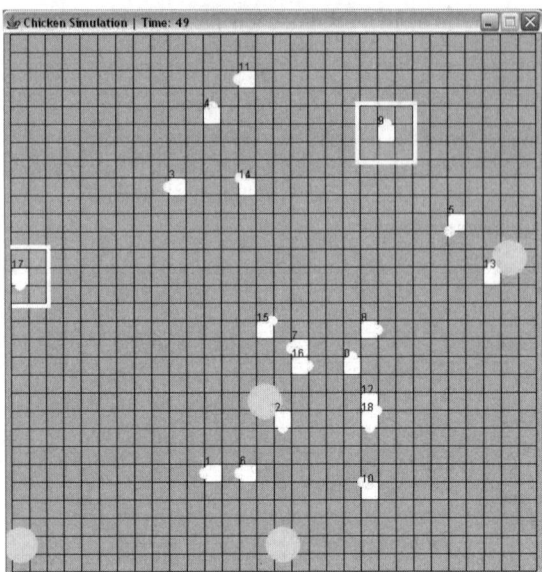

Fig. 5. Scenario with 20 chickens and 4 feeding dishes

the emergence indicators of the collected three chicken attributes. Also, the data analyser uses the x- and y-coordinates to determine the coordinates of cluster centroids. For calculation we use the K-Means clustering algorithm. The results of the data analyser (three emergence indicators and the available cluster centroids) are passed to the aggregator. The aggregator forwards the situation parameters, composed of the computed data analyser values and the number of killed chickens, to the controller.

To close the observer/controller loop and to obtain first experimental results we have implemented a controller with reduced functionality, which includes basically a simple mapping and a mechanism to prevent overshooting of control actions. The controller waits for a fixed duration before interfering again. Minimising the number of killed chickens is defined as the global goal, given by the developer, and leads to the following control logic: If the emergence indicator of the x-coordinates is ≥ 0.42, a noise signal with fixed intensity and fixed duration is applied around the computed cluster centroid to frighten the chickens and disperse the cluster. We plan to automate this control action with more degrees of freedom and to integrate a learning process to adapt intensity and duration of the noise signal.

For the duration of noise control some fields of the playground around the cluster centroid and within a fixed radius are highlighted with a noise flag. A chicken reacts to this flag and uses this information within its decisions of movement and behaviour during the duration of noise control. A chicken changes its status to frightened, if it percieves noise, and moves one field per simulation step outside the fields that are highlighted with noise. All chickens try to maximise

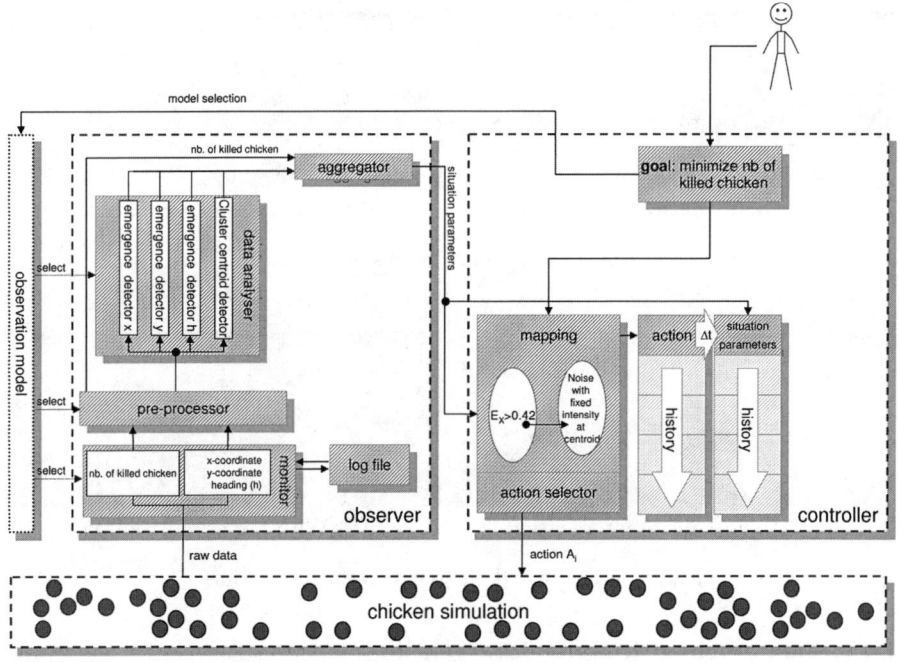

Fig. 6. The generic architecture applied to the chicken scenario

the distance between the wounded chicken and themselves. They move as fast as possible to fields which are not highlighted, and change their status to disturbed.

4.3 Observation of Emergence

Typical values of the relative emergence indicator of the x-coordinates can be seen in Fig. 7(a). The figure shows the trend for one run of simulation. We can observe the recurrence of a cycle that constitutes of three phases: The cluster formation phase (the curve increases from 0.25 to 0.42 or higher), a cluster phase (the values oscillate on a high level), and a dispersion phase (the curve decreases from a high level to a low one). The interpolated values show a more exact disjunction of the three phases, as depicted in Fig. 8(a).

During the absence of a wounded chicken the chickens are uniformly distributed over the area. After the appearance of a wounded chicken a cluster is formed after a short delay and the emergence indicator increases. After each cluster phase a chicken is killed (see Fig. 9(a)), and after a short distribution phase a new cycle begins.

From these observations we learn that a control intervention should be triggered when a certain emergence value is exceeded. Tests have suggested that a well working critical emergence threshold is located close to 0.42.

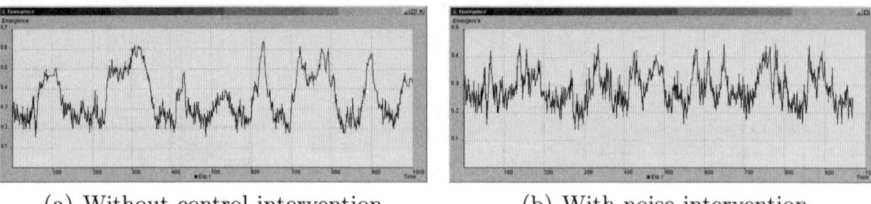

(a) Without control intervention   (b) With noise intervention

Fig. 7. Emergence values over time

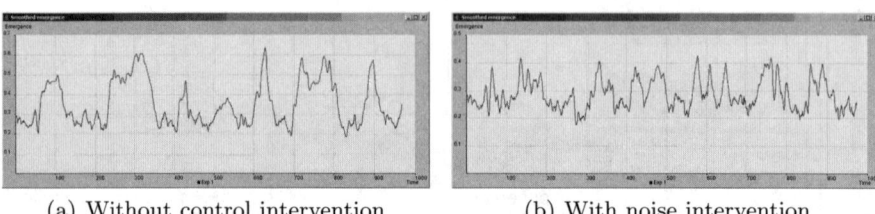

(a) Without control intervention   (b) With noise intervention

Fig. 8. Interpolated emergence values over time

4.4   Emergence with Noise Intervention

The controller interferes only if the observer measures a cluster and the specified threshold is exceeded. Using a freely applicable noise emitter with an interim fixed intensity the chickens are frightened and the cluster disperses. Other control actions to influence the behaviour of the chickens are imaginable, e. g. attracting them with feed in the area around a cluster, but so far not integrated into the simulation. In comparison with Fig. 7(a) the emergence values with noise intervention show no possible separation into three phases any more (see Fig. 7(b)). The interpolated curve in Fig. 8(b) shows this effect in detail. The values are characterised by continuous increasing and decreasing phases. The cluster phase is left out.

   Comparing Fig. 9(a) with Fig. 9(b) we observe a significantly lower death rate and notice that the death rate has decreased due to the control interventions. However, not all interventions have been successful, and for some cases many control interventions have been necessary to deal with the emergent situation

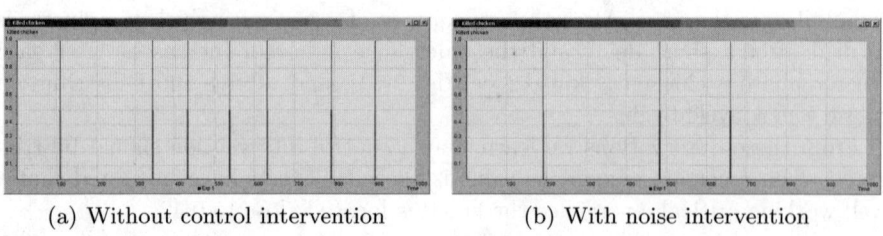

(a) Without control intervention   (b) With noise intervention

Fig. 9. Death rate over time

(e. g. see the occurrence of two successive cluster phases between time 400 and 600). Whether a controller decision is successful or not depends on the energy status of the wounded chicken. And the odds that a chicken will heal during the duratioon of a controller intervention depend on how many other chickens run away randomly in the same direction as the wounded chicken does, so that they can attack it again after getting out of the intervention area and after the disturbance time has elapsed.

5 Conclusion and Outlook

In this paper we have given a short introduction to OC and have summarised the main goals of the OC community, namely to tame particularly increasing complexity in technical systems. To cope with the possibility of emerging global behaviour as a result of bestowing upon the systems some life-like characteristics, the paper outlines the idea of an observer/controller architecture. This architecture allows for organising several aspects of the system behaviour in an autonomous way, independent of an explicit external interference that keeps a system alive and running. The SuOC adapts to changes in its environment in order to acquire robustness and the ability to overcome breakdowns.

We have adapted the generic approach of the observer/controller architecture to a scenario of a swarm of simple robots, which present a macroscopic behaviour that depends only on local rules. The robots are attracted by "injured" ones and frightened by noise.

Providing feedback and decision capabilities to this technical scenario we have shown that the unwanted emergent behaviour, clustering of robots around injured agents, can be observed and prevented automatically with respect to a global objective function. We have used an entropy-based measurement method to observe the described order pattern. For controlling the robots we have implemented a simple method that changes the environment and affects their local behaviour indirectly.

In this paper we presented first steps towards a complete observer/controller loop applied to a concrete test scenario. Our simulation results validate the idea of the proposed generic observer/controller architecture, and we could show advantages and potentials of controlled self-organisation in technical scenarios. Without control actions the robots will meet in a cluster, hinder or attack each other, and the system might break down. The controller extends the life of the robots and keeps on running the system.

Our future work will focus on the following challenges:

– Endow the controller with adaptation capability. The controller should have the possibility to evaluate the success of its interventions and to adapt the fitness and the parameters of the used rules depending on the global objective function. This suggests the ability of online learning. The controller should also be able to generate new rules with adequate parameters (e. g. the controller should learn the correlation between the increasing of the

x-emergence over a critical value of 0.42 and the death of a chicken as shown in Sect. 4.4). Adaptation over the time will optimise the controller's behaviour and the guided process of self-organisation.

- Endow the simulation with other control possibilities. At the moment the controller has only the possibility to control the swarm of robots with noise. In analogy to the introduced chicken behaviour other control actions are possible, e. g. the controller can spread some feed around the cluster to attract the robots in another direction.
- To make the scenario more complex and to enhance the task of the observer/controller architecture we plan to integrate distributed and collective learning to the robots. Robots with local adaptation will show a more complex – and thus challenging – behaviour.

Acknowledgment: We gratefully acknowledge the financial support by the German Research Foundation (DFG) within the priority program 1183 "Organic Computing". We are especially indebted to Fabian Rochner, Leibniz Universität Hannover, and Holger Prothmann, Universität Karlsruhe (TH), for their valuable suggestions.

References

1. Schmeck, H.: Organic Computing – A new vision for distributed embedded systems. In: Proceedings of the 8th IEEE International Symposium on Object-Oriented Real-Time Distributed Computing (ISORC 2005), IEEE Computer Society (2005) 201–203
2. Di Marzo Serugendo, G., Gleizes, M.P., Karageorgos, A.: Self-organisation and emergence in MAS: An overview. Informatica **30**(1) (2006) 45–54
3. Di Marzo Serugendo, G., Foukia, N., Hassas, S., Karageorgos, A., Kouadri Mostéfaoui, S., Rana, O.F., Ulieru, M., Valckenaers, P., Van Aart, C.: Self-organisation: Paradigms and applications. In Di Marzo Serugendo, G., Karageorgos, A., Rana, O.F., Zambonelli, F., eds.: Postproceedings of the Engineering Self-Organising Applications (ESOA 2003) Workshop at the Second International Joint Conference on Autonomous Agents & Multi-Agent Systems (AAMAS 2003). Volume 2977 of LNAI., Springer (2004) 1–19
4. Mnif, M., Müller-Schloer, C.: Quantitative emergence. In: Proceedings of the 2006 IEEE Mountain Workshop on Adaptive and Learning Systems (IEEE SMCals 2006). (2006) 78–84
5. Dorigo, M., Maniezzo, V., Colorni, A.: Ant system: Optimization by a colony of cooperating agents. Technical Report 26 1, IEEE Transactions on Systems (1996)
6. Müller-Schloer, C., Sick, B.: Emergence in Organic Computing systems: Discussion of a controversial concept. In Yang, L.T., Jin, H., Ma, J., Ungerer, T., eds.: Proceedings of the 3rd International Conference on Autonomic and Trusted Computing (ATC 2006). Volume 4158 of LNCS., Springer (2006) 1–16
7. Müller-Schloer, C.: Organic Computing: On the feasibility of controlled emergence. In Orailoglu, A., Chou, P.H., Eles, P., Jantsch, A., eds.: Proceedings of the 2nd IEEE/ACM/IFIP International Conference on Hardware/Software Codesign and System Synthesis (CODES+ISSS 2004), ACM (2004) 2–5

8. Richter, U., Mnif, M., Branke, J., Müller-Schloer, C., Schmeck, H.: Towards a generic observer/controller architecture for Organic Computing. In Hochberger, C., Liskowsky, R., eds.: INFORMATIK 2006 – Informatik für Menschen! Volume P-93 of GI-Edition – Lecture Notes in Informatics (LNI)., Köllen Verlag (2006) 112–119
9. Sterritt, R.: Autonomic Computing. Innovations in systems and software engineering **1**(1) (2005) 79–88
10. Branke, J., Mnif, M., Müller-Schloer, C., Prothmann, H., Richter, U., Rochner, F., Schmeck, H.: Organic Computing – Addressing complexity by controlled self-organization. In: Proceedings of the 2nd International Symposium on Leveraging Applications of Formal Methods, Verification and Validation (ISoLA 2006). (2006)
11. Jäger, M.: Kooperierende Roboter: Gemeinsame Erledigung einer Reinigungsaufgabe. KI **18**(2) (2004) 59–60

Autonomous Learning of Load and Traffic Patterns to Improve Cluster Utilization*

Andrew Sohn[1], Hukeun Kwak[2], and Kyusik Chung[2]

[1] Computer Science Department, New Jersey Institute of Technology,
Newark, NJ 07102, U.S.A.
[2] School of Electronic Engineering, Soongsil University, 511 Sangdo-dong, Dongjak-gu,
Seoul 156-743, Korea
sohn@cs.njit.edu, {gobarian, kchung}@q.ssu.ac.kr

Abstract. Adaptive clustering aims at improving cluster utilization for varying load and traffic patterns. Locality-based least-connection with replication (LBLCR) scheduling that comes with Linux is designed to help improve cluster utilization through adaptive clustering. A key issue with LBLCR, however, is that cluster performance depends much on a single threshold value that is used to determine adaptation. Once set, the threshold remains fixed regardless of the load and traffic patterns. If a cluster of PCs is to adapt to different traffic patterns for high utilization, a good threshold has to be selected and used dynamically. We present in this report an adaptive clustering framework that autonomously learns and adapts to different load and traffic patterns at runtime with no administrator intervention. Cluster is configured once and for all. As the patterns change the cluster automatically expands/contracts to meet the changing demands. At the same time, the patterns are proactively learned that when similar patterns emerge in the future, the cluster knows what to do to improve utilization. We have implemented this autonomous learning method and compared with LBLCR using published Web traces. Experimental results indicate that our autonomous learning method shows high performance scalability and adaptability for different patterns. On the other hand LBLCR-based clustering suffers from performance scalability and adaptability for different traffic patterns since it is not designed to obtain good threshold values and use them at runtime.

1 Introduction

Improving the utilization of computing resources [7] is a critical part of building and managing computational infrastructure. Invariably, computational demands on a cluster of PCs change at runtime. Some machines will be under utilized while others over utilized. Numerous approaches have been reported to date to solve this utilization problem from various different perspectives, including adaptive infrastructure, utility computing, virtualization, process migration, adaptive clustering, etc.

---

* Hukeun Kwak and Kyusik Chung are supported in part by the Basic Research Program of the Korea Science & Engineering Foundation grant No.R01-2006-000-11167-0.

P. Lukowicz, L. Thiele, and G. Tröster (Eds.): ARCS 2007, LNCS 4415, pp. 224–239, 2007.

The N1 Grid initiative by Sun is an example of such technology that can elastically manage computing resources on the fly [18]. Utility computing and adaptive enterprise being deployed by HP are centered on the concepts of treating computing resources as definable and controllable units such as electricity or water [10]. In recent years, IBM has been actively working on their on-demand computing, which attempts to maximize utilization of computing resources as well as automatically detect faults and subsequently heal them [8]. The Gray-box system by Wisconsin [2] and the black-box model by HP [12] help automatically manage the performance of computing services. Process migration is designed to solve the utilization problem. A particular process that causes overload can be migrated to an idle machine. Or, other processes on that machine can be migrated to another one so that all the resources can be made available to the resource-intensive process. Projects such as Wisconsin's Condor [15] and OpenMosix [17] along with Checkpointing [4] have been deployed to migrate processes in various real settings. Virtualization enables multiple "guest" operating systems or *virtual machines* to run on a "host" operating system that runs on a physical machine [1]. If a particular physical machine in a cluster is overloaded, moving the overloaded guest operating system to a less loaded physical machine can help solve the utilization problem. The open source community has been working on Xen, User-mode Linux (UML), Bochs, Plex86, etc. Commercial products include IBM360, VMware, Microsoft's Virtual PC, Sun Containers, etc. Hardware supports for virtualization include Intel's VT and AMD's Pacifica through multi-core processors.

Adaptive clustering aims at improving cluster utilization by changing the number of physical machines for different traffic patterns in the cluster [20]. Initially, a cluster of servers is set up for various services and a subset of servers, or *server set*, is designated for a particular service. If the demand for a particular service exceeds the *threshold* of a server set, the front-end load balancer will automatically recruit another server from the cluster that may be a part of other server set(s), resulting in expansion. When the demand recedes, the particular server set will contract, releasing the newly recruited server to the cluster. Locality-based least connection with replication (LBLCR) scheduling that comes with Linux is designed to adapt to different traffic patterns in cluster [9]. Linux Virtual Server (LVS)-based load balancer [14] distributes client requests to back-end servers based on the number of *connections* between the LVS load balancer and the back-end servers. The decision as to which server to send the request is made based solely on the number of connections, or *threshold*, regardless of the implications of each connection. If the current traffic is beyond the predetermined threshold, LBLCR scheduling calls for an expansion of server set for that particular service. This threshold value is chosen at the time of initial cluster configuration. Once set, it stays the same regardless of the traffic patterns and the cluster condition. Since the performance of cluster hinges on this single value, the system administrator must carefully choose this value through extensive experiments, searching the Web, or ask other experienced administrators.

It is precisely the purpose of this study to solve this single threshold problem. We introduce a learning method that allows the cluster to autonomously adapt to different traffic and load patterns at runtime with *no* administrator intervention. Our approach is in two fold. First, we use both CPU usage and number of connections to accurately assess the entire cluster usage, instead of connections only as done in LBLCR. Second, threshold value is autonomously determined and updated at runtime for different load and traffic patterns with no human intervention.

The report is organized as follows. Section 2 presents background materials on LBLCR scheduling. In Section 3, we explain our approach in detail. In section 4, we provide our experimental environment and results using publicly known Web traces. In Section 5, we analyze our results and compare with LBLCR and simple-minded Round-Robin. The last section concludes our work.

2   Adaptive Clustering with LBLCR Scheduling

An experimental environment consists of N_C client machines, a front-end LVS-based load balancer, a cluster of N_S backend servers, and a Web server. Clients send requests to the LVS. The LVS-based load balancer distributes requests to one of the servers [5]. The load balancer can be mirrored to provide high availability. Initially, a cluster is set up for various services. There will be multiple server sets, each of which will be designated for a particular service.

If the demand for a particular service exceeds the *threshold* of a server set, the front-end load balancer will automatically recruit another server from the cluster, resulting in server set expansion. When the demand recedes, the server set will contract, releasing the newly recruited server to the cluster.

Locality-based least connection with replication (LBLCR) scheduling [11] is designed to adapt to new traffic patterns. When a request is received from the client, the LVS load balancer checks if this is the first time or repeated. For the first time requests LVS determines a target server. The decision is based on the number of connections between the LVS and a back-end server. To be more precise, the LVS computes the connectivity $V = C_{active} \times A_{weight} + C_{inactive}$ for all servers, where C_{active} is the number of active connections and $C_{inactive}$ is the inactive number of connections between the LVS and the server. LVS selects a server with the lowest V regardless of the nature of connections. For repeated requests, LVS directs them to the server that has already been assigned earlier. Suppose that a certain service/page/site/content is popular and there are many requests destined to the same server. The target server will soon be overloaded and some requests may not be served in a reasonable time period or may even be dropped. To prevent this potential server overload, LVS uses LBLCR scheduling to increase the number of servers for this particular service. First, LVS determines if the target server is overloaded. The decision is based on comparing C_{active} and H_{target}, where C_{active} is the number of active connections between LVS and the target server and H_{target} is the *predetermined* threshold for the target server. If $C_{active} > H_{target}$, it is clear that the target server is deemed overloaded. LVS initiates an adaptation process, which entails two steps: first, LVS computes the connectivity V for all servers. Second, LVS selects the one with the lowest connectivity. This additional server along with the overloaded server forms a server set of two to service the "hot" pages/contents [34]. Adaptive clustering has just taken place to meet the increased demand.

LBLCR scheduling presents two challenges. *First*, decisions are made based on the number of connections between the load balancer and each backend server. No other information is taken into consideration. A connection asking for a page of 1KB plain text is treated the same as a connection asking for an image of 1 MB that may require compression and decompression. *Second*, the threshold for each server is fixed at the time of server configuration. This is the value used to determine whether another server should be brought in to the server set. If the threshold is set low, the frequency of expansion will increase, causing much overhead of decision-making and expansion. As a result, the overall effectiveness will suffer since the cluster will spend much time on overhead. If the threshold is set high, the frequency of cluster expansion will decrease, causing some servers overloaded while others idling. Threshold is central to adaptation. It must be carefully chosen with much experiments and efforts. However, finding threshold at the time of setup is not simple. Doing so requires extensive and lengthy experiments, searching the Web, or even asking experienced administrators for good values. Suppose that the right threshold is determined, updating them is another hurdle that has yet to be overcome because it requires human intervention. It is clear that the threshold be *autonomously* determined and updated at runtime to adapt to different traffic patterns with no human intervention. This is exactly what we shall present in the following section.

3 Autonomous Learning

3.1 Learning Environment Overview

Figure 1 shows the new load balancer with three tables: *usage status*, *pattern matrix*, and *learning index*. Raw resource usage data are classified into a predetermined number of regions. The usage status table normalizes raw data to find a load and traffic pattern. The pattern matrix records load and traffic patterns. The values of each entry in the matrix uniquely define the nature of each pattern. The learning index table prescribes how learning should take place. While the details of these three tables will be presented shortly, in what follows we present a brief overview of the entire procedure.

Given the three tables, we perform the following procedure to realize our autonomous learning:

(a) Measure periodically the CPU usage and number of connections for all backend servers,
(b) Normalize the raw data to find two indices to the pattern matrix,
(c) Identify a pattern in the pattern matrix M using the two indices,
(d) Detect pattern changes and capture them,
(e) Set the thresholds for new patterns and measure the throughputs based on the learning index,
(f) Pick the threshold with the highest throughput and adjust autonomously.

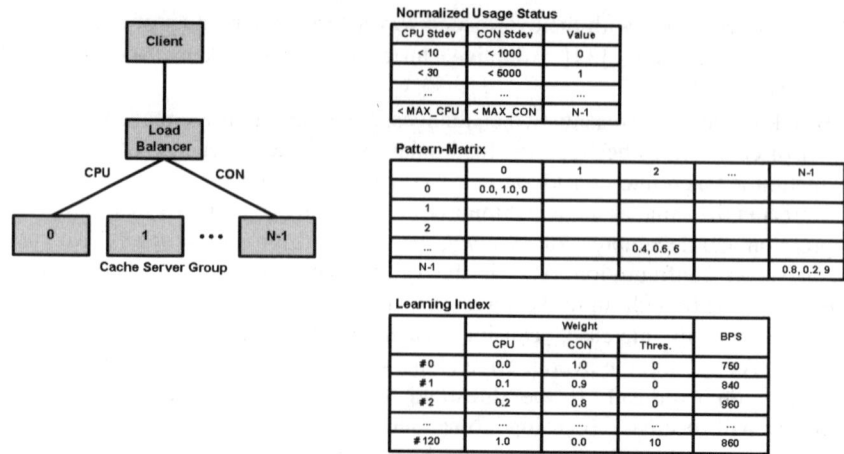

Fig. 1. The overall organization

3.2 Normalizing Load and Traffic Patterns

Load balancer periodically collects raw CPU usage and number of connections from the backend servers. CPU usage is computed using *vmstat* idle time while a connection refers to a client request established between the front-end LVS and a backend server. These raw values are normalized using standard deviation to find their relative importance in detecting unusual traffic patterns. Standard deviation indicates how far a particular CPU usage or a number of connections is away from the mean. Using the standard deviations for CPU usage and connections, we can quickly identify how far the current pattern is from the mean, and thus make a decision for the pattern.

Standard deviation for CPU usage is computed from the CPU utilization samples of the N_S backend servers, where each server can have 0 to 100% utilization. The maximum standard deviation of 71 is computed based on the samples taken from *two* servers as follows: $\sqrt{\dfrac{n\sum x^2 - (\sum x_{avg})^2}{n(n-1)}}$, where n is the number of samples, x is a sample value, and x_{avg} is the average of n sample values. While standard deviation should be computed based on n samples drawn from n machines, this definition is not suitable for our study. The main reason is the number of servers, or server set size n, changes at runtime for different traffic patterns due to its elastic and adaptive nature.

Similarly, connections will be normalized based on the connection standard deviation. Each backend server can establish up to 56,462 connections to the front-end LVS server. Or, the front-end LVS server can establish up to 56,462 connections to the backend servers. The maximum standard deviation of 39,925 is computed based on the samples taken from two servers from an individual machine's perspective: $\sqrt{\dfrac{2(0^2 + 56462^2) - (28231 + 28231)^2}{2(2-1)}} = 39,925$.

The tables below show an example for CPU usage normalization and connections through standard deviation. The normalized CPU index is 2 when the CPU usage standard deviation is 24.

Standard deviation for CPU Usage	<= 10	<= 20	<= 30	...	<= 71 (max)
Normalized CPU index	0	1	2	...	7

On the other hand, the normalized connection index is 1 when the number of connections is 7000.

Standard deviation for connection	<= 5000	<= 10000	<= 15000	...	<= 39925 (max)
Normalized connection index	0	1	2	...	7

The two indices will result in an 8x8 pattern matrix of 64 entries, each of which represents a pattern of load and traffic. This normalization process determines the granularity of traffic patterns and in turn sensitivity of learning. If, for example, the standard deviation for CPU usage is divided into every 5 and connections into every 2500, there will be 16 indices for CPU and 16 indices for connection, resulting in a 16x16 pattern matrix of 256 patterns. While they will help capture the subtle nature of various patterns, more and finer patterns will also entail higher overhead and be susceptible to noises. We'll revisit this issue with experimental results.

3.3 Identifying Load and Traffic Patterns

A pair of normalized CPU and connection indices uniquely defines a pattern of load and traffic. Using these indices, we construct a pattern matrix M to define all possible traffic patterns. Table 1 shows an example 8x8 matrix with 64 patterns. Each entry has three values: CPU weight W_{cpu}, connection weight W_{con}, and threshold H, where $W_{cpu} = W_{con} = 0.0, ..., 1.0$, $W_{cpu} + W_{con} = 1$, and $H = 0, ..., 10$. Each weight indicates its relative importance with respect to the other weight when making distribution decisions while threshold controls adaptive activities.

Table 1. Pattern matrix M. Each entry consists of CPU weight W_{cpu}, connection weight W_{con} and tolerance threshold H, where $W_{cpu} = W_{con} = 0.0 ... 1.0$, $W_{cpu} + W_{con} = 1$, and $H = 0 ... 10$.

Normalized CPU index	Normalized connection index					
	0	1	2	3	...	7
0				...	...	
1		...		...	...	
2		...			...	
3			0.4, 0.6, 7			
4				0.3, 0.7, 5		
...						
7				...	...	

Consider entry $(3,2) = (0.4, 0.6, 7)$, where the CPU weight is 0.4, the connection weight 0.6, and the threshold 7. The entry indicates that when traffic pattern $(3,2)$ is encountered, the number of connections will be more important than the CPU usage by 20%. The threshold value of 7 indicates that a new server should be recruited if the load of a server set is beyond 7, as explained shortly. The fact that an entry exists indicates that the particular pattern has been encountered in the past. If an entry does not exist, no such pattern has emerged yet.

3.4 Expanding/Contracting Server Set Boundary

Adaptation takes place when a set of servers is overloaded. Suppose a client request is received and the load balancer finds that this request has been served in the past since there is a set of servers assigned for this request. Since this traffic pattern has been known, the pattern matrix will have a corresponding entry with W_{cpu}, W_{con}, and H. The load balancer determines which server among the server set to send the request to.

Let i be the normalized CPU usage index and j be the normalized connections index. These two indices uniquely define a traffic pattern in the pattern matrix. Let $M_{i,j} = (W_{cpu}, W_{con}, H)$. First, the load balancer computes the tolerance factor F for each server s in the target server set $S = \{s_0...s_{k-1}\}$ as follows:

$$F_i = s_{i,cpu} \times W_{cpu} + s_{i,con} \times W_{con}, i = 0,...,k-1$$

where $s_{i,cpu}$ is the target server CPU usage and $s_{i,con}$ is the number of connections between the target server s_i and the LVS load balancer, and W_{cpu} and W_{con}, are the weights specified in $M_{i,j}$.

Second, the load balancer selects the target server with the minimum tolerance factor F_{min} among the server set S.

Third, the load balancer compares F_{min} with H. If $F_{min} > H$, it is deemed that the target server set S is overloaded. An expansion of the server set S thus commences.

Expanding a server set entails rechecking, target selection and expansion. Rechecking ensures if the target server set has indeed been overloaded for a reasonable period of time. In our case, we use T_S seconds. If the overload has indeed been persistent for over T_S seconds, the load balancer computes the load tolerance factor of all backend servers in the cluster and recruits the one that has the lowest tolerance factor. It should be noted that the selected server might also be a part of other server set(s) that are underutilized at the time of recruitment. The client request will be subsequently sent to the newly recruited backend server. Contracting a server set is similar to expansion. The load of the server that was just recruited is monitored for T_S seconds. If the load stays below the threshold, the server will be released from the server set to which this server belongs. The server selected for release might belong to one to many other server sets, each of which services various client requests.

3.5 Learning New Patterns

Learning a new pattern takes place when the corresponding entry does not exist in the pattern matrix M. We have developed two types of learning: *incremental* and *leap*. Incremental learning is designed to take advantage of the known traffic patterns and their associated weights and thresholds while leap learning is for patterns that are completely new. The key to this separation, not surprisingly, is the history of the patterns encountered thus far.

Suppose a request is received. A server will be selected and a pattern matrix entry will be identified using the normalized CPU usage and connections. Suppose further that the entry does not exist, which warrants learning, but any of the four immediate neighbors (up, down, left, and right) exists in the pattern matrix. These neighbors would provide a possible clue as to how the new traffic should be handled.

Consider again the pattern matrix shown in Table 1. Suppose that the current traffic points to the entry $M_{4,2}$, which does not exist. However, there are two immediate neighbors, $M_{3,2} = (W_{cpu}, W_{con}, H) = (0.4, 0.6, 7)$ and $M_{4,4} = (W_{cpu}, W_{con}, H) = (0.3, 0.7, 5)$, both of which have already been defined. The tiebreaker for the two immediate neighbors is lower CPU usage, followed by lower connections because choosing the entry with higher CPU usage can endanger the existing processing requirements. For our case, we choose $M_{4,4} = (0.3,0.7,5)$ as the starting point.

Given $M_{4,4} = (0.3, 0.7, 5)$, we measure the throughputs for the following 11 possible combinations of CPU and connection weights for the predetermined learning period:

(0.8,0.2,5), (0.7,0.3,5), (0.6,0.4,5), (0.5,0.5,5), (0.4,0.6,5),
(0.3,0.7,5),
(0.2,0.8,5), (0.1,0.9,5), (0.0,1.0,5), (1.0,0.0,6), (0.9,0.1,6).

Among the 11 throughputs measured, we pick the combination that gives the highest throughput. If, for example, the combination (0.2, 0.8, 5) gave the highest throughput, this would be assigned as the weights and threshold for the entry $M_{4,2}$. Incremental learning is now completed.

Leap learning. If an entry and its four immediate neighbors are not defined, it is clear that this is a completely new pattern. Leap learning will start from scratch. There are 121 possible pairs to examine since $W_{cpu} = W_{con} = 0.0, 0.1, ..., 0.9, 1.0$ and Leap learning measures the throughputs for all 121 pairs. Among these 121 throughputs is the highest that will be selected for the current pattern. If there are multiple entries that have the same throughput, the tiebreaker will be the one with the smallest number of files since it will incur less data transfer time. The corresponding entry will be subsequently updated with the new values of W_{cpu}, W_{con}, and H.

Re-learning. Entries are not permanent even if they are defined. While the weights for a particular traffic pattern should remain the same for consistent performance, there are other possibilities that require weight changes. For example, machine specifications might have been changed due to an additional memory card or CPU upgrade. In this case, it is essential that the weights be re-learned, or re-adjusted for better and correct performance. This type of relearning is not automatic and can be performed at the time of machine reconfiguration.

4   Experimental Results

4.1   Experiment Environment

To test the proposed approach we have set up an experimental environment. There are four types of servers: clients, LVS-based front-end server, a cluster of backend servers, and a web server. The machine specifications are listed in Table 2.

The client machines generate requests based on Webstone [19] and Surge [3]. Various researchers have used these traces to perform their experiment, including [3] for Surge, [16] for UNC, and [6] for Berkeley. While these traces may not be the best ones, we attempted to use publicly available traces available on the Web. The requests based on these traces consist of text files and images, which determine traffic patterns. The LVS load balancer decides where and how to send requests based on the scheduling and routing policy. Three policies are used in this study, which are LBLCR, Round-Robin, and autonomous learning.

Table 2. Specifications of the machines used in the study

		Hardware		Software	#
		CPU (MHz)	RAM (MB)		
Client/Master		P-4 2260	256	Webstone [19] Surge [3]	10 / 1
LVS		P-4 2400	512	Direct Routing	1
Server	Cache	P-2 400	256	Squid [13]	16
	Distiller			JPEG-6b	
Web Server		P-4 2260	256	Apache	1

If the requested pages are cached in one of the 16 servers, the selected server will send them directly to the client without going through the LVS machine, hence *direct* routing. The LVS will only handle one-way incoming traffic. If the requested pages are not cached, the Web server is called to provide the contents. When the fresh contents are received from the Web server, they are cached (stored) in the selected target server, which will subsequently send directly to the client. Squid is used to cache contents received from the Web server.

When caching fresh pages received from the Web server, Round-robin copies them to *every* server in the cluster while LBLCR and autonomous learning do not. Our method and LBLCR keep the data in only one server among the cluster. There is no copy involved, hence eliminating the cache coherence problem and enabling storage scalability. Round-Robin is used mainly for comparison purposes since it will provide the optimal performance at the expense of storage scalability. As all servers have exactly the same files in RR, no particular distribution strategy is required and there will be no hotspot for popular pages.

Table 3 lists some important variables used in the experiments. Among the variables is the number of requests served per second that is used as the main performance metric for the three methods. Given the machine configuration and various parameters we present our experimental results in terms of overall learning behavior, learning new traffic patterns, sensitivity of learning, and learning frequency. Performance is measured in terms of number of requests serviced per second.

Table 3. Key variables in the experiments

Variables	Values
Number of client machines (N_C)	10
Number of servers (N_S)	1, 2, 4, 8, 16
Distribution policy	RR, LBLCR, Self-Learning
LBLCR-Threshold	63, 126, 252
Learning-Time	1, 3, 5, 10 sec
Status table (stdev-con x stdev-CPU)	10 x 5, 10 x 10, 30 x 30, 50 x 50, 100 x 100
Learning Weights (con-CPU-weight x threshold)	11 x 11 = 121, 6 x 6 = 36
Traffic patterns drawn from Web traces (Types of client requests)	Image + HTML (19 files, weights) Surge : Image + HTML (50 or 100 files, weights) Real web traces : UNC, Berkeley (100 files, weights)
Weight selection criterion	bit per second (BPS)
Number of simultaneous connections per client to a server through LVS	63
Performance indicator	Request serviced per second
Active-connection weight (A_{weight})	50 (the overhead of processing active connections is fifty times higher than that of inactive connections on average [11])
Rechecking time for expanding/contracting server set boundary (T_S)	3 seconds

4.2 Learning Behavior

Figure 2 shows the overall behavior of our approach for different traces. The x-axis shows time in minutes while the y-axis shows number of requests served per second. Each experiment was run for at least an hour to identify consistent performance across different traces.

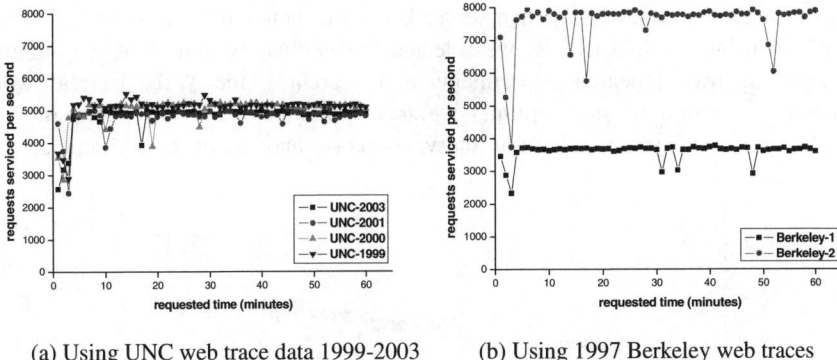

(a) Using UNC web trace data 1999-2003   (b) Using 1997 Berkeley web traces

Fig. 2. Overall learning behaviors

As we observe from the figure, the total number of requests served is inconsistent in the beginning. Each run typically starts out with the performance close to the best but dips in the ensuing minutes. Some runs show significant performance degradation in the first two to three minutes. It is this time period that traffic pattern(s) are learned and the weights and thresholds adjusted. After approximately five minutes, the cluster becomes stable, showing consistent performance throughout the remaining period. The plots based on Berkeley traces in Figure 2(b) show several dips while consistently maintaining similar overall performance. The reason for these occasional dips is to learn new patterns. When the same or similar patterns are presented, the cluster demonstrates essentially the same performance. However, learning new traffic patterns requires some overhead as indicated in the dips.

4.3 Learning New Patterns

The experimental results below are designed to demonstrate how the system handles new traffic patterns. Initially, the pattern matrix is empty. Table 4 lists the traces that clients send. First, traffic patterns 1-5 are used to initiate preliminary learning of traffic patterns and some of the results are shown in Figure 3. This initial learning fills five entries.

Table 4. Traffic patterns using the Web traces

	Initial traffic patterns					New traffic patterns					
	1	2	3	4	5	6	7	8	9	10	11
Type	Image	Html	Image +Html	Surge-50	Surge-100	UNC-2003	UNC-2001	UNC-2000	UNC-1999	Berk.-1	Berk.-2
# req	1000	1000	1000	1000	1000	1000	500	1000	500	1000	500

The second step of the experiment is to send six new traffic patterns 6-11 of the above table. As seen from Figure 3, the very first traffic pattern causes some initial fluctuation in throughput. As soon as learning settles and the weights are selected, our approach shows performance comparable to Round Robin and LBCLR.

It should be noted that the performance of round-robin and LBCLR are the optimal performance. In Round-Robin each server keeps the same copy of all files. There is no file distribution. In LBCLR, we selected the *optimal* weights that we obtained through our own repeated experiments and searching the Web. In real world situations, selection of such optimal weights would take days to weeks or even months with extensive administrator intervention or it may not even be practical.

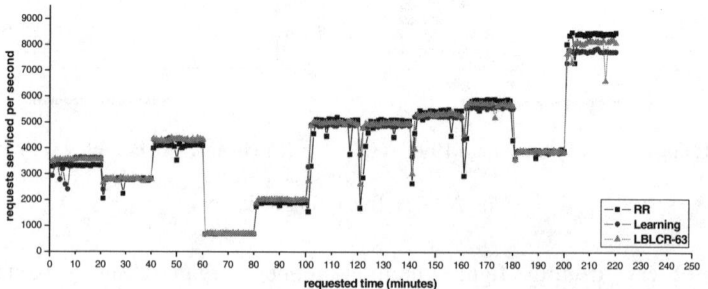

Fig. 3. Learning new traffic patterns

As the time progresses, similar traffic patterns (patterns 2-5) are efficiently processed with little overhead. This is apparent from the figure up to 100 minutes. Each dip of the curve indicates another similar traffic pattern. When new traffic patterns emerge starting at the 100-minute mark, all three methods show fluctuations.

The last new traffic pattern (Berkeley1997-2) distinguishes the three methods, where RR performs better than LBCLR by 4 % and learning by 8%. While the difference is not substantial, the main reason for this minor performance difference stems from the fact that the front-end load-balancer LVS becomes a bottleneck for a large number of requests. LVS has reached its capacity of processing approximately 8,000 client requests per second.

4.4 Learning Frequency

Figure 4 shows the impact of learning frequency on performance. The figure demonstrates that performance is proportional to frequency. Figure 4(b) supports this proportional behavior. At the frequency of learning in every second, the performance converges to the maximum in approximately 5 minutes. On the other hand, it takes approximately 12 minutes to reach the maximum performance when learning takes place every 3 seconds. It is further evident that the convergence takes approximately 42 minutes when the learning frequency is 10 seconds. However, it is not always advantageous to learn as often as a second. The cost for this is overhead. First of all, statistics have to be measured and communicated to the load balancer every second. Second, the computing overhead associated with learning itself will become visible for the load balancer as frequency increases. This increase in overhead can be seen in the figure.

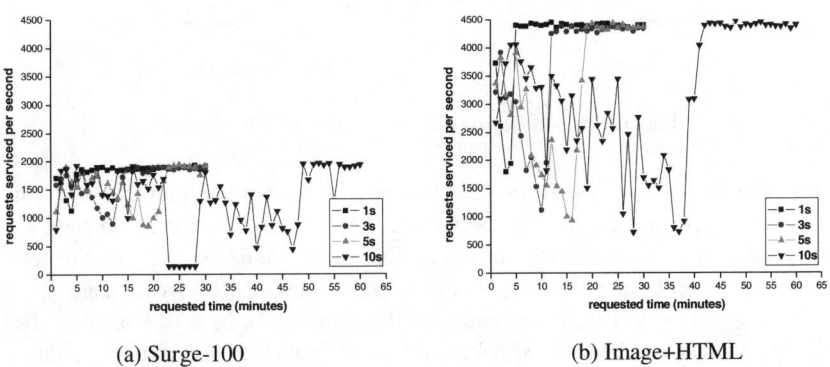

(a) Surge-100   (b) Image+HTML

Fig. 4. Impact of the frequency of learning on performance

Consider again Figure 4(b). We find that as frequency decreases, unit convergence rate actually decreases, indicating that learning became effective faster. However, when learning takes places in every 10 seconds, the unit convergence rate increases again demonstrating that learning is now less effective. In summary, learning often does not necessarily give better performance because it incurs additional overhead of collecting data, analyzing, and making routing decisions. On the other hand, learning in every 10 seconds also is not a good practice either since it tends to ignore new traffic patterns that require immediate attention. It is therefore essential that learning frequency be changed at runtime to adapt to different traffic patterns.

5 Performance Comparisons

5.1 Overall Comparison

Comparisons are made against Round-Robin, no learning, and LBLCR with minor variations. Round Robin is the reference point for our comparisons since it provides the optimal performance. Figure 5 shows the relative performance of three methods. Several variations of each method are introduced to highlight the individual methods. RR is fixed, as there can be no variations. LBLCR has the three variations of 63, 126, and 252. LBLCR-63 uses the threshold of 63, which we find is the optimal policy based on our experiments and searching the Web.

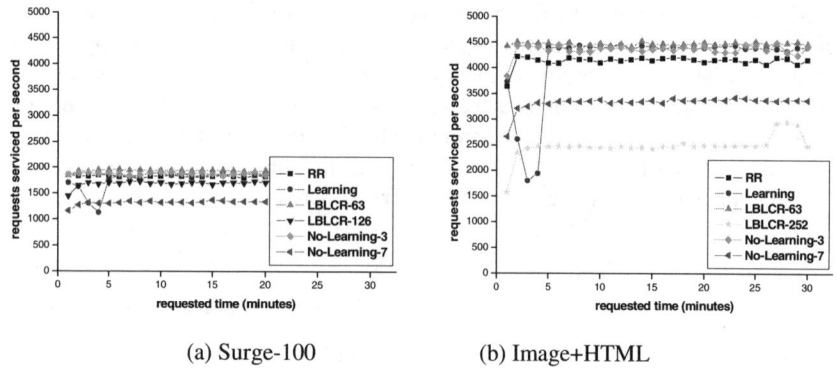

(a) Surge-100   (b) Image+HTML

Fig. 5. Comparisons of the three methods with variations

Our method has two variations: No-learning-3 and No-learning-7. No-learning-3 refers to no adaptation but the threshold set to 3 out of 10. Low threshold causes more adaptive activities while high threshold does less often. Various parameters for learning are listed below: learning frequency 1 second, sensitivity of learning 10x5, learning range 11x11=121, three pattern matrix entries (0, 0) = (0.8,0.2,0.3), (0,1) = (0.6,0.4,0.3), (1,0) = (0.9,0.1,0.3), where the three values are connection weight, CPU weight, and threshold, respectively.

Learning clearly outperforms no learning in both plots. No-learning-3 actually shows performance comparable to learning. The main reason is that the threshold is set low so that server sets can be formed more easily and often, resulting in good performance. However, this frequent formation of server sets comes at the expense of frequent file copying within a server set. No-learning-7, on the other hand, shows poor performance, a clear indication that the threshold is set too high to form server sets.

LBLCR also shows different performance depending on connection weight. While Figure 5(a) shows approximately a 10% performance difference among the LBLCR variations, Figure 5(b) shows that the performance difference between LBLCR-63 and LBLCR-252 is large, approximately 50%. This large difference in performance indicates that weight selection is critical in overall performance since once set the weights will remain the same until manually changed. However, finding optimal weights is painstaking and time-consuming. This intervention entails time consuming steps, including monitoring system performance, determining poor performance, alerting the administrator, finding good weights that would give better performance than the current, having the administrator update the weight, etc.

The fact that learning takes several minutes in the beginning can be critical for high performance applications. One solution to this boundary condition is to spread the learning over the entire period instead of front-loading. For example, the cluster starts with LBLCR settings and every several minutes learning would take place. Since learning is now spread over time, the clients may not notice the difference. This incremental learning, when implemented properly, will avoid such significant performance drop that is seen in the first several minutes.

5.2 Scalability

It is critical to study if adding more machines will improve performance. Scalability is measured in terms of speedup, which is defined as the execution time on P servers compared to that on 1 server. Figure 6 shows the speedup for two traces, Surge-100 and Image+Html.

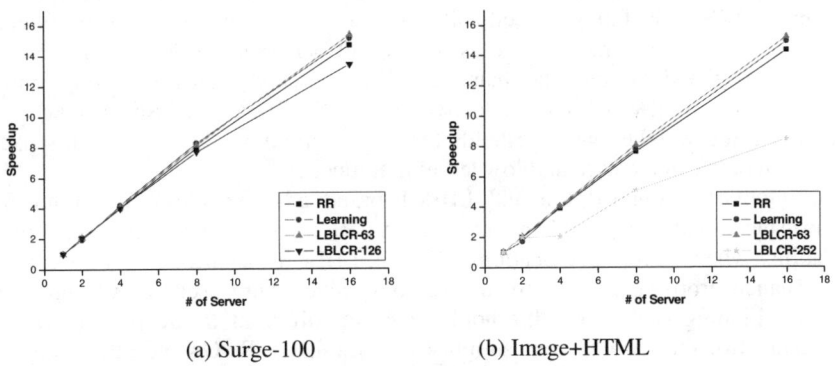

(a) Surge-100   (b) Image+HTML

Fig. 6. Scalability

The figure demonstrates that the scalability of our learning method is essentially linear indicating that adding more machines will proportionally provide more performance. As expected, RR shows performance comparable to our method because every server using RR has the same copy of data.

On the other hand, we note in Figure 6(b) that LBLCR-252 shows a significance drop in scalability while LBLCR-63 maintains the highest scalability. The key threshold value chosen for LBLCR-252 at the time of machine configuration was apparently not good. Since this poor performance will become apparent after some trial and errors, the system administrator will inevitably have to find a better value such as LBLCR-63 either through various experiments or searching the Web. This is the fundamental motivation for our study. Our approach removes the system administrator or human intervention from the loop by autonomously learning the key parameters and putting them to work with no manual and time-consuming search.

6 Conclusions

We have presented in this report an adaptive clustering framework that autonomously learns and adapts to different traffic patterns to improve cluster utilization with no human intervention. Cluster is configured once and for all at the time of initial setup. Autonomous learning handles the rest. Each traffic pattern is represented in a pair of normalized CPU usage and number of connections. The normalized values of 0 to n-1 are used to construct an $n \times n$ pattern matrix. Each entry of the pattern matrix holds three values of CPU weight, connection weight, and threshold. The LVS-based load balancer uses these three values to determine if the target machine is overloaded. When the target machine is overloaded, a new server is brought in to a server set from the cluster. When a new pattern emerges which does not exist, it is this set of three values that will be autonomously learned by using the existing patterns in the pattern matrix.

To test our approach, we have set up an environment consisting of a cluster of 16 servers, an LVS load-balancer, 10 client machines, and a Web server. Client machines generated requests based on the publicly available Web traces: Berkeley [6], UNC [16], Surge [3], Webstone [19], etc. Experimental results have demonstrated that the performance of our learning method is comparable to or even better than that of the manual LBLCR scheduling. Specifically, we have found that learning too often, or every second, is counterproductive as the overall performance decreased due to the increased overhead. On the other hand, learning too many patterns including spurious ones turned out also counterproductive as the load balancer spent much time readjusting the weights and thresholds for minor variations. Our approach strikes a balance in between the high and low learning frequency.

In summary, Round-Robin and LBLCR-based adaptive clustering suffer from performance scalability and administration. RR has shown little performance scalability when server specifications changed. LBLCR has also shown a significant performance drop when suboptimal threshold values were used for varying traffic patterns. Finding such good threshold values for different traffic patterns requires time and efforts by the system administrator for LBLCR. On the other hand, our learning-based adaptive clustering has demonstrated performance scalability regardless of the changes in traffic patterns and machine configuration while it is all done autonomously.

References

1. G. M. Amdahl, G. A. Blaauw, F. P. Brooks, Architecture of the IBM System 360, IBM Journal of Research and Development, Vol. 8, No. 2, April 1964, pp. 87 - 101.
2. Andrea C. Arpaci-Dusseau and Remzi H. Arpaci-Dusseau, Information and Control in Gray-Box Systems, 8th ACM Symposium on Operating Systems Principles, 2001, pp. 43-56.

3. P. Barford and M. Crovella, Generating Representative Web Workloads for Network and Server Performance Evaluation, In Proc. ACM SIGMETRICS Conf., Madison, WI, Jul. 1998.

4. M. Bozyigit, M. Wasiq, User-Level Process Checkpoint and Restore for Migration, Vol. 35, No. 2 (2001) 86 - 96

5. Y. Chang and J. Chen, Designing an enhanced PC cluster system for scalable network services, 19th International Conference on Advanced Information Networking and Applications (2005) 163-166

6. M. Christiansen, K. Jeffay, D. Otta and F. Smith, Tuning RED for Web traffic, IEEE/ACM Transactions on Networking, Vol. 9, No. 3 (2001) 249-264

7. C. Ernemann, V. Hamscher, and R. Vahyapour, Benefits of global grid computing for job scheduling, 5th IEEE/ACM International Workshop on Grid Computing (2004) 374-379

8. A. G. Ganek and T. A. Corbi, The dawning of the autonomic computing era, IBM System Journal, Vol. 42, No. 1 (2003) 5 - 18

9. Y. Hong, J. No, and I. Han, Evaluation of fault-tolerant distributed Web systems, 10th IEEE International Conference on Advanced Information Networking and Applications (2005) 163-166

10. HP Grid and Utility Computing, http://devresource.hp.com/drc/topics/utility_comp.jsp

11. Job Scheduling Algorithms in Linux Virtual Server, http://www.linuxvirtual-server.org/docs/scheduling.html.

12. M. Karlsson and M. Covell, Dynamic Black-Box Performance Model Estimation for Self-Tuning Regulators, 2nd International Conference on Autonomic Computing (ICAC'05) (2005) 172-182

13. W. Liao and P. Shih, Architecture of proxy partial caching using HTTP for supporting interactive video and cache consistency, 11th International Conference Computer Communications and Networks (2002) 216-221

14. Linux Virtual Server Project, http://www.linuxvirtualserver.org/.

15. M. Litzkow, T. Tannenbaum, J. Basney, M. Livny, Checkpoint and Migration of UNIX Processes in the Condor Distributed Processing System, Technical Report #1346, University of Wisconsin-Madison Computer Science Department (1997)

16. D. Lu, Y. Qiao, P. Dinda and F. Bustamante, Modeling and Taming Parallel TCP on the Wide Area Network, Proceedings of 19th IEEE International Parallel and Distributed Processing Symposium (2005)

17. S. McClure, R. Wheeler, MOSIX: How Linux Clusters Solve Real World Problems, in Proc. 2000 USENIX Annual Tech. Conf., San Diego, CA. (2000) 49 - 56

18. N1 Grid: Managing n computers as 1, Sun Microsystems, http://wwws.sun.com/software/solutions/ n1/

19. V. Olaru and W. Tichy, Request distribution-aware caching in cluster-based Web servers, 3rd IEEE International Symposium on Network Computing and Applications (2004) 311-316

20. C. Yang and M. Luo, Building an adaptable, fault tolerant, and highly manageable web server on clusters of non-dedicated workstations, IEEE International Conference on Parallel Processing, (2000) 413-420

Parametric Architecture for Function Calculation Improvement

María Teresa Signes Pont, Juan Manuel García Chamizo,
Higinio Mora Mora, and Gregorio de Miguel Casado

Specialized Processor Architectures Lab (I2RC-DTIC)
University of Alicante, 03690, San Vicente del Raspeig, Alicante, Spain
{teresa,juanma,hmora,demiguel}@dtic.ua.es
http://www.dtic.ua.es/spa-lab/index.html

Abstract. This paper presents a new approach to the problem caused by the exploding needs of computing resources in function calculation. The proposal argues for increasing the computing power at the primitive processing level in order to reduce the number of computing levels required to carry out the calculations. This trade-off is developed within the limits of function evaluation by substituting the usual primitives, namely sum and multiplication, by a unique weighted primitive that can be tuned for different values of the weighting parameters. All function points are carried out by successive iterations of the primitive. A parametric architecture implements the design. The case of combined trigonometric functions involved in the calculation of the Hough transform (HT) is analyzed under this scope. It provides memory and hardware resource saving as well as speed improvements, according to the experiments carried out with the HT.

1 Introduction

Information processing in conventional CPUs is carried out by successive stages with an increasing computing power as they go away from the initial primitive level implemented by the hardware circuitry. The increasing requirements of computation-intensive applications lead to a growth of the number of computing levels or to the replication of equivalent computing stages. In both cases, a scalability problem may appear. CPUs implement primitive operations, such as sum and multiplication, to perform a sequential execution in the electronic circuits. This physical basis supports a hierarchical organization of computing levels characterized by different languages and operations. Therefore, these operations can be expressed in these languages with more or less success, according to their suitability. A high level language consists of a set of machine language instructions that involve a great amount of sums and shifts for the digital electronic circuitry. There are several well-known methods which have succeeded in improving calculations: pipelining [1] [2] and anticipation [3] [4] can reduce time delay whereas resource sharing can lower hardware requirements [5].

In this paper, an improvement of the computing capabilities of the primitive processing level is pursued in order to contain the growth of the number of

P. Lukowicz, L. Thiele, and G. Tröster (Eds.): ARCS 2007, LNCS 4415, pp. 240–253, 2007.
© Springer-Verlag Berlin Heidelberg 2007

derivation levels of the calculations. As a consequence, it can be expected that modeling, formalization and calculation tasks become easier. The new primitive is a weighted sum of two operands which performs function evaluation by means of successive iteration. The generic definition of the new primitive can be achieved by a two dimensional table in which the cells store combinations of the weighting parameters. This evaluation method suits for a great amount of functions, particularly when the evaluation requires a lot of computing resources. It also allows implementation schemes which offer a good balance between speed, area saving and accuracy.

Proposed in 1962, the Hough Transform (HT) has become a widely used technique in image segmentation: plane curve detection [6], object recognition [7], air picture vectorization [8], 3D image reconstruction [9], industrial quality inspection [10], biomedical applications [11] [12], quasar reckoning [13], OCR [14], etc. The HT is very suitable because of its robustness, although the great amount of temporary and spatial resources that requires has moved it away from real time applications. This way, the investigation efforts in HT have dealt with the design of fast algorithms and parallel or ad-hoc architectures. As the HT consists of function evaluation using arithmetic operations different algorithmic approaches have been developed: piece-lineal [15], combinatory [16], binary [17], adaptive [18] and fast [19]. There are also implementations of the CORDIC algorithm for applications that demand high speed and precision, such as digital signal and image processing and algebra [20] [21]. However, their drawback is the lower degree of regularity and parallelism capabilities when comparing with the traditional algorithm. The parallelism that underlies in the traditional algorithm allows the implementation of architectures with shared or distributed memory (lineal array, mesh, hypercube and binary tree) as well as systolic ones [22].

The paper is structured in six parts. Following the introduction, Section 2 defines the weighted primitive. Section 3 presents the fundamental concepts of the evaluation method based on the use of the weighted primitive and outlines its computing relevance. A motivational example illustrates the method. In Section 4, an implementation based on Look-up tables is discussed and an estimation of the time delay calculation and space occupation is provided. Section 5 is entirely devoted to the application of the method to the calculation of the HT. A comparison with two different hardware approaches together with another one with a software approach is provided. Finally, Section 6 summarizes results and presents the concluding remarks.

2 Definition of a Weighted Primitive

This section presents the definition of the weighted primitive and outlines its computational relevance.

The weighted primitive is an operation $\otimes$ defined in the following way:

$$
\begin{aligned}
&\otimes : \mathbb{R} \times \mathbb{R} \longrightarrow \mathbb{R} \\
&(a, b) \longrightarrow a \otimes b = \alpha a + \beta b \\
&(\alpha, \beta) \in \mathbb{R}^2
\end{aligned}
\tag{1}
$$

The operation $\otimes$ can also be defined by means of a two-input table. Figure 1 defines the operation for the particular case of integer values, in binary sign-magnitude representation (the sign is the more left bit, 0 for positive values, 1 for negative values) and for $k = 1$ (k stands for the number of significant bits in the representation, in bold characters). The arguments have been represented in binary and decimal notation and the results are referred in a generic way, as combinations of the parameters α and β.

a $\otimes$ b	01 = 1	10 = 00 = 0	11 = -1
01 = 1	$\alpha + \beta$	α	$\alpha - \beta$
10 = 00 = 0	β	0	$-\beta$
11 = -1	$-\alpha + \beta$	$-\alpha$	$-\alpha - \beta$

Fig. 1. Definition of the operation $\otimes$ for $k = 1$

The same operation can be represented for greater values of k (see Figure 2 for $k = 2$). Central cells are equivalent to those of Figure 1. The number of cells in a table is $(2(k+1) - 1)^2$ and it only depends on k. The cells are organized as concentric rings centered in 0. It can be noticed that increasing the value of k causes a table growth arranged in peripheral rings. The number of rings increases at a rate of 2^k when k increases one unit. The smallest table is defined for $k = 1$ but the same information about the operation $\otimes$ is provided for any value of k.

a $\otimes$ b	011= 3	010= 2	001= 1	100=000= 0	101= -1	110= -2	111= -3
011= 3	$3\alpha+3\beta$	$3\alpha+2\beta$	$3\alpha+\beta$	3α	$3\alpha-\beta$	$3\alpha-2\beta$	$3\alpha-3\beta$
010= 2	$2\alpha+3\beta$	$2\alpha+2\beta$	$2\alpha+\beta$	2α	$2\alpha-\beta$	$2\alpha-2\beta$	$2\alpha-3\beta$
001= 1	$\alpha+3\beta$	$\alpha+2\beta$	$\alpha + \beta$	α	$\alpha - \beta$	$\alpha-2\beta$	$\alpha-3\beta$
100=000= 0	3β	2β	β	0	$-\beta$	-2β	-3β
101= -1	$-\alpha+3\beta$	$-\alpha+2\beta$	$-\alpha + \beta$	$-\alpha$	$-\alpha - \beta$	$-\alpha-2\beta$	$-\alpha-3\beta$
110= -2	$-2\alpha+3\beta$	$-2\alpha+2\beta$	$-2\alpha+\beta$	-2α	$-2\alpha-\beta$	$-2\alpha-2\beta$	$-2\alpha-3\beta$
111= -3	$-3\alpha+3\beta$	$-3\alpha+2\beta$	$-3\alpha+\beta$	-3α	$-3\alpha-\beta$	$-3\alpha-2\beta$	$-3\alpha-3\beta$

Fig. 2. Definition of the operation $\otimes$ for $k = 2$

The operation $\otimes$ is performed when the arguments (a, b) address the table and the result is picked up from the corresponding cell. The first argument (a) addresses the row whereas the second (b) addresses the column. When the precision of the arguments n is greater than k, these must be fragmented in $k-$sized fragments in order to perform the operation. Therefore, t double accesses are necessary to complete the t cycles of a single operation (remember $n = k \cdot t$). A single operation requires picking up from the table so many partial results as fragments are contained in the argument. The overall result is obtained by adding t partial results, according to their position.

3   Function Evaluation Method Based on a Weighted Primitive

Usually, under the scope of the extended machine model with primitives such as sum and multiplication, function evaluation is provided by higher and higher levels in the machine according to the increasing computational complexity of the function which is going to be calculated. Nevertheless, we can easily imagine the same calculation performed at a lower computing level by other primitives, whenever the new primitives intrinsically assume part of the complexity. This approach is considered in this paper as far as it may be the way to perform the calculation of functions with both algorithmic and architectural benefits, as well as time delay and area costs saving.

Our inquiry for a primitive operation that bears more computing power than the usual primitives sum and multiplication points towards the operation $\otimes$. As shown in Section 2, this new primitive is more generic (sum and multiplication are particular cases of it) and, as it will be shown, the recursive development of the operation $\otimes$ offers quite different features which turn to be much more meaningful than the formal combination of both operations. This issue has crucial consequences because function evaluation is performed with no more difficulty than iteratively applying a simple operation defined by a two-input table.

In order to implement the evaluation of a given function Ψ we propose to approximate it by a function F defined as follows:

$$
\begin{aligned}
& F_0, \\
& F_{i+1} = F_i \otimes G_i, \\
& i \in \mathbb{N}; \forall i, (F_i, G_i) \in \mathbb{R}^2
\end{aligned}
\tag{2}
$$

The first value of the function is given (F_0) and the next values are calculated by iterative application of the recursive equation. The approximation capabilities of function F can be understood as the equivalence of the respective sets of real values $\{\Psi(i)\}$ and $\{F_i\}$. The definition of the mapping between the two sets causes function Ψ to be quantized with a quantization step h. The independent variable is denoted $x + ih$, where x is the initial real value and $i \in \mathbb{N}$ can take successive increasing values. The mapping implies three initial conditions to be fulfilled. They are:

1. x (initial Ψ value) is mapped to 0 (index of F), that is to say $\Psi(x) \equiv F_0$.
2. Any quantization step h will be referred as the unity in the weighted primitive formulation ($h \equiv 1$).
3. The previous assumptions allow not having to discern between i (index belonging to the independent variable) and i (iteration number), that is to say: $\Psi(x + ih) = \Psi(0 + i1) = \Psi(i) \equiv Fi$.

So, the mapping of the function Ψ by the recursive function F succeeds in the approximation by means of the normalization process defined in 1), 2) and 3). It can be noticed that the function F is not unique because different mappings can be done.

As an example of the approximation capabilities of the method, the calculation scheme for an exponential function is illustrated in Figure 3.

$$\Psi(t) = 1 - e^{-0.9t}$$

Approximation: $F_0 = 1; \alpha = 0.42; \beta = 0.582; \forall i, G_i = 1.$

$$F_{i+1} = 0,42Fi + 0,582,$$

with $\Psi(0) \equiv F_0 = 0$ and $h = \Delta t \equiv \Delta i = 1$.

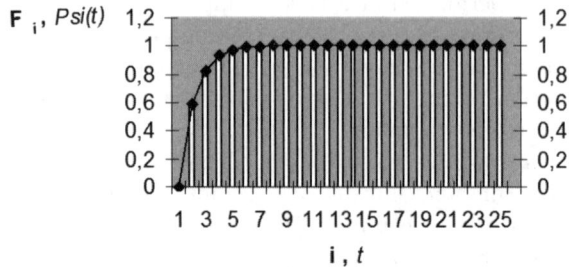

Fig. 3. Approximation of $\Psi(t)$ (bars) by recursive function F (dots). error of 0.33%.

Any calculation of $\{F_i\}$ is performed with computational complexity $O(N)$ whenever $\{G_i\}$ is known or carried out with the same (or less) complexity. It can be outlined that the interest of mapping the function F is limited to the fulfillment of this condition. This fact outlines at least two different computing issues. The first develops new function evaluation upon the previous, that is to say, when the function F has been calculated, it can play the role of G in order to generate a new F function. This spreading scheme provides a lot of increasing computing power with linear cost. The second scheme deals with the crossed calculation of the functions F and G, that is to say G is the auxiliary function involved in the calculation of F as well as F is the auxiliary function for the calculation of G. In addition to the linear cost, the crossed calculation scheme provides time delay saving as both functions are calculated simultaneously.

4   Architecture

As mentioned in Section 3, the two main computing issues lead to different architectural counterparts. The development of a new function evaluation upon the previous one in a spreading calculation scheme is carried out by the processor presented in Figure 4. The second scheme deals with the crossed paired calculation of the functions F and G, that is to say G is the auxiliary function involved in the calculation of F as well as F is the auxiliary function for the calculation of G (see Figure 5).

The proposed implementation uses a Look-up table (LUT) that contains all partial products $\alpha A_k + \beta B_k$. These A_k, B_k are few bits chunks of the current input data F_i and G_i. On every cycle the LUT is accessed by the A_k and B_k coming from the shift registers. Then the partial products are taken out of the

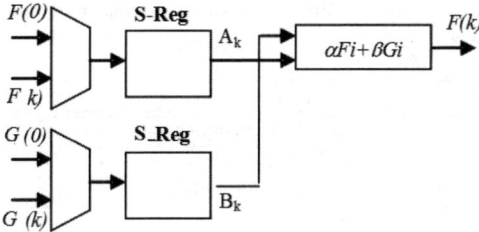

Fig. 4. Arithmetic processor for the spreading calculation scheme

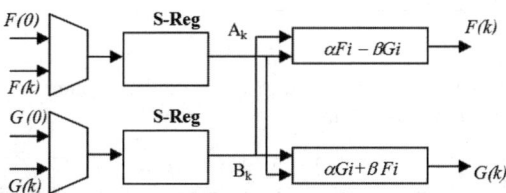

Fig. 5. Arithmetic processor for crossed paired evaluation

cells (partial products in the LUT are the hardware counterpart of the weighted primitives presented in Figures 1 and 2). The overall partial product $\alpha F_i + \beta G_i$ is obtained by adding all the shifted partial products corresponding to all fragment inputs A_k, B_k of F_i and G_i, respectively. At the following iteration the new calculated F_{i+1} value and the next G_{i+1} value are multiplexed and shift registered before they access the LUT, in order to repeat the addressing process. The processor shown in Figure 4 is different from the one of Figure 5 in what concerns to the function G. The values of G are obtained in the same way as for F but the LUT for G is different from the LUT for F.

In order to perform a comparison of computing resources, an estimation of the area cost and time delay of the proposed architectures is presented here. The model used for the estimations is taken from [23], [24] and [25]. The unit τ_a represents the area of a complex gate. The complex gate is defined as the pair (AND, XOR) which provides a meaningful unit for implementing the most basic computing device: the one bit full-adder. The unit τ_t is the delay of this complex gate. This model is very useful because it provides a direct way to compare different architectures, irrespectively from their implementation. As an example, the area cost and time delay for 16 bits and one-bit fragmented data are estimated for both processors (Table 1).

If the fragments of the input data are greater than one bit the occupied area and the time delay of access of the LUT vary. The relationship between area, time delay and fragment length k, for 16 bits data is shown in Table 2 for the processor of Figure 5.

Table 2 outlines the fact that the LUT area increases exponentially at a rate of k, and represents an increasing portion of the overall area as k increases.

Table 1. Arithmetic processor estimations of area cost and time delay for 16 bits one-bit fragmented data

Hardware devices	Occupied area	Time delay
Multiplexer	$0,25 \cdot 2 \cdot 16\tau_a = 8\tau_a$	$0,5\tau_t$
Shift Register	$0,5 \cdot 16\tau_a = 8\tau_a$	$15\tau_t \cdot 0,5\tau_t = 7,5\tau_t$
LUT	$40\frac{\tau_a}{Kbit} \cdot 16 \text{ bits} \cdot 16 \text{ cells} = 10\tau_a$	$3,5\tau_t \cdot 16 \text{ acc.} = 56\tau_t$
Register	$0,5 \cdot 16\tau_a = 8\tau_a$	$1\tau_t$
Reduction structure $4:2$ + adder	$4\tau_a + 16\tau_a = 20\tau_a$	$3 \text{ red} \cdot 3\tau_t + \log 16\tau_t = 13\tau_t$
Arith. processor Fig. 4	$70\tau_a$	$78\tau_t$
Arith. processor Fig. 5	$108\tau_a$	$78\tau_t$

Table 2. Relationship between area, time delay and fragment length k, for 16 bits data for the processor shown in Figure 5

	$k=1$	$k=2$	$k=4$	$k=8$	$k=16$
LUT area	$20\tau_a$	$80\tau_a$	$\approx 2K\tau_a$	$\approx 500K\tau_a$	$> 1G\tau_a$
LUT area vs overall area	$\frac{20\tau_a}{108\tau_a} = 0,18$	$\frac{80\tau_a}{168\tau_a} = 0,47$	$\frac{2048\tau_a}{2136\tau_a} = 0,96$	$> 0,99$	$> 0,99$
LUT acc. time	$56\tau_t$	$28\tau_t$	$14\tau_t$	$7\tau_t$	$3\tau_t$
LUT acc. time vs overall proc. time	$\frac{56\tau_t}{78\tau_t} = 0,72$	$\frac{28\tau_t}{50\tau_t} = 0,56$	$\frac{14\tau_t}{36\tau_t} = 0,39$	$\frac{7\tau_t}{29\tau_t} = 0,24$	$\frac{3\tau_t}{25\tau_t} = 0,12$

The access time for the LUT decreases linearly at a rate of $1/k$. The percentage of access time versus overall processing time decreases slowly at a rate of $1/k$. The trade-off between area and time has to be defined according to the final application. The proposed architecture has also been tested in the XS4010XL-PC84 FPGA. Time delay estimation in usual time units can also be provided assuming $\tau_t = 1$ns.

A complete study of the error is still under development and numerical results are not available yet. Nevertheless, some important traits can be outlined. The recursive calculation accumulates absolute error as the number of iterations increases. This drawback can be faced by both decreasing the number of iterations and considering techniques for error compensation. A trade-off must be found between the accuracy of the evaluation (related to the number of calculated values) and the increasing calculation error. Parallelization provides a mean to deal with this problem by defining computing levels. Between two consecutive values, $F_i F_{i+1}$ obtained at a computing level p a new calculation can take place at level $p + 1$ to carry out a set of new values by different parameters α, β and G. The improving process can extend on more levels. It can be outlined that the processors involved in these calculations configure a tree-structured architecture. Figure 6 shows an example of a four-level structure organization.

In a more generic way, when the calculation of N values for function F is pursued, the organization of the calculation can be set as follows, assuming that

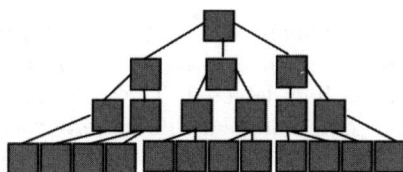
1st Level: 1 processor calculates 3 values

2nd Level: 3 processors calculate 2x3=6 values

3rd Level: 6 processors calculate 2x6=12 values

4th Level: 12 processors calculate ...

Fig. 6. Four level processing tree-structure

$N = N_1 \cdot N_2 \cdot \ldots \cdot N_p$, where N_i stands for the number of values to be calculated at a given level i. If the error for the calculation of one value is assumed to be ε, the maximum error affects the values at the last level and is $(N_1 + N_2 + \ldots + N_p)\varepsilon$. It can be minimized when the sum $(N_1 + N_2 + \ldots + N_p)$ is minimized, that is to say when all N_i in the sum are prime numbers. The sequential calculation would imply a greater error $N\varepsilon$. The time delay calculation follows a similar evolution scheme as the error. Assuming T is the time delay for one value calculation, the overall time delay is $(N_1 + N_2 + \ldots + N_p)T$. The minimization condition is the same as for the error, so the sequential calculation would imply a greater time delay NT. For the occupied area, the precise structure of the tree in what concerns the depth (number of computing levels) and the number of branches (number of calculated values per processor) is quite relevant to the result. That is to say the distribution of N_i defines some improving tendencies. The number of processors P in the tree-structure can be bounded as shown in 3:

$$P = 1 + N_1 + N_1 N_2 + N_1 N_2 N_3 + \ldots + N_1 N_2 N_3 \ldots N_p < 1 + (p-1)\frac{N}{N_p}. \quad (3)$$

P increases as the number of computing levels p increase but the growth can be contained if N_p is the maximum value of all the N_i, that is to say, in the last computing level $p - 1$ the number of values calculated per processor is the highest. Summarizing the main ideas:

- The parallel calculation has benefits on error calculation and time delay whereas sequential calculation improves in area saving.
- A trade-off must be established between the time delay, occupied area and the approximation accuracy (considering the number of calculated values).

5   Calculation of the Hough Transform

The geometric primitive detection using the HT implies three stages: image outline creation by using an edge detector, application of the HT to each point of the image and a voting process in the Hough domain in order to extract the geometric primitives. If the geometric primitive to be detected is a straight line, the HT transforms each point $P(x, y)$ in the Cartesian domain in a point (ρ, θ) in the Hough domain, and vice versa. So, the Hough domain is complete and unique for $0 \leq \rho < \Pi$ line representation. The Hough domain can be interpreted as a voting grid. Each point in the Cartesian domain votes for a set of lines that intersect it and that stand for a grid point (ρ, θ). A local maximum point in the voting grid represents the best adjusted line detected. The grid point's increments $\Delta\rho$ and $\Delta\theta$ establish both distance and angular difference between lines in the Cartesian domain, respectively. The HT is a robust technique since the voting process is not affected by isolated noise points because wrong votes do not affect the local maximum. The HT also manages successfully line occlusion problems, because the distance between points is not relevant. The parametric space is quantized in N_θ levels, from 0 to Π and N_ρ levels, from ρ_{min} to ρ_{max}. The HT calculates the ρ values for all the angles in $[0, \Pi[$ and for every pixel in the image. The direct calculation has a complexity $O(N^2)$ and the global amount of operations is $N^2 \cdot N_\theta$. If $[0, \Pi[$ is considered as $[0, \Pi/2[U[\Pi/2, \Pi[$, the HT for every pixel (x_i, y_j) in the image can be written as:

$$
\begin{aligned}
(\rho_I)_i &= x_i \cdot \cos\theta_i + y_j \cdot \sin\theta_i, \quad 0 \le \theta_i < \tfrac{\Pi}{2}. \\
(\rho_{II})_i &= y_j \cdot \cos\theta_i - x_i \cdot \sin\theta_i, \quad \tfrac{\Pi}{2} \le \theta_i < \Pi.
\end{aligned} \tag{4}
$$

If:

$$
\begin{aligned}
\theta_k &= \theta_{k-1} + \Delta\theta, \\
\cos\theta_k &= \cos(\theta_{k-1} + \Delta\theta), \\
\sin\theta_k &= \sin(\theta_{k-1} + \Delta\theta), \\
\cos\Delta\theta &= \alpha, \quad \sin\Delta\theta = \beta.
\end{aligned} \tag{5}
$$

When substituting 5 in 4 we have that:

$$
\begin{aligned}
(\rho_I)_i &= \alpha(\rho_I)_{i-1} + \beta(\rho_{II})_{i-1}, \\
(\rho_{II})_i &= \alpha(\rho_{II})_{i-1} + \beta(\rho_I)_{i-1}, \\
&\text{with } \alpha^2 + \beta^2 = 1.
\end{aligned} \tag{6}
$$

It appears that $(\rho_I)_i$ and $(\rho_{II})_i$ can be cross-evaluated by applying twice the Equation 2, using $G_q = (\rho_I)_i$ when evaluating $F_q = (\rho_{II})_i$ and using $G_q = (\rho_{II})_i$ when evaluating $F_q = (\rho_I)_i \cdot (\rho_I)_0$ and $(\rho_{II})_0$ should be initialized with the value of the coordinates of each pixel in the image.

Two fast HT architectures based on CORDIC are considered for comparison with our method: a pipelined reconfigurable implementation [26] and a parallel one [27]. Area and delay are compared but error is only treated for the first one. In addition, a comparison with a software implementation [26] is provided.

5.1   Comparison with Pipelined CORDIC

In the first proposal [26], a HT using 16-bit fixed-point arithmetic, 12-iteration CORDIC is implemented using a Xilinx XS4010XL-PC84 FPGA for fast prototyping. The HT using pipelined CORDIC with serial scale factor compensation uses 83% or 333 CLBs out of 400 of the XS4010XL FPGA. This implementation can be clocked at more than 40MHz with a computational complexity of $O(N2)$ for a NxN image. At this frequency, a $128x128$ binary image with 128 discrete angles ($\Delta\theta = 1.40625^\circ$) takes 0.0262 seconds to transform one image.

In the pipelined CORDIC implementation, the global error $(E = 2N2^{-(n-1/2)} + 2^{-M} \cdot n)$ falls with the number of bits of the fractional part M and grows with the number of iterations n, when $n > 16$.

According to the evaluation model presented by [23] [24] [25] and to the characteristics of the Xilinx XC4000 FPGA devices (a CLB consists of one LUT-3, two LUT-4 and 2 latches), Table 3 shows the area estimation for pipelined CORDIC, data precision is 16 bits and reduction structure chosen is $3:2$.

Table 3. Area estimation for the pipelined CORDIC

Pipelined CORDIC	Num CLBs=333	Complex gates
LUT-3	$1 \cdot 333 = 333$	$333 \cdot 2^3 \cdot 2^4 \cdot 40 \frac{\tau_a}{Kbit} = 1665\tau_a$
LUT-4	$2 \cdot 333 = 666$	$2 \cdot 333 \cdot 2^4 \cdot 2^4 \cdot 40 \frac{\tau_a}{Kbit} = 6660\tau_a$
Latches	$2 \cdot 333 = 666$	$2 \cdot 333 \cdot 0.5 \cdot 2^4 \cdot \tau_a = 5328\tau_a$
Overall		$13653\tau_a$

With respect to the area, the comparison between our method and the pipelined CORDIC has been achieved by observing Tables 1, 2 and 3. As shown in Table 4, it appears that our method occupies an increasing area with k, but our implementations are better than the CORDIC ones up to $k = 8$. For delay estimation, the HT calculation involves $64 \cdot 128 \cdot 128$ iterations and each iteration has $16/k$ cycles. We assume that $\tau_t \approx 1$ns in the XS4010XL-PC84 FPGA. Table 4 shows that only for $k = 16$ our method can provide equal time delay than the pipelined CORDIC, which takes 0.0262s to perform the calculation. Unfortunately, the occupied area would not be acceptable.

Table 4. Delay and area estimates for our proposal for the sequential implementation

Fragment size	Time delay (ms)	Area (τ_a)
k=1	$128 \cdot 128 \cdot 12 \cdot (22 + 56)\,\tau_t = 15.33$	$9 \cdot 2 \cdot (44 + 10)\,\tau_a = 972$
k=2	$128 \cdot 128 \cdot 12 \cdot (22 + 28)\,\tau_t = 9.83$	$9 \cdot 2 \cdot (44 + 40)\,\tau_a = 1512$
k=4	$128 \cdot 128 \cdot 12 \cdot (22 + 14)\,\tau_t = 7.08$	$9 \cdot 2 \cdot (44 + 1024)\,\tau_a = 19224$
k=8	$128 \cdot 128 \cdot 12 \cdot (22 + 7)\,\tau_t = 5.70$	$9 \cdot 2 \cdot (44 + 262444)\,\tau_a = 4724784$
k=16	$64 \cdot 128 \cdot 128 \cdot (22 + 3)\,\tau_t = 4.91$	$9 \cdot 2 \cdot (44 + 17179869184)\,\tau_a > 1G$

It can be shown that our method has a better performance in what concerns to time delay and occupation area. With respect to the error, the pipelined CORDIC increases its error when the number of calculated values increases. Our implementation has a bounded value for the tree-structured architecture. If ε is the error obtained after one iteration, then the maximum error of this structure has an upper bound of 12ε.

5.2   Comparison with a Parallel CORDIC

The second proposal [27] is a parallel implementation of the CORDIC algorithm in order to calculate the HT. The computation of the HT for a NxN image with a single CORDIC processor requires $N^3/2$ cycles, assuming that in eacheval-uation two values for parameter ρ are obtained. The computation time can be reduced by introducing parallelism. Three possible approaches exist, namely parallelization for the pixels in the image, for the angle θ or both simultaneously. The latter requires $N^3/2$ CORDIC processors, that is, one processor per pixel per angle. The evaluation of the transform takes only the time of one CORDIC operation: n cycles for radix 2, $n/2 + n/4$ cycles for mixed radix $2 - 4$ and $n/2$ for radix 4 (n is the data precision), but the hardware is considerably increased. Also, some conflicts occur in the voting process because the results obtained by the processors for the same angle θ can vote over the same element in the Hough space at the same time. The introduction of parallelism only for the pixel calculation requires N^2 processors, one for each pixel. The number of CORDIC operations is $N/2$ plus a latency, which depend on radix. There are also conflicts in the voting process. A solution which does not produce voting conflicts is the parallelization of the angles. In this case a processor per angle is needed, in which all the pixels of the image are processed sequentially. The total number of processors is $N/2$ and the number of cycles for the evaluation of the transform is N^2 plus a latency and one pixel is processed in each cycle.

The implementation considered in [27] uses a 12-bit-precision CORDIC processor with 10 stages (6 are the standard iteration-stages, 1 for the compensation of the scaling factor and 3 for performing the scaling). Each stage consists of two registers, two multiplexes and two adders/subtracters. The standard stage needs 24 bits for each angle in the ROM. The time delay is $0,819$ms.

For transforming a $128x128$ image with 128 rotation angles, in the angle parallelization case, 64 processors are needed and $128 \cdot 128$ cycles are performed. The area of the $n-$bit register can be estimated as $0.5 \cdot n\tau_a$. The area of the multiplexors depends on the number of input vectors v and on their size n. The associated area is about $0.25 \cdot v \cdot n\tau_a$. The area estimates are shown in Table 5.

In order to speed up the time delay and contain the error, our method can parallelize the execution with a similar tree-structured architecture as for the pipelined CORDIC. $N = 128x128$ are the initial pixel values and 64 new values need to be calculated per each pixel value. In order to approximate the parallel CORDIC performance (0.819 ms and $31580\tau_a$), the following implementations can suit:

- $k = 1$, 16 identical tree-structures performing 210 calculations each one. So, time delay will be 0.9581ms and area $15552\tau_a$.

Table 5. Area estimation for the parallel CORDIC implementation

Parallel CORDIC	Quantity	Complex gates
Registers	$20 \cdot 64$	$20 \cdot 64 \cdot 0.5 \cdot 12\tau_a = 7880\tau_a$
Multiplexers	$20 \cdot 64$	$20 \cdot 64 \cdot 0.5 \cdot 12\tau_a = 7880\tau_a$
Adders/Subtracters	$20 \cdot 64$	$20 \cdot 64 \cdot 12\tau_a = 15760\tau_a$
LUT tables	$64 \cdot 24$ bits	$64 \cdot 24 \cdot 40\frac{\tau_a}{Kbit} = 60\tau_a$
Overall		$31850\tau_a$

- $k = 1$, 32 identical tree-structures performing 29 calculations each one. So, time delay will be 0.4790ms and area $31104\tau_a$.
- $k = 2$, 16 identical tree-structures performing 210 calculations each one. So, time delay will be 0.6144ms and area $24192\tau_a$.

If ε is the error, the maximum error caused by this structure is 12ε after one iteration. The referred work [27] does not provide data on the error performed by the implementation presented. It can be noticed that the presented implementation achieves a satisfying trade-off between area, time delay and error.

5.3 Comparison with a Software Implementation

Reference [26] provides a comparison with a software implementation. A typical software program that performs the HT was written for comparing the performance of a microprocessor and that of a FPGA implementation. The program performs 128 $\rho-$value calculations using *sine* and *cosine* functions for $128x128 = 16384$ pixels. The following code shows the calculation:

```
register int i,j;
register double k;
start = clock();
register double result = 0;
for (i=0; i<128;i++){
     for (j=0;j<128;j++){
          for (k=0; k<PI; k+=PI/128 { }
     }
}
finish=clock();
```

This program was built and run in Debug mode with Microsoft Visual C++ 6.0 on a PIII 667 MHz with 133 MHz FSB and 512 KB cache, Windows 2000 PC. The results show it can process 16384 pixels in 0.921s. This is about 35 times slower than the throughput of a comparable pipelined CORDIC HT and 11 times slower than our proposal (see Table 3 for $k = 1$). The performance advantages made available by an FPGA's ability for custom computation are clear from this comparison.

6   Conclusion

This paper has presented an architectural approach to the scalability problem caused by the exploding requirements of computing resources in function calculation approaches. Fundamentals of the method claim that the use of a more complete primitive, namely a weighted sum, converts the calculation of the function values into a recursive operation defined by a table. The strength of the method is concerned with the fact that the operation to be performed is the same for the evaluation of different functions (elementary or not). Therefore, only the table must be changed because it holds the features of the concrete evaluated function in the parameter values. The method provides a linear computational cost when some conditions are fulfilled. The comparison with other methods outlines the advantages of this approach, at least for calculations that involve a lot of elementary operations, as is the case of combined trigonometric functions. The sequential and parallel architectures allow achieving a satisfying trade-off between time delay and occupied area as well as encouraging partial results in what concerns to error contention.

References

1. Schwarz, E.M.: Rounding for Quadratically Converging Algorithm for Division and Square Root. Proc. ASILOMAR'96, (1996) 600–603
2. Beaumont-Smith, A.; Burgess, N. and Lefrere, S.: Reduced Latency IEEE Floating-point Standard Adder Architectures. Proc. ARITH'99, (1999) 35–43
3. Schmookler, M.S. and Nowka, K.J.: Leading zero anticipation and detection-a comparison of methods. Proc. ARITH'01, (2001) 7–12
4. Lang, T. and Bruguera, J.D.: Floating-Point Multiply-Add-Fused with Reduced Latency. IEEE Trans. on Computers, Vol. 53, No. 8, (2004) 988–1003
5. Tan, D; Danysh, A and Liebelt, M.: Multiple-precision fixed-point vector multiply-accumulator using shared segmentation. Proc. ARITH'03, 2003 12–19
6. Muamar, H.K and Nixon, M.: Tristage HT for multiple ellipse extraction. IEEE Proc. Part E: Computer and Digital Techniques, Vol. 138, No. 1, (1991) 27–35
7. Haule, D.D and Malowany, A.S.: Object Recognition using fast adaptive HT. Proc PACRIM'89, (1989) 91–94
8. Silva, I.: Vectorization from aerial photographs applying the HT method. Proc. SPIE, Vol. 1395, No. 2, (1990) 956–963
9. Yamazawa, K.; Yagi, Y. and Yachida, M.: 3d Line Segment Reconstruction by Using Hyperomni Vision and Omnidirectional Hough Transforming. Proc. ICPR'00, Vol. 3, (2000) 487–490
10. Bariani, M.; Cucchiara, R.; Mello, P. and Piccardi, M.: Exploiting symbolic learning in visual inspection. LNCS, Vol. 1280, (1997) 223–234
11. Dong, F.; Clapworthy, G.J. and Krokos, M.: Volume Rendering of Fine Details Within Medical Data. Proc. VIS'01, (2001) 387–394
12. Tezmol, A.; Sari-Sarraf, H.; Mitra, S. and Gururajan A.: Customized HT for Robust Segmentation of Cervical Vertebrae from X-Ray Images. Proc. SSIAI'02, (2002) 224–228
13. Huang, L.Y.; Hu, Z. and Sun, F.M.: A New Automatic Quasar Recognition Technique Based on PCA and the HT. Proc. ICPR'00, (2000) 2499–2502
14. Sural S. and Das, P.K.: A genetic algorithm for feature selection in a neuro-fuzzy OCR system. Proc. ICEDAR'6, (2001) 987–991
15. Koshimizu, H. and Numada, M.: On fast HT method PLHT based on piecewise-linear Hough function. J. System Computer in Japan, Vol. 21. No. 5, (1990) 62–73
16. Ben-Tzvi, D. and Sandler, M.: A Combinatorial HT. J.P. Recognition Letters, Vol. 11, (1990) 167–174,
17. da Fontura, L. and Sandler, M.B.: A binary HT and its efficient implementation in a systolic array architecture. J.P. Recognition Letters, Vol. 10, (1989) 329–334
18. Walther, J.S.: A unified algorithm for elementary functions. Proc. Spring Joint Computers Conf., (1971) 379–385

19. Li, H.F.; Lavin, M.A. and Le Master, R.J.: Fast HT: a hierarchical approach. J. Computer Vision Graphics Image Processing, Vol. 36, (1986) 139–161
20. Hu, Y.H.: CORDIC-based VLSI architectures for Digital Signal Processing. IEEE Signal Processing Magazine, Vol. 7, (1992) 16–35
21. Shankar, R.V. and N. Asokan.: A parallel implementation of the HT method to detect lines and curves in pictures. Proc. MWSCAS'32, (1990) 321–324
22. Chuang , H.Y.H. and Li, C.C. A systolic processor for straight line detection by modified HT. IEEE Conf. on Computer Architecture for Pattern Analysis and Image Database Management, (1995) 300–304
23. Ercegovac, M. and Lang, T.: Division and Square root: Digit-Recurrence, Algorithms and Implementations. Klüwer Academic Pub., (1994)
24. Piñeiro, J-A.; Bruguera, J.D. and Muller. J.M.: Faithful powering computation using table look-up and a fused accumulation tree. Proc. ARITH'01, (2001) 40–47
25. Piñeiro, J-A.; Ercegovac, M.; Bruguera, J.D.: High-Radix Logarithm with selection Rounding. Proc. ASAP'02, (2002) 101–110
26. Deng Dixon, D.S and El Gindy H.: High speed Parametrizable HT using reconfigurable hardware. Proc. ACM VIP, 2001 51–57
27. Bruguera, J.D. and Guil, N.: CORDIC-based parallel/pipelined architecture for the HT. J. of VLSI Signal Processing, Vol. 12, No. 3, (1996) 207–221

Design Space Exploration of Media Processors: A Generic VLIW Architecture and a Parameterized Scheduler

Guillermo Payá-Vayá, Javier Martín-Langerwerf,
Piriya Taptimthong, and Peter Pirsch

Institute of Microelectronic Systems
University of Hannover
Appelstr.4, 30167 Hannover Germany
{guipava,jamarlan,pirsch}@ims.uni-hannover.de

Abstract. This paper presents a new environment for exploring and optimizing VLIW architectures for multimedia applications. The environment consists of a generic VLIW architecture, in which virtually all characteristics can be changed, and an assembler with the corresponding parameterized scheduler based on an enhanced version of the list scheduling algorithm. A novel partitioned register file architecture is proposed and analyzed with this environment. This is performed using a highly time consuming task of the H.264 video decoder application. Performance improvements of up to 67% can be achieved when running this application on different architecture configurations.

1 Introduction

Nowadays, continuous improvements in algorithm research require increasingly sophisticated multimedia operations, demanding a rising amount of processing power. Current media processors cannot meet the tremendous demand on multimedia hardware without an adaptation to special processing characteristics. Current video applications, e.g. MPEG-4 or H.264, are pushing the limits of existing media processors for high definition quality.

Due to the capability of exploiting the high degree of inherent parallelism in multimedia applications, VLIW architecture approaches have received great interest in the last ten years [1]. VLIW architectures execute multiple operations within a single long instruction word. Therefore, multiple parallel functional units have to be implemented in order to allow this concurrency. In general, the performance of a VLIW architecture can be further improved by subword parallelism [2,3], instruction set extensions [4], specialized functional units and instruction level parallelism [5]. In contrast to other advanced processors, VLIW architectures require instruction scheduling to be performed at compile time to assemble the long instruction words.

Multiple architectural options should be taken into account before designing a media processor. For example, increasing the amount of instruction level

P. Lukowicz, L. Thiele, and G. Tröster (Eds.): ARCS 2007, LNCS 4415, pp. 254–267, 2007.

parallelism (ILP) requires more registers to hold the operands and results of those instructions. This leads to a complex register file with a high number of read/write ports to allow the concurrent access. On the other hand, large multi-port memories introduce performance degradation [6]. Therefore, partitioned register files should be considered for this kind of processor architectures.

Existing design space exploration techniques comprise, e.g., software-based performance estimation [7]. Therewith, fast results can be obtained, but they do not have the possibility to perform architecture-specific code optimizations. Other techniques obtain more precise results by actually simulating different applications on a parameterized VLIW cluster-based architecture [8].

In order to explore and optimize a VLIW architecture for a specific group of applications, a generic VLIW architecture is desirable, in which virtually all characteristics can be changed. This paper presents a design space exploration environment that comprises a parameterized VLIW pipeline architecture simulator, together with an aggressive code scheduler based on the list scheduling algorithm [9]. For this, a scheduler that considers all these characteristics is mandatory. Using this environment and the motion compensated prediction algorithm specified on the H.264 video coding standard [10] as a reference application, the performance of a generic VLIW architecture for different configurations has been studied.

This paper is organized as follows. Section 2 presents the generic VLIW architecture. Section 3 introduces some basic concepts of code scheduling, and gives details of a parameterized code scheduler. After that, in Section 4, the results of a H.264 decoding task run on different architecture configurations are shown. Finally, conclusions are presented in Section 5.

2   A Generic VLIW Architecture

In this section, we present a generic VLIW architecture that covers the basic requirements for media processing. This architecture is based on the experience obtained on previous works [11]. Two important characteristics have been considered. A fully parameterized architecture is defined, which is easily extendable by adding or modifying functional units. Additionally, a novel register file structure avoids a high number of required ports when the number of parallel operations increases.

For this generic VLIW architecture, a parameterized simulator was written in OpenVera [12] and standard C. This reference model is used, not only to build a stand-alone simulator, but also to perform an IP verification of the generic VLIW architecture hardware description. The use of OpenVera allows easy testbench creation and structured RTL interfacing [13]. Figure 1 shows an overview of the design space exploration environment.

2.1   Vector Unit Structure

The basic structure of this VLIW architecture, hereafter called **Vector Unit** (VU) (see Figure 2), has been specially designed for efficient processing of blocks

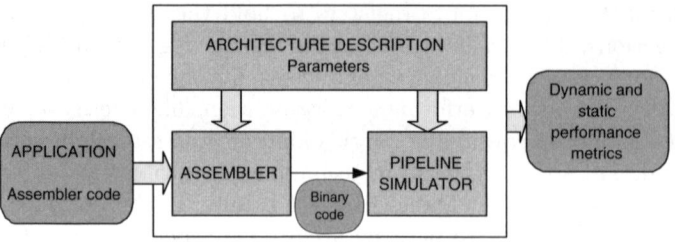

Fig. 1. Design Space Exploration Environment

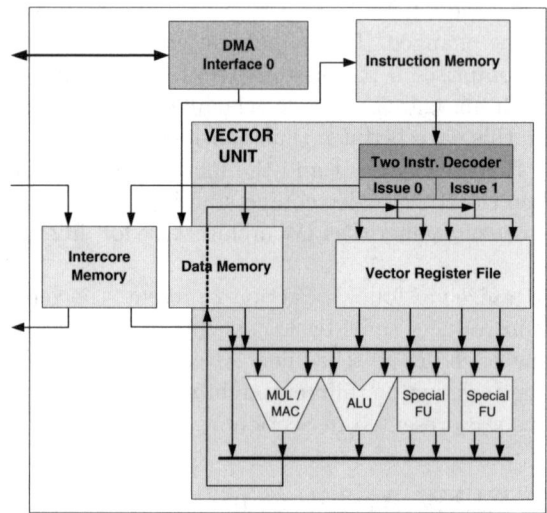

Fig. 2. Vector unit architecture

of data or macroblocks normally used in video coding algorithms. The VU comprises a flexible datapath controlled by a 64-bit dual-issue VLIW (two 32-bit operations). Moreover, the datapath implements subword parallelism (e.g. for a 64-bit datapath, operands are splitable into one 64-bit subword, two 32-bit subwords, four 16-bit subwords or eight 8-bit subwords) in almost every functional unit.

The control path is divided initially into 5 different basic stages: Instruction Fetch (IF), Instruction Decode (DE), Register Access (RA), Execution (EX) and Write back (WB). The EX stage, as shown in Figure 3, can be subdivided into more stages depending on the characteristics, e.g. latency, of the functional units.

The VU implements a configurable global address map, which allocates a 64-bit dual port instruction memory with configurable size. There is also a dual port data memory with configurable size. Each memory is accessed by the processor itself and by a DMA unit, which performs data transfers in background between

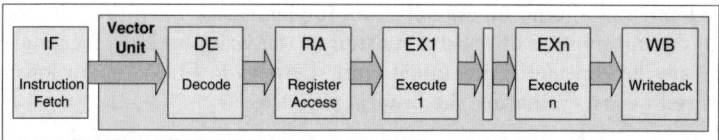

Fig. 3. Vector Unit pipeline

memories and external slaves through a configurable bus system. Eventually, if this processor is integrated in a multi-core system, dual port intercore memories can be added to allow shared memory communication.

2.2 Specialized Instructions and Functional Units

The VU implements an orthogonal instruction set that contains several extensions for typical video processing computations, e.g. data formatting, different kinds of rounding, multiply-and-accumulate operations [3]. To facilitate adding new instructions and functional units, most instructions use a common bit-pattern (see Figure 4). Move and flow operations use different patterns. Access to memory or special registers, e.g. address registers for indirect addressing mode or configuration registers located in the functional units, are performed by move operations.

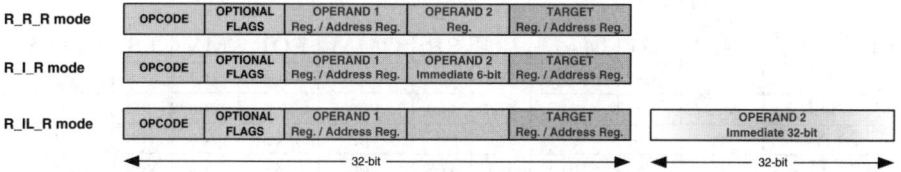

Fig. 4. Instruction types

Functional units are represented by a set of characteristics, e.g. latency, which is expressed in number of execution stages, and type and number of operands, which are related with the number of read and write ports required in the register file. An instruction example is the multiply-and-accumulate (MAC) operation in a R_R_R instruction (see Figure 4). It requires 2 read registers that contain the data to be multiplied, 2 read registers that contain the value to be added to the multiplication result and 2 write registers that store the double-sized MAC result. Additionally, each instruction has different variants depending on the functional units, e.g. condition read and set, overflow and saturation modes, or resolutions to fix the subword parallelism type. Table 1 shows the implemented functional units and some of their characteristics.

Another interesting feature is that a VU can execute all kind of operations in either of its two issues. Therefore, two operations that execute the same function

Table 1. Details about the functional units implemented: operations performed, latency required and number of register written in the write-back (WB) stage. MV, LS and FLOW are not considered functional units. Latency in FLOW operations indicates cycles required before evaluating the branch.

Functional Units	Operations performed	Latency (typical)	Write Registers
AU	*Arithmetic*	1	1
MAC	*Multiplication and Multiply and Accumulation*	2	1 or 2
LU	*Logic*	1	1
SR	*Shift and Round*	1	1
CMM	*Clip, Max and Min*	1	1
FOR	*Data formatting*	1	1
MV	*Move*	1 or 3	0 or 1
LS	*Load and Store*	1	0 or 1
FLOW	*Control flow*	2 or 3	0 or 1

can only be scheduled in the same instruction, if the needed functional unit has been replicated for the desired architecture configuration. This operation pairing is represented in a matrix (see Table 2).

Table 2. Operation pairing in a vector unit (typical configuration: no functional unit replicated). (*) This pairing is only possible if a MV operation does not access memory.

Issue 1-0	NOP	AU	MAC	LU	SR	CMM	FOR	MV	LS	FLOW
NOP	yes	yes	no	yes	yes	yes	yes	yes	yes	yes
AU	yes	no	no	yes	yes	yes	yes	yes	yes	yes
MAC	yes	yes	no	yes	yes	yes	yes	yes	yes	yes
LU	yes	yes	no	no	yes	yes	yes	yes	yes	yes
SR	yes	yes	no	yes	no	yes	yes	yes	yes	yes
CMM	yes	yes	no	yes	yes	no	yes	yes	yes	yes
FOR	yes	yes	no	yes	yes	yes	no	yes	yes	yes
MV	yes	yes	no	yes	yes	yes	yes	yes*	no*	yes
LS	yes	yes	no	yes	yes	yes	yes	no*	no	yes
FLOW	yes	yes	no	yes	yes	yes	yes	yes	yes	no

2.3   Multiple Vector Unit and Partitioned Register File

In some applications, using a higher degree of parallelism results in considerably more performance. This can be achieved by increasing the number of VUs to two or even three within the architecture as shown in Figure 5. The pipeline configuration is mostly the same (see Figure 6), with the exception of the IF stage, which is now common for all VUs. The other stages are independent for each VU. In this new architecture, every VU has its own data memory and DMA, allowing multiple transfers in parallel (if allowed by the bus system).

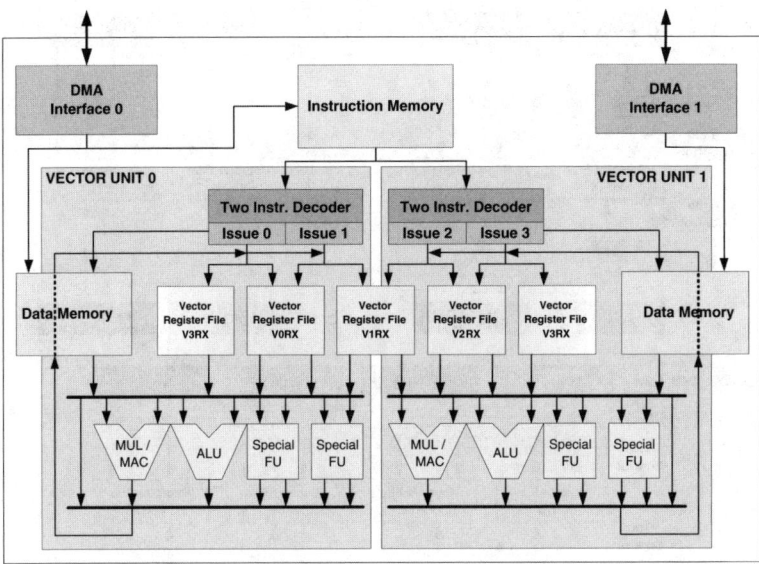

Fig. 5. Dual VU architecture

The use of multiple VUs implicates a redesign of the register file structure. On the one hand, having a unique register file for two VUs (with two issues each) results in implementing a 16 read and 8 write ports memory, which is not practicable. On the other hand, having separated register files for each VU (i.e. cluster [14]) results in inefficient inter-VU data transfers. Therefore, the register file structure for multiple VUs can be implemented in a circular manner as shown in Figure 5. This new register file structure increases the operation parallelism allowing to parallelize operations that require same functional units and same input data. Input data is not duplicated, instead it is stored in the register file located between VUs. This structure occupies less area and improves timing performance of the hardware implementation.

Depending on the required number of ports, the register file structure is configurable. For example, there are some functional units that, due to the output

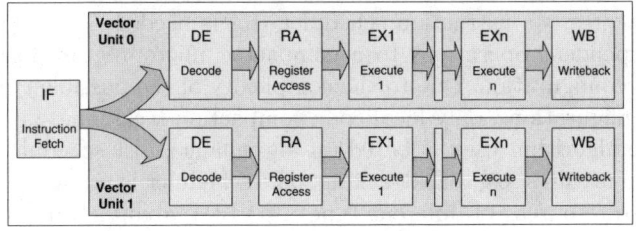

Fig. 6. Dual VU pipeline architecture

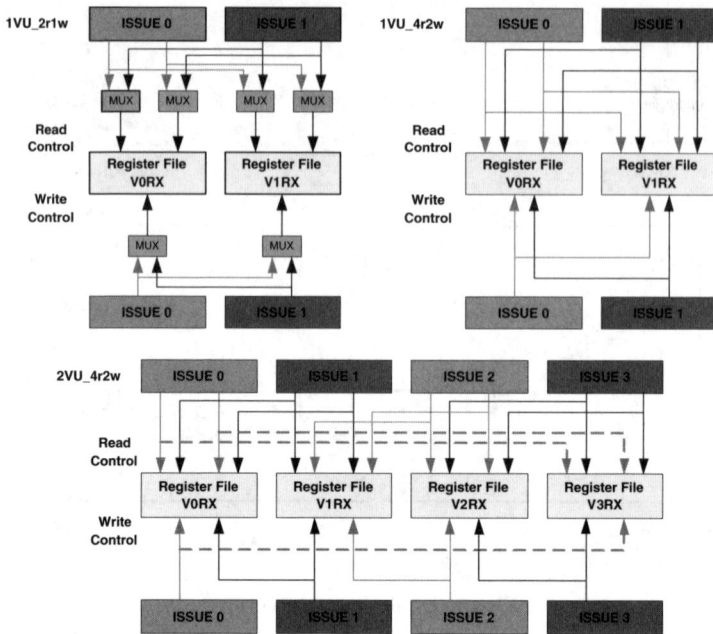

Fig. 7. Register file control signals for different configurations (dot-lines for better visualisation)

resolution, require storing the results in two registers (see section 2.2). This means that 3 write ports, even 4 in case of duplicating these functional units, should be available in a register file. Different register file structures for single and dual VU are shown in Figure 7.

Certainly, to properly compare the performance on different configurations of a different application, a parameterized scheduler that efficiently exploits the available hardware resources of each configuration is mandatory.

3   Parameterized Scheduler

In order to exploit ILP of an application and utilize hardware resources efficiently, an aggressive instruction scheduler [15] is needed. The scheduler reorganizes independent operations to a compacted microprogram for the chosen architecture configuration. Due to the complexity of the possible generic VLIW processor configurations, only local code compaction techniques [9] will be considered. The algorithm used is based on an enhanced list scheduling including sophisticated features, e.g. backtracking. The scheduler is parameterized to utilize hardware resources of different types of VLIW architecture configurations efficiently.

The assembler, written in C, divides an input assembly application program into collections of consecutive branch-free operation sequences, called straight line microcodes (SLMs) or basic blocks. After that, the scheduler analyzes data dependencies of all operations in a SLM and reorganizes operations that neither depends on nor conflicts with other operations to form a sequence of instructions. Finally, the assembler compiles sequences of instructions of each SLM to the compacted microprogram. This general process of local code compaction is shown in Figure 8.

Register allocation is an issue that influences the quality of the compacted microprogram regardless of the architecture of the processor. It is obvious that, for the same hardware configuration, compaction efficiency gets worse if any register is used too often, producing redundant data dependencies. Register allocation is performed manually with good results, since scheduler output can be used as a guideline for register reallocation. However, it is planned to automate register allocation in the future versions of the scheduler.

3.1   Scheduler Structure

The scheduler operating mode depends on the properties of the operations and the underlying hardware architecture. Relevant operation properties are the utilized functional unit, accessed registers, number of read and write ports of the register file and latency. Significant architecture information is, for example, number of vector units, types and number of functional units, number of read and write ports of vector register files and their configuration.

To parameterize the scheduler, its implementation should be loosely bound to the hardware architecture and operations. One approach is storing architecture description and operations in tables. For instance, the operation set table contains the functional units, the type of operands and the latency required by all possible operations. When new operations are introduced or the number of stages in the pipeline is extended, this table has to be modified and the scheduler will be able to handle those new operations. The operation pairing table (see Table 2) is used to reflect the number of functional units. The vector register files table specifies the vector register file configuration and the number of ports available (see Figure 7). This approach is also useful when describing special parts of the architecture, like indirect address register files that have two access modes depending on the operation addressing mode used (direct or indirect), operations that implicitly access special registers (most of them effect control flow of the program) and operations having variable latency depending on their operands.

3.2   Enhanced List Scheduling Algorithm

The scheduling process is shown in Figure 9. The first step is a data dependency analysis. The output of this process is a data dependency graph (DDG) which can be represented in a matrix form. The DDG provides information about the instruction order to preserve semantic equivalence with the input program. The list scheduling is a particular tree searching algorithm with heuristic functions,

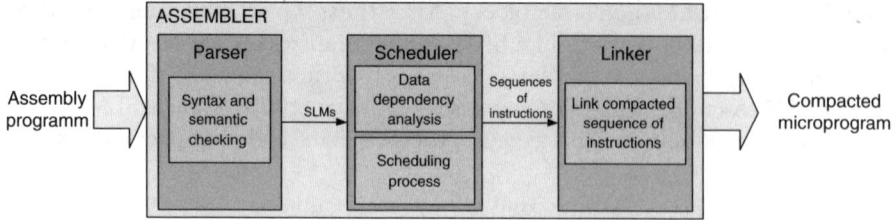

Fig. 8. Assembler and local code compaction

e.g. critical path heuristic. The major advantages of list scheduling are its moderate complexity and its acceptable performance. However, drawbacks of list scheduling algorithm are lacking the ability to undo decisions that led to worsen results and missing heuristic functions to evaluate hardware resources. The enhanced list scheduling overcomes the mentioned drawbacks using backtracking to undo inappropriate decisions. Additional heuristic function is used to allow hardware resource management.

A detailed explanation about the enhancements developed for the list scheduling is not focus of this paper and has therefore been postponed for a future publication.

4   Design Space Exploration Example

In this section, performance statistics for a specific application on different configurations of the generic VLIW architecture are presented. Architectural decisions should be combined in the future with physical parameters, i.e. area and timing, to obtain detailed evaluation results.

4.1   H.264 Motion Compensated Prediction

H.264/AVC is the latest video coding standard, and provides gains in compression efficiency of up to 50 % compared to previous standards [16]. This compression efficiency requires the use of powerful processors. For example, the H.264/AVC baseline decoder is two to three times more complex than an H.263 baseline decoder in a Pentium 3 platform [17]. For H.264/AVC decoding, motion compensated prediction is one of the most time consuming routines.

H.264 motion compensated prediction (MC) uses displacement vectors accuracy of a quarter of a picture element (quarter-pel). These fractional-pel displacements are generated by interpolating sub-pel positions. Therefore, a 6-tap FIR filter has to be applied horizontally and vertically to generate half-pel. Quarter-pel positions are obtained by averaging the luminance signal at full- and half-pel positions [10]. Each motion vector can specify 8 different situations depending on the sub-pel position (Full, Half or Quarter) and the directions (Horizontal and Vertical). Chrominance signal is generated by averaging.

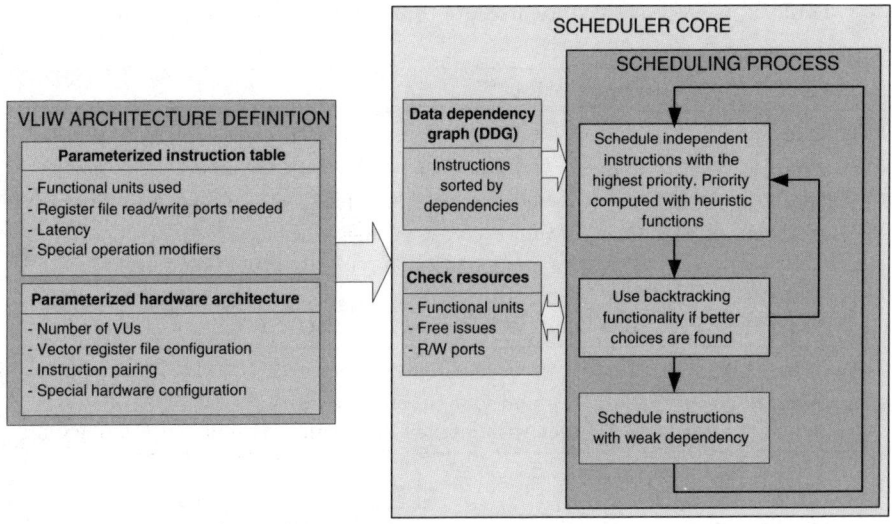

Fig. 9. Scheduler core and scheduling process

A full operational 4x4 block motion compensated prediction routine has been developed. Two different versions of the same code are available and optimized for an architecture with one and two VUs, respectively.

4.2   Performance Measures

Using the previous application as an input, a static performance measure has been applied. The measure is based on the maximum number of system clock cycles required (see Figure 10) for computing a motion compensated prediction of a 16x16 block. Analyzing the results, affirmations concerning the achievable performance on different architecture configurations can be made (see Table 3). All architecture configurations have a 64-bit datapath and, if no otherwise specified, the register file size is 32x64 and the latency of the functional units correspond to the values from Table 1.

The proposed register file structure used on the 1VU_4r2w configuration requires the same number of cycles as the 1VU_8r4w configuration with only one register file. On the other hand, the 1VU_2r1w configuration requires more cycles because of penalties introduced by the MAC operations, which requires storing the results on two registers. Therefore scheduling other instructions together with the MAC operation is impossible. Finally, the architectural decision will be a trade-off between achievable performance, programming flexibility and chip area. The proposed register file structure allows dramatical gate count reduction due to reduced number of read/write ports needed without performance degradation.

Duplicating functional units reduces hardware resource conflicts and allow more flexible scheduling. For example, by duplicating the SR and CMM units,

Table 3. Generic VLIW architecture configurations used in the example

Name	Description
1VU_2r1w	One VU and two register files with 2 read / 1 write ports.
1VU_4r2w	One VU and two register files with 4 read / 2 write ports.
1VU_8r4w	One VU and one reg. file (64x64) with 8 read / 4 write ports.
1VU_4r2w_2SR_2CMM	One VU and two register files with 4 read /2 write ports. Two SRs and two CMMs replicated functional units.
1VU_4r2w_Lat	One VU and two register files with 4 read / 2 write ports. MAC unit with latency 3. AU and SR units with latency 2.
2VU_4r2w	Two VUs and four register files with 4 read / 2 write ports.
2VU_4r2w_Lat	Two VUs and four register files with 4 read / 2 write ports. MAC unit with latency 3. AU and SR units with latency 2.

the scheduling efficiency, i.e. the percentage of no-NOP instructions in the code, increases up to 80% (see Table 4). Although more execution stages introduce a high latency, performance gains can be achieved since operation frequency could be significantly higher. This architectural modification can aggravate scheduling due to data depencencies.

Finally, using a second VU decreases the scheduling efficiency down to 60%, but also decreases the amount of cycles required by the H.264 task. This performance improvement should be taken with care, because more area for hardware implementation is needed.

In contrary to synthetic sequences, motion compensated prediction is not needed for every macroblock. Moreover, not always the most time consuming mode, HH (see Figure 10), is used. By processing real movie sequences, dynamic performance can be measured (see Figure 11). Assuming a 10 frames buffer, the following results are obtained. While 1VU_2r1w configuration requires an average of 3.61M cycles per frame, the 2VU_4r2w configuration only requires 2.15M cycles. Therefore, a hardware implementation of the latter configuration allows real time processing (720x576@30fps) at only 65 MHz.

Table 4. Application code size (in 64 or 128 bit word for one VU or two VU respectively) for the different generic VLIW architecture configurations and scheduling efficiency (percentage of no-NOP operations in the complete code)

	1VU 2r1w	1VU 4r2w	1VU 8r4w	1VU 4r2w_2SR_2CMM	1VU 4r2w_Lat	2VU 4r2w	2VU 4r2w_Lat
Size	631	567	567	555	608	466	507
Compaction	70.4%	78.3%	78.3%	80.0%	73.1%	62.0%	57.0%

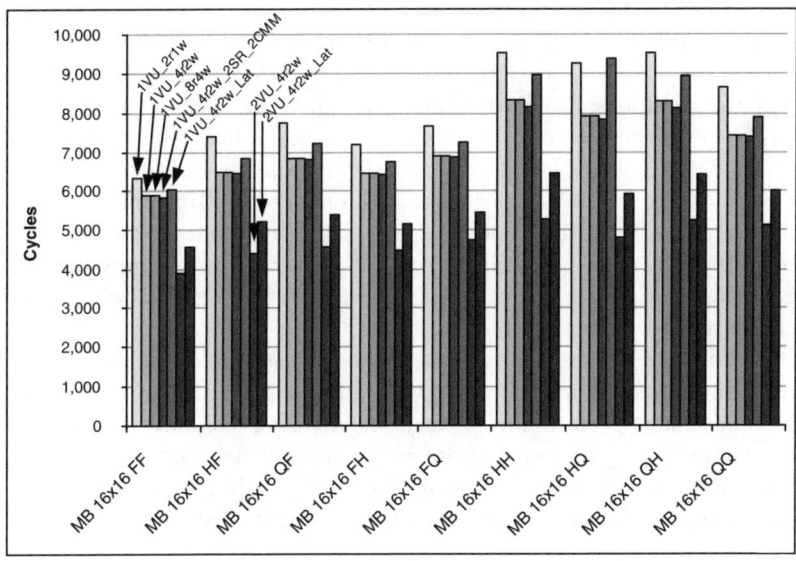

Fig. 10. Number of system clock cycles required by the H.264 motion compensated prediction task for different generic VLIW architecture configurations (see Table 3)

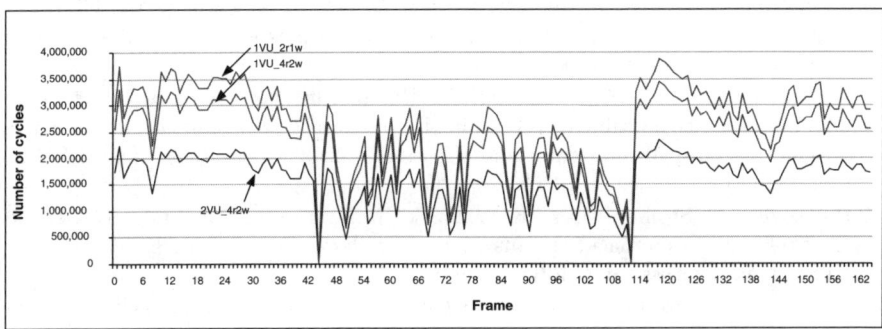

Fig. 11. Number of system clock cycles per frame used for the motion compensation routine in an H.264 video decoder application. Input is a commercial movie sequence with 720x576 frame size.

5 Conclusions

An environment for exploring and optimizing VLIW architectures for multimedia applications has been presented. Based on previous work, a generic VLIW architecture has been implemented as an executable specification, written in an object-oriented language. An assembler, with the corresponding scheduler, has been also developed. Both, model and assembler, are configurable in terms of architectural parameters, like type and number of functional units, word size and number of register file ports, memory sizes, etc. Therefore, any application can be run on different hardware configurations and hereby assist the hardware developer at exploring the design space.

The motion compensated prediction algorithm of the H.264 video coding standard has been implemented for this architecture and scheduled for different configurations. The application code size and the dynamic performance for different architecture configurations, which consist of one Vector Unit and a vector register file structure with different sizes and number of ports, demonstrate the performance of the novel partitioned register file structure presented in this paper. More precisely, there is no difference in scheduling efficiency between a 4r2w and a 8r4w register file configuration.

The flexibility of the architecture model and the scheduler allowed assembling and simulating the chosen application with different architecture configurations. Therewith, dynamic performance measurements using real movie sequences have been obtained. These present a performance improvement of up to 67%, when comparing the configuration consisting of one VU with a 2r1w register file and the configuration consisting of two VUs with a 4r2w register file.

References

1. Dasu, A., Panchanathan, S.: A survey of media processing approaches. Circuits and Systems for Video Technology, IEEE Transactions on **12**(8) (2002) 633–645
2. Fridman, J.: Sub-word parallelism in digital signal processing. Signal Processing Magazine, IEEE **17**(2) (2000) 27–35
3. Lee, R., Fiskiran, A., Shi, Z., Yang, X.: Refining instruction set architecture for high-performance multimedia processing in constrained environments. Application-Specific Systems, Architectures and Processors, 2002. Proceedings. The IEEE International Conference on (2002) 253–264
4. Berekovic, M., Stolberg, H.J., Kulaczewski, M.B., Pirsch, P., Müller, H., Runge, H., Kneip, J., Stabernack, B.: Instruction Set Extensions for MPEG-4 Video. J. VLSI Signal Process. Syst. **23**(1) (1999) 27–49
5. Espasa, R., Valero, M.: Exploiting instruction- and data-level parallelism. Micro, IEEE **17**(5) (1997) 20–27
6. Kneip, J., Berekovic, M., Pirsch, P.: An algorithm-hardware-system approach to VLIW multimedia processors. In: Multimedia Signal Processing, 1997., IEEE First Workshop on. (1997) 433–438
7. Gong, J., Gajski, D., Narayan, S.: Software estimation using a generic-processor model. In: European Design and Test Conference, 1995. ED&TC 1995, Proceedings. (1995) 498–502
8. Fisher, J., Faraboschi, P., Desoli, G.: Custom-fit processors: letting applications define architectures. In: Microarchitecture, 1996. MICRO-29. Proceedings of the 29th Annual IEEE/ACM International Symposium on. (1996) 324–335
9. Landskov, D., Davidson, S., Shriver, B., Mallett, P.W.: Local Microcode Compaction Techniques. ACM Comput. Surv. **12**(3) (1980) 261–294
10. ISO/IEC: Coding of Audiovisual Objects - Part 10: Advanced Video Coding. ISO/IEC 14496-10:2003 (2003)

11. Stolberg, H.J., Berekovic, M., Friebe, L., Moch, S., Flügel, S., Mao, X., Kulaczewski, M.B., Klusmann, H., Pirsch, P.: HiBRID-SoC: A Multi-Core System-on-Chip Architecture for Multimedia Signal Processing Applications. In: DATE '03: Proceedings of the conference on Design, Automation and Test in Europe, Washington, DC, USA, IEEE Computer Society (2003) 189–194
12. Synopsys: Vera User Guide. (2003) version 6.0.
13. Haque, F., Khan, K., Michelson, J.: The Art of Verification with VERA. Verification Central (2001)
14. Lapinskii, V., Jacome, M., de Veciana, G.: Application-specific clustered VLIW datapaths: early exploration on a parameterized design space. Computer-Aided Design of Integrated Circuits and Systems, IEEE Transactions on **21**(8) (2002) 889–903
15. Taptimthong, P.: Design, Implementation and Verification of an Assembler Translation Program for a VLIW Processor Model. Master's thesis, Institute of Microelectronic Systems. University of Hannover (2006)
16. Ostermann, J., Bormans, J., List, P., Marpe, D., Narroschke, M., Pereira, F., Stockhammer, T., Wedi, T.: Video coding with H.264/AVC: tools, performance, and complexity. Circuits and Systems Magazine, IEEE **4**(1) (2004) 7–28
17. Horowitz, M., Joch, A., Kossentini, F., Hallapuro, A.: H.264/AVC baseline profile decoder complexity analysis. Circuits and Systems for Video Technology, IEEE Transactions on **13** (July 2003) 704– 716

Modeling of Interconnection Networks in Massively Parallel Processor Architectures

Alexey Kupriyanov[1,*], Frank Hannig[1,*], Dmitrij Kissler[1,*], Jürgen Teich[1,*],
Julien Lallet[2,§], Olivier Sentieys[2,§], and Sébastien Pillement[2,§]

[1] Department of Computer Science 12, Hardware-Software-Co-Design,
University of Erlangen-Nuremberg, Germany
{kupriyanov, hannig, kissler, teich}@cs.fau.de
[2] IRISA/R2D2, University of Rennes, France
{lallet, sentieys, pillemen}@irisa.fr

Abstract. In this paper, we present a new concept for modeling of interconnection networks in the field of massively parallel processor embedded architectures. The main focus of the paper is on two interconnection concepts, namely, *interconnect-wrapper* and *DyRIBox* definitions of reconfigurable interconnection networks. We compare both interconnection concepts against each other and formally prove their equality. Both concepts allow to model many different reconfigurable inter-processor networks efficiently. Furthermore, we point out how to define the interconnect using an architecture description language for massively parallel processor architectures called *MAML*. Finally, we demonstrate the pertinence of our approach by modeling and evaluation of different reconfigurable interconnect topologies.

1 Introduction

The desire for more mobility and the enthusiasm for ubiquitous electronic gadgets on the one hand side and the unbowed progress in semiconductor industry on the other hand are driving forces in the market of embedded digital systems. These application specific systems have to fulfil different performance, cost, and power requirements. In addition, changing standards or add-ons as unique selling point of a product demand for more flexible solutions. Thus, generic highly parameterizable architecture templates in terms of IP-cores (intellectual property) have become more and more important when building such so-called Systems-on-a-Chip (SoC).

In the domain of application specific instruction set processors (ASIP), there exist some holistic design methodologies which consider the simultaneous development of architectures, simulators, and compilers. Central link between the different design aspects are Architecture Description Languages (ADL). In the following we only refer a few of the most known ADLs. At the ACES laboratory of the University of California, Irvine, for example, the architecture description language EXPRESSION [6] has

---

* Supported in part by the German Science Foundation (DFG) in project under contract TE 163/13-1.

§ This work has been supported by the French-German Reasearch Network Programm P2R.

P. Lukowicz, L. Thiele, and G. Tröster (Eds.): ARCS 2007, LNCS 4415, pp. 268–282, 2007.
© Springer-Verlag Berlin Heidelberg 2007

been developed. From an EXPRESSION description of an architecture, the retargetable compiler EXPRESS and the cycle-accurate simulator SIMPRESS can be automatically generated. The Trimaran system [16] has been designed to generate efficient VLIW code. It is based on a fixed basic architecture (HPL-PD) being parameterizable in the number of registers, the number of functional units, and operation latencies. Parameters of the machine are specified in the description language HMDES. The ADL LISA [14] is the basis for a retargetable compiled approach aiming at the generation of fast simulators for microprocessors even with complex pipeline structures. Finally, we refer to the language MAML which has been developed in the BUILDABONG project [5]. MAML is used for the efficient architecture/compiler co-generation of ASIPs and VLIW processor architectures.

Beside the classical usage of DSPs for dataflow dominant digital signal processing several architectures based on massively parallel processor arrays are emerging. Many academic coarse-grained reconfigurable arrays have been developed [7] and, since a while, more and more commercial ones are being developed like the D-Fabrix [3], the DRP from NEC [13], the PACT XPP [2], or Bresca from Silicon Hive (Philips) [15]. All of the above described ADLs only target at the design of ASIPs and VLIW processors. Only very few approaches are known that consider also coarse-grained processor arrays. For instance, DRAA [11] is a generic reconfigurable architecture template which can represent a wide range of coarse-grained reconfigurable arrays or the Adres [12] reconfigurable architecture template which is described by an architecture description in XML. Furthermore in [1,12], the authors use their ADL in order to explore different interconnect topologies. In this context the contributions of our paper can be summarized as follows:

1. We introduce two generic reconfigurable interconnect methodologies for parallel processor architectures (Section 2),
2. By graph and formal languages theory we proof the equivalence of the two models (Section 2.3),
3. We extend our ADL MAML in order to model the two interconnect concepts (Section 3),
4. We perform a case-study for different well-known interconnect topologies and evaluate the cost for the flexibility being able to switch from one topology to another by dynamic reconfiguration (Section 4).

2   Modeling of Interconnection Networks

Since design time and cost are critical aspects during the design of processor architectures it is important to provide efficient modeling techniques in order to evaluate architecture prototypes without actually designing them. In the scope of the methodology presented here, we are looking for a flexible reconfigurable interconnect architecture in order to find out trade-offs between different interconnect topologies for a given set of applications.

Generally, a massively parallel processor architecture is defined by an array of processing elements $P = \| \ p_{ij} \ \|$, $i = [1, M]$, $j = [1, N]$ and its interconnect as shown in Fig. 1 (a). Usually, a centralized approach is a classical and convenient way to model

(a)   (b)

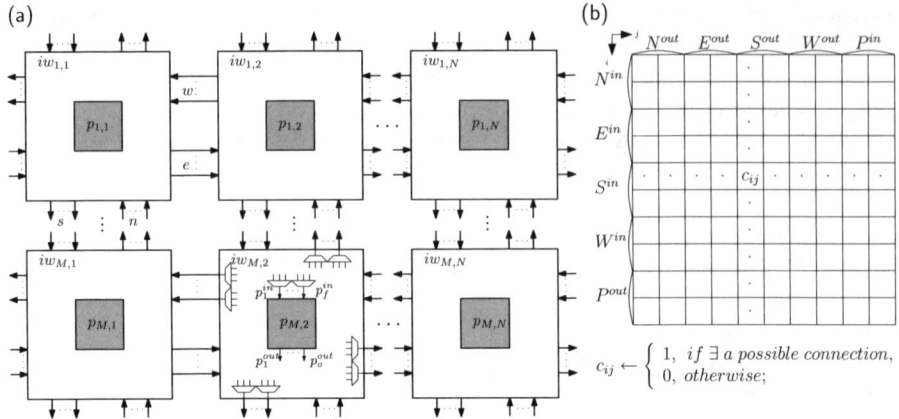

Fig. 1. In (a), a structure of parallel processor architecture with interconnect modeled by the use of interconnect-wrappers (IW). In (b), a configuration of an IW - interconnect adjacency matrix.

the interconnection of the processing elements (PE). In this case, the processors are connected to one or more central switch-nodes, which provide the reconfigurable interconnect. Whereas, a decentralized approach may be more reasonable in case of homogeneous architectures. In the following, we present a modeling concept for distributed reconfigurable interconnect, compare it to the classical approach, and formally prove their equality. This will show an ability of the decentralized approach to model an arbitrary dynamically reconfigurable interconnect topology. But first, we introduce a decentralized concept for modeling reconfigurable interconnection based on so-called *interconnect-wrappers* and a classical centralized interconnection approach called *DyRIBox*.

2.1   Modeling of Interconnection Network Using Interconnect-Wrappers

In order to model and specify a reconfigurable interconnect topology, a PE *interconnect-wrapper* (IW) [9, 10] concept is introduced. An interconnect-wrapper describes the ingoing and outgoing signal ports of a processing element. Each interconnect wrapper has a constant number of inputs and outputs on each of its sides which are connected to the inputs and outputs of neighbor IW instances. An arbitrary massively parallel processor array can be defined as a tuple $A = (P, IW)$ with an array of PEs $P = \parallel p_{ij} \parallel$ and an array of interconnect-wrappers $IW = \parallel iw_{ij} \parallel$, $i = [1, M]$, $j = [1, N]$. Each processing element is represented by the sets of ingoing and outgoing ports $p = (p^{in}, p^{out})$, where $p^{in} = \{p_1^{in}, \dots, p_f^{in}\}$, $p^{out} = \{p_1^{out}, \dots, p_o^{out}\}$. An interconnect wrapper is represented as a rectangle around a PE and consists of the input and output ports on the northern, eastern, southern, and western side of it as shown in Fig. 1 (a). However, we require the input ports and the output ports on the opposite sides of an IW (i.e., northern inputs and southern outputs) to have equal bitwidths. Also, the number of them must be the same. Introduction of this condition proves the correct interconnection between neighbor IW instances. The condition can be completely satisfied by the introduction of *directed interconnect channels*. Each directed interconnect channel represents a pair of one input and one output port on the opposite sides of the interconnect wrapper with a certain common bitwidth. The direction of the channel is determined by the position of the output port. For example, if we consider a pair of *northern input* and *southern output* IW ports, then the direction of corresponding interconnect channel is southward. The interconnection between all interconnect-wrappers of the processor array is correct if and only if each IW has equivalently directed interconnect channels. An interconnect-wrapper is defined by a quintuple $iw = (CS, CN, CE, CW, C)$ with southward channels $CS = \{cs_1, \ldots, cs_s\}$, northward channels $CN = \{cn_1, \ldots, cn_n\}$, eastward channels $CE = \{ce_1, \ldots, ce_e\}$, and westward channels $CW = \{cw_1, \ldots, cw_w\}$. The configuration of an interconnect-wrapper is defined by $C = \| c_{ij} \|$. C is the so-called *interconnect adjacency matrix* (IAM) (see Fig. 1 (b)). By the *configuration of an IW*, the definition of the mapping of the possible connections between the ports of an interconnect wrapper and a processor element is meant. Therefore, the particular ports of an IW should be considered instead of interconnect channels (the pair of ports). The rows of IAM represent the input ports of an IW, except the last few rows (dependent on the number of the PEs output ports) which represent the output ports of the PE. The columns represent the output ports of an IW, except the last few columns (dependent on the number of the PEs input ports) which represent the input ports of the PE. The matrix contains the values c_{ij} which are equal to "1" if there exists a possible connection between input and output ports and equal to "0" otherwise. The last rows and columns of IAM represent the port mapping between PE and IW ports. The positions of input PE ports are interchanged with the positions of the output PE ports in the IAM. This allows to avoid the configuration of such incorrect connections as a connection between IW input and PE output or a connection between PE input and IW output. If many input signals are allowed to drive a single output signal a *multiplexer* (see Definition 1) with appropriate number of input signals is generated.

Definition 1 (Multiplexer). *A multiplexer f is a boolean function where $f : \mathbb{B}^{n+m} \to \mathbb{B}$ with inputs $I_f = \{x_0 \ldots x_{n-1}\}$ and control signals $S_f = \{s_0 \ldots s_{m-1}\}$ (with $m = \lceil log_2(n) \rceil$ and n inputs): $f(x_0 \ldots x_{n-1}, s_0 \ldots s_{m-1}) = x_{u(s_0 \ldots s_{m-1})}$, where $u(s_0 \ldots s_{m-1})$ is the interpretation of the positive binary number $s_0 \ldots s_{m-1}$.*

The inputs of a multiplexer are connected to the corresponding IW input signals and the output to the corresponding IW output signal. The control signals for such generated multiplexers are stored in configuration registers and can therefore be changed dynamically. By changing the values of the configuration registers in an interconnect-wrapper component, different interconnect topologies can be implemented and changed dynamically at run-time. Configuration registers and reconfiguration mechanisms are not in the scope of this paper.

In order to be able to compare the interconnect-wrapper concept to others, a parallel processor architecture with IW-based interconnect is represented as a directed graph $G^{IW}(V, E)$. Initially, the processor architecture is given by a *netlist* (see, Definition 2).

Definition 2 (Netlist). *A netlist $N = (V, F)$ is a set V of logic elements and a set F of nets interconnecting the elements $v \in V$ with each other. It defines a circuit in terms of*

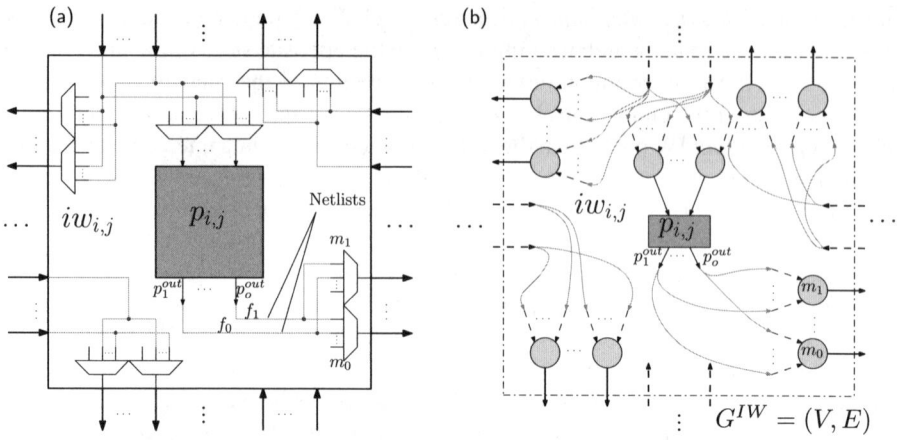

Fig. 2. Representation of parallel processor architecture netlist. In (a), a netlist of IW $iw_{i,j}$, and in (b), its netgraph representation.

basic logic elements. A unidirectional[1] net $f \in F$, which interconnects $n + m$ elements will be represented through $f = (\{v_1, \ldots, v_n\}, \{u_1, \ldots, u_m\})$, where $v_1, \ldots, v_n \in V$ are source nodes and $u_1, \ldots, u_m \in V$ are target nodes of net f.

In Fig. 2 (a), a netlist of interconnect-wrapper $iw_{i,j}$ is shown. For example, net f_0 is given by $f_0 = (\{p_1^{out}\}, \{m_0, m_1\})$. Nodes named p_i^{out} denote output ports of processing elements whereas nodes named m_i denote the multiplexer elements.

A netlist can be seen as a hypergraph, where the elements and nets are vertices and edges, respectively. This hypergraph can be transformed into a graph by the introduction of a *netgraph* concept.

Definition 3 (Netgraph). *A netgraph $G^{IW} = (V, E)$, $E \subseteq V \times V$ is a directed graph containing two disjoint sets of vertices $V = V_p \cup V_m$, representing the processing elements or processor output ports V_p and multiplexer elements V_m of a given netlist $N = (V, F)$. Netlist interconnections are represented by directed edges $e = (v_1, v_2) \in E$.*

In Fig. 2, an IW netlist and its netgraph $G^{IW} = (V, E)$ are shown. The subset of multiplexer elements $V_m = \{m_0, m_1, \ldots\}$ is shown as circles and the subset of processing elements $V_p = \{p_{i,j}\}$ is represented by rectangles. In order to transform a given netlist N into a netgraph G all elements of the netlist must be analyzed first and, according to the element's type (processing element or multiplexer), they are either included in the subset V_p or in the subset V_m. All of the nets are represented as directed edges $e \in E$ of a netgraph G^{IW}. In case when a net contains a $n : m$ connection, it is transformed into $n \times m$ directed edges of the netgraph, (in Fig. 2 (a), for f_0 the case of a $1 : 2$ connection is represented which is transformed into edges (p_1^{out}, m_0) and (p_1^{out}, m_1) of the graph G^{IW} in Fig. 2 (b)).

---

[1] A bidirectional net can be modeled by two unidirectional nets.

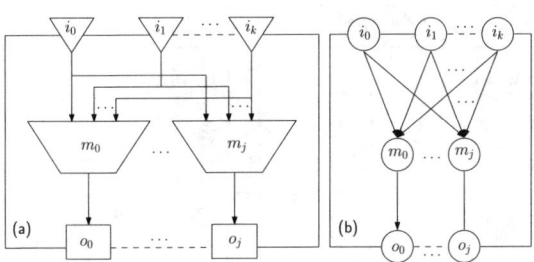 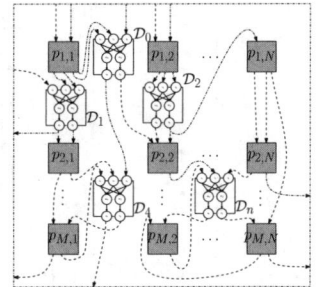

Fig. 3. DyRIBox with multiplexer and node view

Fig. 4. Example of a heterogeneous architecture using DyRIBoxes for interconnect modeling

2.2 Modeling of Interconnection Network Using DyRIBoxes

The second methodology to model reconfigurable parallel processor networks is used to permit the centralized description of dynamically reconfigurable processors. In this purpose, the interconnections have to be flexible (to be able to modify the connections "on the fly"), parameterizable in reconfiguration time and size. In this section, we give a brief description of the "**Dy**namically **R**econfigurable **I**nterconnections **Box**" (DyRIBox) concept which assumes all these constraints.

The basic elements of a DyRIBox are multiplexers (see Definition 1) connected in a way that any input can be connected to the desired output.

The DyRIBox formal description based on multiplexer formal description. A DyRIBox is a one level switching network of multiplexers. A DyRIBox $\mathcal{D}$ (see Fig. 3 (a)) can be described as an oriented acyclic graph $\mathcal{D} = (V, E)$, where V is a number of nodes and $E \subset V \times V$ is a number of oriented edges (see Fig. 3 (b)). The set of nodes V contains a set of inputs $\mathcal{I}(\mathcal{D})$, a set of outputs $\mathcal{O}(\mathcal{D})$, and a set of multiplexers $\mathcal{M}(\mathcal{D})$:

- $deg^-(v) = 0, deg^+(v) > 0$, when $v \in \mathcal{I}(\mathcal{D})$
- $deg^-(v) = 1, deg^+(v) = 0$, when $v \in \mathcal{O}(\mathcal{D})$
- $deg^-(v) = |\mathcal{I}(\mathcal{D})|, deg^+(v) = 1$, when $v \in \mathcal{M}(\mathcal{D})$,

where deg^- and deg^+ are input and output degrees of a node, respectively.

Each edge $e \in E$ which comes in a multiplexer node $v \in M(\mathcal{D})$ is an input $i_v(e) \in \{0, \dots deg^-(v) - 1\}$. A configuration of a DyRIBox is the following representation χ : $M(\mathcal{D}) \to \mathbb{N}$, with for all multiplexer nodes $v \in M(\mathcal{D}) : 0 \leq \chi(v) \leq deg^-(v) - 1$ and $\chi(v)$ is a configuration for v.

A route taken across a DyRIBox $\mathcal{D} = (V, E)$ is described as an oriented route $R = (v_0, e_0 \dots v_r, e_r, v_{r+1})$ in the graph (V, E) with $v_i \in V$ and $e_i \in E$ where $v_0 \in I(\mathcal{D}), v_{r+1} \in O(\mathcal{D})$ and for $i = \{1 \dots r\}$ is $v_i \in M(\mathcal{D})$. The length of R is define as $l(R) \overset{def}{=} r$. The source of the route R is defined as $S(R) = v_0$ and the destination as $D(R) = v_{r+1}$. The route R is defined on a DyRIBox as soon as a χ exist with: $\forall i = \{1 \dots r\} \, S(\chi(v_i)) = v_{i-1}$ and $v_{r+1} = succ(v_r)$. The predecessor of v_i through the input of $\chi(v_i)$ is v_{i-1}. The

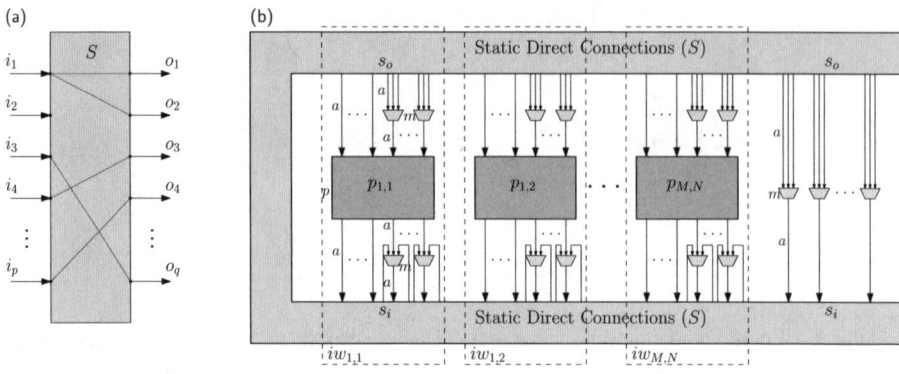

Fig. 5. Parallel processor architecture general case representation. In (a), an example of a static direct connector is shown, in (b) the general case of interconnect-wrapper interconnection is presented.

configuration K_R from the route R is defined as all configurations from the DyRI-Box $\mathcal{D}$. This provides R to be configurable on $\mathcal{D}$. It is also possible to configure two routes R_1 and R_2 on one DyRIBox $\mathcal{D}$ using two separate multiplexer nodes with $R_1 = (v_0, e_0 \ldots v_r, e_r, v_{r+1})$ and $R_2 = (v'_0, e'_0 \ldots v'_s, e'_s, v'_{s+1})$. R_1 and R_2 are compatible when the two nodes are completely disconnected: $v_i \neq v'_j$ ($\forall i \in \{1 \ldots r\}$ and $\forall j \in \{1 \ldots s\}$).

A massively parallel processor array can be described as an oriented acyclic graph $\mathcal{A} = (\mathcal{D}, P, E)$, where E is the set of oriented edges, $\mathcal{D}$ is the set of DyRIBoxes used, and P is the set of PEs with $\mathcal{D} = \|\mathcal{D}_n\|$ ($n \in \mathbb{N}^+$) and $P = \|\mathcal{P}_{ij}\|$ ($\forall i \in \{1 \ldots M\}$ and $\forall j \in \{1 \ldots N\}$). Considering DyRIBox inputs $\mathcal{I}(\mathcal{D})$, DyRIBox outputs $\mathcal{O}(\mathcal{D})$, PE inputs $\mathcal{I}(\mathcal{P})$, and PE outputs $\mathcal{O}(\mathcal{P})$, all the following connections are feasible (in Fig. 4): $e = (v \in \mathcal{O}(\mathcal{D}), v \in \mathcal{I}(\mathcal{D}))$, $e = (v \in \mathcal{O}(\mathcal{D}), v \in \mathcal{I}(\mathcal{P}))$, $e = (v \in \mathcal{O}(\mathcal{P}), v \in \mathcal{I}(\mathcal{D}))$, and $e = (v \in \mathcal{O}(\mathcal{P}), v \in \mathcal{I}(\mathcal{P}))$ with $e \in E$.

2.3 Comparison of Interconnect-Wrapper and DyRIBox Concepts

In the following, we compare both interconnection concepts. In fact, the DyRIBox concept allows for modeling of all possible interconnect topologies ranging from linear arrays, meshes, multistage hierarchical networks, butterflies, trees, etc. A special case of a DyRIBox network is a full crossbar, when the number of DyRIBox inputs and outputs is equal. In this section, we prove that interconnect-wrapper and DyRIBox interconnect definitions are equivalent.

In order to generalize the parallel processor architecture with configurable interconnect and to separate the dynamically reconfigurable interconnection from static connections we introduce an interconnect object called *static direct connector* (SDC).

Definition 4 (Static Direct Connector). *A static direct connector (SDC) $S = (S^{in}, S^{out})$ is an interconnection black-box or a tuple with an infinite set of input ports $S^{in} = \{i_1, \ldots i_p\}$ and infinite set of output ports $S^{out} = \{o_1, \ldots o_q\}$, where $p \to \infty$, $q \to \infty$. The input ports are statically connected to the output ports inside of SDC.*

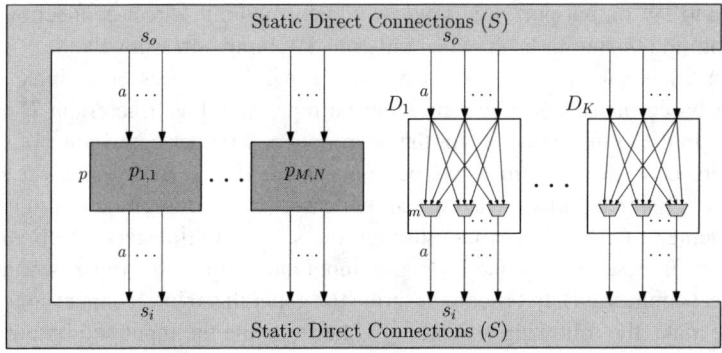

Fig. 6. A general case of DyRIBox Interconnection

$$(a)$$
$$L^{IW}(G^{IW})$$

$$(b)$$
$$L^{DB}(G^{DB})$$

$G^{IW} = (\ \{S,A,M\}, \{s_i, s_o, a, p, m\}, P^{IW}, S)$

$G^{DB} = (\ \{D,E,F\}, \{s_i, s_o, a, p, m\}, P^{DB}, D)$

$P^{IW} = \{\ S \to s_o As_i S,$

$P^{DB} = \{\ D \to s_o Es_i D,$

$\qquad S \to s_o ApAs_i S,$

$\qquad E \to a,$

$\qquad A \to a,$

$\qquad E \to aFa,$

$\qquad A \to aMa,$

$\qquad F \to m,$

$\qquad M \to m,$

$\qquad F \to p,$

$\qquad S \to \varepsilon\}$

$\qquad D \to \varepsilon\}$

Fig. 7. Representation of general case interconnect as formal language. In (a), a definition of formal language $L^{IW}(G^{IW})$ for IW is shown, in (b), a definition of formal language $L^{DB}(G^{DB})$ for DyRIBox is presented.

An SDC provides a formal description of a static, i.e. not configurable, interconnection network. A graphical representation of an SDC is shown in Fig. 5 (a).

As shown in Section 2.1, a generic massively parallel processor architecture with IW-based interconnect structure can be represented as a netgraph $G^{IW} = (V, E)$ (see Definition 3). Due to Definition 4, all edges E of this directed graph can be represented as an SDC. Therefore, a generic parallel processor architecture with IW-based interconnect topology can be considered as shown in Fig. 5 (b). There are only few types of routes or signal paths through the interconnect-wrappers $\{iw_{1,1}, \ldots, iw_{M,N}\}$ possible. All of them are shown in Fig. 5 (b): (i), a direct route through the processing element inside of the IW, (ii), a route through the multiplexer directly connected to one of the PE inputs (this is the case, when more than one line are connected to the input of the PE), (iii), a route through the PE inside of the IW, where one of the PE outputs is directly connected to the multiplexer (this is the case, when more than one line are connected to one of the IW output ports), (iv), a route through the multiplexer directly connected to one of the PE inputs, where one of the PE outputs is directly connected to another multiplexer, and (v), a direct route through the multiplexer inside of the IW not passing the processing element through (this is the case, when IW input ports are directly connected to IW output port; this route is transformed into direct connection without passing through the multiplexer, when only one IW input port is involved).

Section 2.2 shows that a generic massively parallel processor architecture with DyRIBox-based interconnect structure can be represented as a netgraph $\mathcal{D} = (V, E)$. Therefore, in the same manner as for the interconnect-wrapper-based concept, an architecture with DyRIBox definition can be represented with an SDC as shown in Fig. 6. The following signal paths are shown in this case: (i), a direct route through the processing element, (ii), a direct route through the set of multiplexers which compound a DyRIBox (in case of only one DyRIBox input and output, this route is transformed into a direct connection). In order to describe the generalized interconnect concepts formally, we make the following assumptions: Let s_i, s_o be the input and output ports of the static direct connector, respectively. Let a be a direct connection, let p be a route through the PE, and let m be a route through the multiplexer.

Let G^{IW} and G^{DB} be formal grammars [8]. Now, the definitions of IW and DyRIBox interconnect can be considered as two formal languages $L^{IW}(G^{IW})$ and $L^{DB}(G^{DB})$, respectively [8]. The definition of the formal languages is given in Fig. 7. The grammar G^{IW} for language L^{IW} is defined by the finite set $\{S, A, M\}$ of nonterminal symbols, by the alphabet of terminal symbols $\{s_i, s_o, a, p, m\}$, and by the set P^{IW} of production rules with start symbol S. The nonterminals S, A, and M represent the possible signal paths through the interconnect-wrapper. The grammar G^{DB} for language L^{DB} is defined by the finite set $\{D, E, F\}$ of nonterminal symbols, by the alphabet of terminal symbols $\{s_i, s_o, a, p, m\}$, and by the set P^{DB} of production rules with start symbol D. The nonterminals D, E, and F represent the possible signal paths through the DiRIBox. An ε in both grammars denotes the empty string, i.e., the string of length 0 such that $\varepsilon \notin L^{IW} \wedge \varepsilon \notin L^{DB}$. The grammars G^{IW} and G^{DB} are context-free grammars, as the left hand sides of a their production rules are formed by only a single non-terminal symbol. Moreover, both of the grammars have the same alphabet of terminal symbols $\{s_i, s_o, a, p, m\}$. Therefore, we will prove the equivalence of modeling power of both concepts by proving language equivalence.

Theorem 1. *The PE interconnection networks based on IW are equivalent to the PE interconnection networks based on DyRIBox:* $L^{IW}(G^{IW}) \equiv L^{DB}(G^{DB})$.

Proof. Two formal languages are equivalent if their grammars are equivalent. The equivalence of the grammars must be shown in both directions $L^{DB}(G^{DB}) \subseteq L^{IW}(G^{IW}) \wedge L^{IW}(G^{IW}) \subseteq L^{DB}(G^{DB})$ in order to prove the equivalence of the languages.

First, the relation $L^{DB}(G^{DB}) \subseteq L^{IW}(G^{IW})$ will be shown. Obviously, the following production rules are similar: $M \to m \sim F \to m$, $A \to a \sim E \to a$, $A \to aMa \sim E \to aFa$, $S \to \varepsilon \sim D \to \varepsilon$, and $S \to s_oAs_iS \sim D \to s_oEs_iD$. The similarity of production rules $S \to s_oApAs_iS$ and $F \to p$ remains questionable. Consider the production rule $S \to s_oApAs_iS$. After applying of a rule $A \to aMa$ on it, the derivation $S \to s_oaMapaMas_iS$ occurs. Let's introduce the equivalent transformation $T^{IW} : a \simeq as_is_oa$. This transformation is equivalent because it physically means just a splitting of a direct connection a into two direct connections which are connected through the input (s_i) and output (s_o) ports of an SDC (see Fig. 8). Applying the transformation T^{IW} on the terminals a staying next to the terminal symbol p, the following production rule is obtained: $S' = T^{IW}(S \to s_oaMapaMas_iS) : S' \to s_oaMas_is_oapas_is_oaMas_iS$. The subsequence of

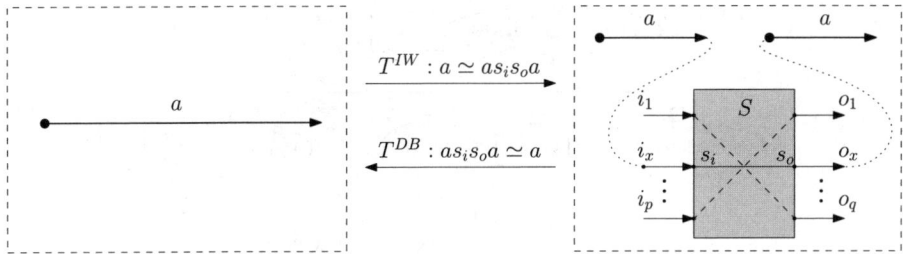

Fig. 8. The equivalent transformations $T^{IW} : a \simeq as_i s_o a$ and $T^{DB} : as_i s_o a \simeq a$ are presented

symbols $s_o a M a s_i$ in S' occurs twice and corresponds to the rule $S \to s_o A s_i S$ which is a start symbol of G^{IW}. Therefore, L^{IW} will not be changed by the elimination of this sub-sequence. Thus, S' takes the following form: $S' \to s_o a p a s_i S$. We get the following set of production rules: $\{S \to s_o A s_i S, A \to a, A \to a M a, M \to m, S \to \varepsilon, S' \to s_o a p a s_i S\}$. After integration of S' into S the set of production rules can be rewritten: $\{S' \to s_o A' s_i S', A' \to a, A' \to a M' a, M' \to m, M' \to p, S' \to \varepsilon\} \simeq P^{DB}$. Thus, $L^{DB}(G^{DB}) \subseteq L^{IW}(G^{IW})$.

Now, the relation $L^{IW}(G^{IW}) \subseteq L^{DB}(G^{DB})$ will be shown. Obviously, the following production rules are similar: $F \to m \sim M \to m$, $E \to a \sim A \to a$, $E \to aFa \sim A \to aMa, D \to \varepsilon \sim S \to \varepsilon$, and $D \to s_o E s_i D \sim S \to s_o A s_i S$. The similarity of production rules $F \to p$ and $S \to s_o A p A s_i S$ remains questionable. Let's include the production rule $D \to s_o E s_i s_o E s_i s_o E s_i D$ in P^{DB}. Language L^{DB} is not changed by this operation because $D \to s_o E s_i s_o E s_i s_o E s_i D \equiv D \to s_o E s_i D$. This is an obvious derivation after three times applying the recursive production rule $D \to s_o E s_i D$. After applying the rule $E \to aFa$ on the new included production rule, we obtain $D \to s_o a F a s_i s_o a F a s_i s_o a F a s_i D$. Let's introduce the equivalent transformation $T^{DB} : as_i s_o a \simeq a$. This transformation is equiv-alent because it physically means replacing of a connection through the input (s_i) and output (s_o) ports of SDC with a direct connection a (see Fig. 8). Applying the trans-formation T^{DB} on the subsequences of terminals $as_i s_o a$, the following production rule is obtained: $D' = T^{DB}(D \to s_o a F a s_i s_o a F a s_i s_o a F a s_i D) : D' \to s_o a F a F a F a s_i D$. The sub-sequence of symbols aFa can be replaced by the nonterminal symbol E using the pro-duction rule $E \to aFa$: $D' \to s_o EFE s_i D$. Applying the rule $F \to p$, D' can be rewritten: $D' \to s_o E p E s_i D$. We get the following set of production rules: $\{D \to s_o E s_i D, E \to a, E \to aFa, F \to m, F \to p, D \to \varepsilon, D' \to s_o E p E s_i D\}$. After integration of D' into D the set of production rules can be rewritten: $\{D' \to s_o E' s_i D', E' \to a, E' \to aF'a, F' \to m, D' \to s_o E' p E' s_i D', D' \to \varepsilon\} \simeq P^{IW}$. Thus, $L^{IW}(G^{IW}) \subseteq L^{DB}(G^{DB})$. □

According to Theorem 1, the set of interconnection networks that can be expressed us-ing the interconnect-wrapper definition (distributed interconnection concept) is equivalent to those that can be modeled using the DyRIBox definition (centralized in-terconnection concept).

3 Modeling of Interconnect Structures Within MAML

In order to allow the specification of reconfigurable interconnect topologies in mas-sively parallel processor architectures we use the *MAchine Markup Language* (MAML)

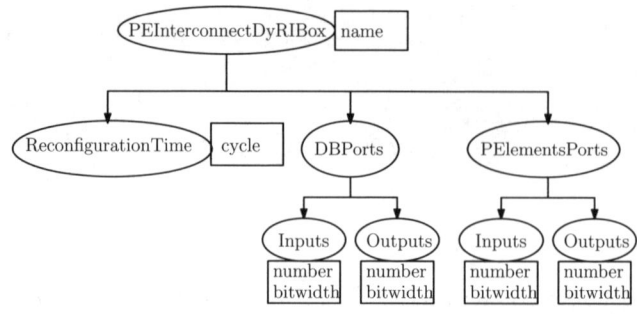

Fig. 9. The <PEInterconnectDyRIBox> elements

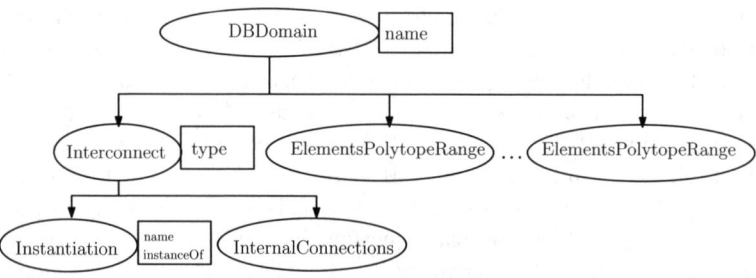

Fig. 10. The <DBDomain> elements

[4, 10]. MAML is based on the XML notation and is used for describing architecture parameters required by possible mapping methods such as partitioning, scheduling, functional unit and register allocation. Moreover, the parameters extracted from a MAML architectural description can be used for interactive visualization and simulation of the given processor architecture. For the MAML description of interconnect-wrappers we refer to [9, 10].

MAML description of DyRIBox and DyRIBox domain. The structure of the DyRI-BOX element description is shown in Fig. 9. In MAML, a DyRIBox is represented by the <PEInterconnectDyRIBox> element. It contains the attribute name which specifies the name of the DyRIBox which is going to be described. It also contains the following set of description subelements: <ReconfigurationTime>, <DBPorts>, and <PElementsPorts>. The subelement <ReconfigurationTime> specifies the number of cycles needed to dynamically reconfigure the complete DyRIBox. This subelement is initialized by giving a value to the cycle parameter. The subelement <DBPorts> specifies the number of ports directly connected to others DyRIBoxes. The number of Inputs, Outputs and their bitwidth are specified by the attributes. The subelement <PElementsPorts> specifies the number of ports directly connected to the ports of processing elements. The number of Inputs, Outputs and their bitwidth are specified by the attributes.

The DyRIBox domain description. A DyRIBox domain <DBDomain> is typically used for describing the interconnections of heterogeneous architectures. The structure of the DyRIBOX domain description is shown in Fig. 10. The element <DBDomain> is followed by the attribute name which specifies the name of the DyRIBox domain which is going to be described. It also contains a set of subelements:

– <Interconnect>
– <ElementsPolytopeRange>

<Interconnect> specifies how the interconnections are going to be described by giving a type. The type can be manual when all connections are explicitly described, or mesh or honeycomb or tree when the connections are automatically generated as a mesh form or honey comb form or tree form. In case that <Interconnect> is a manual type, the subelement <Instantiation> specifies which DyRIBox is being used and follows the subelement <InternalConnections> which describe the internal interconnections of the domain ports by ports one after the other.

<ElementsPolytopeRange> specifies a range of PEs in shape of a given polytope which belong to the domain which is going to be described. Concerning the interconnections with external resources of the DyRIBox domain, an element <PortMapping> is given at the end of the <ProcessorArray> description.

4   Case-Study

We implemented a highly parameterizable template for the generation of parallel processor arrays in VHDL (*Very High Speed Hardware Description Language*). An example 4×4 processor array was instantiated and tested on a *Xilinx Virtex-II Pro™ xc2vp100* FPGA. To reduce synthesis and mapping time, a very simple configuration was chosen for all processing elements in the array. Each WPPE was configured to contain two

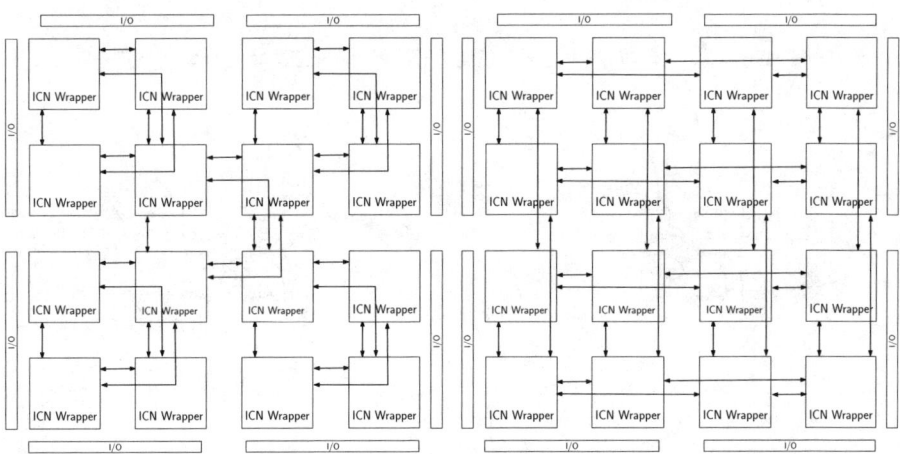

Fig. 11. BFT IW-interconnection scheme embedded in a 4×4 array

Fig. 12. 4D Hypercube IW-interconnection scheme embedded in a 4×4 array

Table 1. Synthesis results for different IW-based interconnection networks in a 4×4 processor array

	Mesh	4D Hypercube	BFT & 4D Hypercube
Equivalent Gates	859482	865941	875412
$\mathrm{MUX}_1^4(16)$	32	32	37
$\mathrm{MUX}_1^8(16)$	0	0	7
Logic LUT	30287	31724	32883
Route-Through LUT	302	244	140
Memory LUT	4768	4768	4768
FPGA Resources	50%	51%	52%

adder modules and one module for the transfer of control signals. The data path width was chosen to be 16 bit. Different interconnection schemes were implemented using interconnect-wrappers. The logical interconnection scheme for the *4D hypercube* topology is depicted in Fig. 12 and for the *butterfly fat tree (BFT)* topology in Fig. 11. The results are shown in Table 1. It can be seen from the synthesis results, that starting from the basic mesh interconnection scheme additional interconnection configurations can be added at relative little hardware cost of roughly 2% in the case of the *4D hypercube* topology. In the case of the 4D hypercube with the additional butterfly fat tree topology the hardware cost amounts to roughly 1%.

We also synthesized the same interconnection topologies for a 4×4 processor array using centralized modeling approach DyRIBox (see Table 2). The logical interconnection scheme for the *4D hypercube* topology is depicted in Fig. 14 and for the *butterfly fat tree (BFT)* topology in Fig. 13. In order to analyze only the interconnection network cost, here a very simple PE architecture was implemented. Each PE contains only the registers on its input and output ports. In order to compare both approaches the IW interconnection cost was derived from the results in Table 1. The stand-alone 6-ports and 4-ports PEs were synthesized and their hardware costs were analyzed. The stand

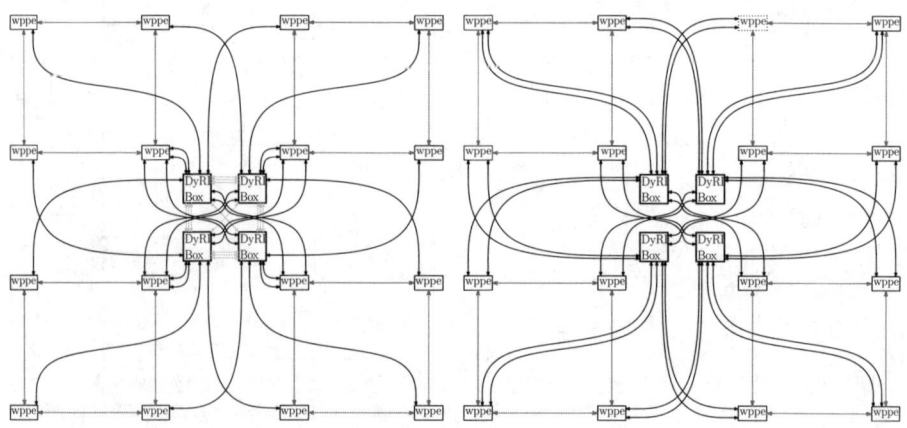

Fig. 13. Butterfly fat tree (BFT) DyRIBox-interconnection scheme

Fig. 14. 4D Hypercube DyRIBox-interconnection scheme

Table 2. Comparison distributed IW approach against centralized DyRIBox concept

	IW (equivalent gates)	DyRIBox (equivalent gates)
Mesh	119258	189051
4D Hypercube	125717	45556
BFT & 4D Hypercube	101848	103070
Intuitive	-	++
Uniform latency	-	+
Homogeneous architecture	+++	-

alone 6-ports PE consumed 54.599 equivalent gates, whereas 4-ports PE occupied only 46.264 equivalent gates. The PE hardware costs were subtracted from the hardware costs for the complete designs in Table 1. The DyRIBox interconnection hardware cost was compared against derived IW interconnection cost. The comparison results are shown in Table 2.

The centralized approach is a classical and convenient way to model the interconnection of the processing elements. In this case, the processors are connected to one or more central switch-nodes, which provide the reconfigurable interconnect. Another advantage of DyRIBox interconnection concept is an uniform latency. Synthesis results show more efficient DyRIBox implementation for *4D hypercube* topology. Whereas, a decentralized approach may be more reasonable in case of homogeneous architectures such as meshes or systolic structures.

5   Conclusions and Future Work

In this paper, we introduced a new concept for modeling of interconnection networks in the field of massively parallel processor embedded architectures. Two interconnection concepts, namely, *intercon-nect-wrapper* and *DyRIBox* definitions of reconfigurable interconnection were formally defined and compared against each other. The equivalence of distributed and centralized interconnection concepts was formally proved which also proved the ability of the *interconnect-wrappers* to efficiently model many possible interconnect topologies. Moreover, we demonstrated the pertinence of our approach by synthesis of a real 4×4 processor array with different reconfigurable interconnect topologies in a case-study.

In the future, we would like to extend MAML in order to model different transport mechanisms like blocking FIFOs, asynchronous data-driven models, or even concepts such as routers in Networks-on-a-Chip (NoC). This would allow for architecture exploration of a vast class of processor arrays.

References

1. N. Bansal, S. Gupta, N. Dutt, A. Nicolau, and R. Gupta. Network Topology Exploration of Mesh-Based Coarse-Grain Reconfigurable Architectures. In *Proceedings Design Automation and Test in Europe (DATE'2004)*, pages 474–479, Paris, France, Feb. 2004.

2. V. Baumgarte, G. Ehlers, F. May, A. Nückel, M. Vorbach, and M. Weinhardt. PACT XPP – A Self-Reconfigurable Data Processing Architecture. *The Journal of Supercomputing*, 26(2):167–184, 2003.

3. Elixent Ltd. www.elixent.com.

4. D. Fischer, J. Teich, M. Thies, and R. Weper. Design Space Characterization for Architecture/Compiler Co-Exploration. In *ACM SIG Proceedings International Conference on Compilers, Architectures and Synthesis for Embedded Systems (CASES 2001)*, pages 108–115, Atlanta, GA, U.S.A., November 2001.

5. D. Fischer, J. Teich, M. Thies, and R. Weper. BUILDABONG: A Framework for Architecture/Compiler Co-Exploration for ASIPs. *Journal for Circuits, Systems, and Computers, Special Issue: Application Specific Hardware Design*, pages 353–375, 2003.

6. A. Halambi, P. Grun, A. Khare, V. Ganesh, N. Dutt, and A. Nicolau. EXPRESSION: A Language for Architecture Exploration through Compiler/Simulator Retargetability. In *Proceedings Design Automation and Test in Europe (DATE'1999)*, 1999.

7. R. Hartenstein. A Decade of Reconfigurable Computing: A Visionary Retrospective. In *Proceedings of Design, Automation and Test in Europe*, pages 642–649, Munich, Germany, Mar. 2001. IEEE Computer Society.

8. J. Hopcroft. *Introduction to Automata Theory, Languages and Computation*. Addison-Wesley Series in Computer Science. Addison Wesley Publishing Company, older edition, Apr. 1979.

9. A. Kupriyanov, F. Hannig, D. Kissler, R. Schaffer, and J. Teich. MAML - An Architecture Description Language for Modeling and Simulation of Processor Array Architectures, Part I. Technical Report 03-2006, University of Erlangen-Nuremberg, Department of Computer Science, Hardware-Software-Co-Design, Mar. 2006.

10. A. Kupriyanov, F. Hannig, D. Kissler, J. Teich, R. Schaffer, and R. Merker. An Architecture Description Language for Massively Parallel Processor Architectures. In *GI/ITG/GMM-Workshop 2006 - Methoden und Beschreibungssprachen zur Modellierung und Verifikation von Schaltungen und Systemen*, pages 11–20, Dresden, Germany, Feb. 2006.

11. J. Lee, K. Choi, and N. Dutt. An Algorithm for Mapping Loops onto Coarse-grained Reconfigurable Architectures. In *Languages, Compilers, and Tools for Embedded Systems (LCTES'03)*, pages 183–188, San Diego, CA, June 2003. ACM Press.

12. B. Mei, A. Lambrechts, D. Verkest, J. Mignolet, and R. Lauwereins. Architecture Exploration for a Reconfigurable Architecture Template. In *IEEE Design and Test of Computers*, pages 90–101, Mar. 2005.

13. M. Motomura. A Dynamically Reconfigurable Processor Architecture. In *Microprocessor Forum*, CA, 2002.

14. S. Pees, A. Hoffmann, and H. Meyr. Retargeting of Compiled Simulators for Digital Signal Processors Using a Machine Description Language. In *Proceedings Design Automation and Test in Europe (DATE'2000)*, Paris, March 2000.

15. Silicon Hive. www.siliconhive.com.

16. Trimaran. http://www.trimaran.org.

Invited Talk: Expanding Software Product Families: From Integration to Composition

Jan Bosch

Nokia, Technology Platforms/Software Platforms,
P.O. Box 407, FI-00045 NOKIA GROUP, Finland
Jan.Bosch@nokia.com
http://www.janbosch.com

Abstract. Software product families have found broad adoption in the embedded systems industry. Product family thinking has been prevalent in this context for mechanics and hardware and adopting the same for software has been viewed as a logical approach. During recent years, however, the trends of convergence, end-to-end solutions, shortened innovation and R&D cycles and differentiation through software engineering capabilities have lead to a development where organizations are stretching the scope of their product families far beyond the initial design. Failing to adjust the product family approach, including the architectural and process dimensions when the business strategy is changing is leading to several challenging problems that can be viewed as symptoms of this approach. This paper discusses the key symptoms, the underlying causes for these symptoms as well as solutions for realigning the product family approach with the business strategy.

Keywords: Software product families, compositionality.

1 Introduction

Mobile phones have, over the last decade, evolved from basic devices aimed primarily at the voice and SMS use cases to a rich set of mobile devices ranging from mobile multi-media computers to mobile enterprise devices. Contemporary mobile devices support a rich set of use cases including taking still and video pictures, playing music, watching television, reading email, instant messaging, navigating, etc. In response to the enormous increase in the features required from mobile devices, the demands on the software present in the mobile device have increased similarly. The developments in the mobile devices industry are illustrative examples of the general trend in embedded systems: the investment in software R&D has increased by an order of magnitude during the last decade.

One can identify three main trends that are driving the embedded systems industry, i.e. convergence, end-to-end functionality and software engineering capability. The convergence of the consumer electronics, telecom and IT industries has been discussed for over a decade. Although many may wonder whether and when it will happen, the fact is that the convergence is taking place constantly. Different from what the name may suggest, though, convergence in fact leads to a portfolio of

P. Lukowicz, L. Thiele, and G. Tröster (Eds.): ARCS 2007, LNCS 4415, pp. 283–295, 2007.

increasingly diverging devices. For instance, in the mobile telecom industry, mobile phones have diverged into still picture camera models, video camera models, music player models, mobile TV models, mobile email models, etc. This trend results in a significant pressure on software product families as the amount of variation to be supported by the platform in terms of price points, form factors and feature sets is significantly beyond the requirements just a few years ago. The second trend is that many innovations that have proven their success in the market place require the creation of an end-to-end solution and possibly even the creation or adaptation of a business eco-system. Examples from the mobile domain include, for instance, ring tones, but the ecosystem initiated by Apple around digital music is exemplary in this context. The consequence for most companies is that where earlier, they were able to drive innovations independently to the market, the current mode requires significant partnering and orchestration for innovations to be successful. The third main trend is that a company's ability to engineer software is rapidly becoming a key competitive differentiator. The two main developments underlying this trend are efficiency and responsiveness. With the constant increase in software demands, the cost of software R&D is becoming unacceptable from a business perspective. Thus, some factor difference in productivity is easily turning into being able or not being able to deliver certain feature sets. Responsiveness is growing in importance because innovation cycles are moving increasingly fast and customers are expecting constant improvements in the available functionality. Web 2.0 [O'Reilly 05] presents a strong example of this trend. A further consequence for embedded systems is that, in the foreseeable future, the hardware and software innovation cycles will, at least in part, be decoupled, significantly increasing demands for post-deployment distribution of software.

Due to the convergence trend, the number of different embedded products that a manufacturer aims to bring to market is increasing. Consequently, reuse of software (as well as of mechanical and hardware solutions) is a standing ambition for the industry. The typical approach employed in the embedded systems industry is to build a platform that implements the functionality common to all devices. The platform is subsequently used as a basis when creating new product and functionality specific to the product is built on top of the platform.

Although the platform model is easy to understand in theory, in practice there are significant challenges. As discussed in [Bosch 00], the platform model is supposed to capture the most generic and consequently the least differentiating functionality. Innovations and differentiating functionality are supposed, over time, to flow from product-specific implementations to the platform. However, in practice one can identify at least two forces that drive innovations directly to the platform. First, in several cases, it is clear that a novel feature or innovation will be required by all or most future products. In this case, there is a clear rationale to implement the new feature or innovation directly in the platform, bypassing the product-specific phase. Second, in the case that a company also licenses the platform to other organizations in the industry, the platform itself needs to be differentiating and contain sufficient novel features to hold or expand its position in the face of competition from other platforms.

Second, the product specific functionality frequently does not respect the boundary between the platform and the software on top of it. Innovations in embedded systems can originate from mechanics, hardware or software. Both mechanical and hardware innovations typically have an impact on the software stack. However, due to the fact that the interface to hardware is placed in device drivers at the very bottom of the stack and the affected applications and their user interface are located at the very top of the stack, changes to mechanics and hardware typically have a cross-cutting effect that causes changes in many places both below and above the platform boundary. A second source of cross-cutting changes is software specific. New products often enable new use cases that put new demands on the software that can not be captured in a single component or application, but rather have architectural impact. Examples include adding security, a more advanced user interface framework or a web-services framework. Such demands result in cross-cutting changes that affect many places in the software, again both above and below the platform boundary.

Software product families have, in many cases, been very successful for the companies that have applied them. Due to their success, however, during recent years one can identify a development where companies are stretching their product families significantly beyond their initial scope. This occurs either because the company desires to ship a broader range of products due to, among others, convergence, or because the proven success of the product family causes earlier unrelated products to be placed under the same family. This easily causes a situation where the software product family becomes a victim of its own success. With the increasing scope and diversity of the products that are to be supported, the original integration-oriented platform approach increasingly results in several serious problems in the technical, process, organizational and, consequently, the business dimension.

The purpose and contribution of this paper is that it analyses the aforementioned problems and to present alternative approaches that are better suited for broad-scoped product families. Both the problem analysis and the proposed alternative approaches are based on experiences from a variety of companies that the author has worked with during the last decade. However, for reasons of confidentiality, no specific references can be provided at this point.

In the remainder of this article, we first present a more detailed assessment of the problems and challenges associated with the traditional, platform-based, integration oriented approach. Subsequently, we discuss five aspects of software product families that are most relevant when broadening the scope of a product family in section 3. We then proceed with presenting two alternative approaches, i.e. hierarchical product families and the compositional-oriented approach, in section 4. Finally, related work is described in section 5 followed by the conclusions of the paper.

2 Problem Statement

This paper discusses and presents the concerns of the integration-oriented platform approach. However, before we can discuss this, we need to first define integration-oriented platform approach more precisely. In most cases, the platform approach is organized using a strict separation between the platform organization and the product organizations. The platform organization has typically a periodic release cycle where the complete platform is released in a fully integrated and tested fashion. The product organizations use the platform as a basis for creating and evolving theirs product by extending the platform with product-specific features.

The platform organization is divided in a number of teams, in the best case mirroring the architecture of the platform. Each team develops and evolves the component (or set of related components) that it is responsible for and delivers the result for integration in the platform. Although many organizations have moved to applying a continuous integration process where components are constantly integrated during development, in practice significant verification and validation work is performed in the period before the release of the platform and many critical errors are only found in that stage.

The platform organization delivers the platform as a large, integrated and tested software system with an API that can be used by the product teams to derive their products from. As platforms bring together a large collection of features and qualities, the release frequency of the platform is often relatively low compared to the frequency of product programs. Consequently, the platform organization often is under significant pressure to deliver as many new features and qualities during the release. Hence, there is a tendency to short-cut processes, especially quality assurance processes. Especially during the period leading up to a major platform release, all validation and verification is often transferred to the integration team. As the components lose quality and integration team is confronted with both integration problems and component-level problems, in the worst case an interesting cycle appears where errors are identified by testing staff that has no understanding of the system architecture and can consequently only identify symptoms, component teams receive error reports that turn out to originate from other parts in the system and the integration team has to manage highly conflicting messages from the testing and development staff, leading to new error reports, new versions of components that do not solve problems, etc.

Although several software engineering challenges associated with software platforms have been outlined, the approach often proves highly successful in terms of maximizing R&D efficiency and cost-effectively offering a rich product portfolio. Thus, in its initial scope, the integration-oriented platform approach has often proven itself as a success. However, the success can easily turn into a failure when the organization decides to build on the success of the initial software platform and significantly broadens the scope of the product family. The broadening of the scope can be the result of the company deciding to bring more existing product categories under the platform umbrella or because it decides to diversify its product portfolio as the cost of creating new products has decreased considerably. At this stage, we have identified in a number of companies that broadening the scope of the software product family without adjusting the mode of operation quite fundamentally leads to a number of key concerns and problems that are logical and unavoidable. However, because of the earlier success that the organization has experienced, the problems are insufficiently identified as fundamental, but rather as execution challenges, and fundamental changes to the mode of operation are not made until the company experiences significant financial consequences.

The problems and their underlying causes that one may observe when the scope of a product family is broadened considerably over time include, among others, those described below:

- **Decreasing complete commonality:** Before broadening the scope of the product family, the platform formed the common core of product functionality. However, with the increasing scope, the products are increasingly diverse in their requirements and amount of functionality that is required for all products is decreasing, in either absolute or relative terms. Consequently, the (relative) number of components that is shared by all products is decreasing, reducing the relevance of the common platform.

- **Increasing partial commonality:** Functionality that is shared by some or many products, though not by all, is increasingly significantly with the increasing scope. Consequently, the (relative) number of components that is shared by some or most products is increasing. The typical approach to this model is the adoption of hierarchical product families. In this case, business groups or teams responsible for certain product categories build a platform on top of the company wide platform. Although this alleviates part of the problem, it does not provide an effective mechanism to share components between business groups or teams developing products in different product categories.

- **Over-engineered architecture:** With the increasing scope of the product family, the set of business and technical qualities that needs to be supported by the common platform is broadening as well. Although no product needs support for all qualities, the architecture of the platform is required to do so and, consequently, needs to be over-engineered to satisfy the needs of all products and product categories.

- **Cross–cutting features:** Especially in embedded systems, new features frequently fail to respect the boundaries of the platform. Whereas the typical approach is that differentiating features are implemented in the product (category) specific code, often these features require changes in the common components as well. Depending on the domain in which the organization develops products, the notion of a platform capturing the common functionality between all products may easily turn into an illusion as the scope of the product family increases.

- **Maturity of product categories:** Different product categories developed by one organization frequently are in different phases of the lifecycle. The challenge is that, depending on the maturity of a product category, the requirements on the common platform are quite different. For instance, for mature product categories cost and reliability are typically the most important whereas for product categories early in the maturity phase feature richness and time-to-market are the most important drivers. A common platform has to satisfy the requirements of all product categories, which easily leads to tensions between the platform organization and the product categories.

- **Unresponsiveness of platform:** Especially for product categories early in the maturation cycle, the slow release cycle of software platforms is particularly frustrating. Often, a new feature is required rapidly in a new product. However, the feature requires changes in some platform components. As the platform has a slow release cycle, the platform is typically unable to respond to the request of the product team. The product team is willing to implement this functionality itself, but the platform team is often not allowing this because of the potential consequences for the quality of the product team.

3   Five Dimensions of Product Families

In this section, we discuss the five dimensions that are of predominant importance in the context of broadening the scope of software product families, i.e. business strategy, architecture, components, product creation and evolution.

3.1   Business Strategy

Based on the author's experience in the domain of software product families, one can identify three predominant business strategies:

- **R&D minimization:** The first argument used by companies moving from product-specific to product family development of software is the reduction of R&D expenditure through the sharing of software artifacts between multiple products.
- **Time-to-market optimization:** Once the organization has successfully adopted a product family approach, the next argument often becomes the decreased time-to-market of new products as these products can share a significant amount of software.
- **Maximizing product family scope:** Once the organization is successfully bringing new products to market with agreeable R&D expenditure and time-to-market, there is often a drive to broaden the scope of the product family. This may be because the approach has proven its success in one product category and the organization is eager to build on this success by expanding it to other product categories. A second scenario is where the organization enjoys success in the market because of its adopted approach and is able to expand into new product categories. This article is concerned with the challenges and consequences of this third stage.

3.2   Architecture

In section 4.1, three, often consecutive, stages of business strategy are discussed. One can also identify four architectural approaches that are partially related to the stages discussed above.

- **Fixed structural architecture:** The first architectural approach, especially suitable for relatively narrow product families, is to specify a complete structural architecture as the basis for the product family. The architecture is the same for all products and variation is primarily captured through variation points in the components.
- **Micro-kernel architecture, optional elements:** An alternative, architecture centric approach is the combination of a micro-kernel, used for all products in the family, and significant set of optional elements that can be included, replaced or excluded depending on the product being derived.

- **Architectural principles guaranteeing compositionality:** The third approach does away with the structural architecture all together and focuses on the architectural principles that components have to satisfy in order to guarantee composability. This approach allows for the richest set of alternative configurations to be derived from the shared product family artefacts.
- **Accidental architecture:** Finally, in an excessive component-oriented approach, the architecture is the result of the opportunistic composition of independently developed components that share no or few architectural principles.

3.3 Components

Although architecture is very important and helps achieve business and operational qualities of software systems, it is of course the components that contain the actual implementations of functionality, features and requirements. Not surprising, however, the relation between the architecture and the components is more intimate than the terms may indicate. Again, we present three approaches to developing, managing and evolving components.

- **Internal integration-oriented components:** The first category is defined by the class of components that have been implemented specifically for a specific architecture that is specified in all or most of its aspects. The components contain variation points to satisfy the differences between different products in the family, but these do not spread significantly beyond the interfaces of the components. Finally, the components are implemented such that they depend on the implementation of other components rather than on explicit and specified interfaces.
- **Internal compositional components:** An alternative approach to implementing components is to develop components against explicitly defined provided, required and configuration interfaces and based on well-defined architectural principles. This approach allows for components that represent relatively independent domains of functionality and that can be freely composed with other components, as long as interfaces are adhered to and principles not violated.
- **External components:** In the era of open-source software, organizations often are able to satisfy a significant part of the requirements of a product through the selection of appropriate components. The resulting collection can often be complemented by commercially available components. This approach specifies an extreme approach, but in practice most product families define some form of common infrastructure consisting of external components. Thus, the latter two approaches are often combined.

3.4 Product Creation

With all the focus on strategy, architecture and component, one would almost forget that the predominant reason for all this work is to cost-effectively and rapidly create a broad set of products. Again, here we can identify alternative approaches.

- **Product-specific code based on pre-integrated platform:** The integration-oriented model presented in section 2 typically assumes a pre-integrated platform that contains the generic functionality required by all or most products in the family. A product is created by using the pre-integrated platform as a basis and adding the product-specific code on top of the platform. Although not necessarily so, often the company is also organized is along this boundary. The approach works very well for narrowly scoped product families, but less well when the scope of the product family is broadening.

- **Composing components in product specific configurations:** An alternative approach is to rely on a composable set of components that can be relatively freely combined by a product team to compose a significant part of the functionality required by the product. The composition of reusable components can be interlaced with product specific functionality at any interconnect between two components as well as, architecturally, on top of the reusable components. This approach is based on the assumption that through enforcing architectural principles on the components relatively free compositionality can be achieved while maintaining system reliability.

- **Opportunistic integration and glueing of external components:** Finally, the third approach that can be pursued is the opportunistic integration and glueing of external components to create a product. Especially companies that, at some point, discover the open source software community and the tens of thousands of ongoing projects experiment with constructing systems through composing open source software components.

3.5 Evolution

The topic that, in my experience, often is the most challenging to manage well is the evolution of shared as well as product-specific software artefacts. Although the creation of an individual product based on reusable components brings benefit to the company, in practice the true benefits and cost appear when the whole machinery consisting of continuous product creation and the evolution and expansion of shared software artefacts is operating in normal mode. This dimension is primarily an organizational one as the evolution of features and requirements anyhow needs to take place from a technical perspective. Below, we discuss different alternatives that we have seen organizations use in practice:

- **Platform organization:** The first model, typically used in an integration-oriented approach is the strong preference towards incorporates new features and requirements into the pre-integrated platform. The reasoning behind this model is that new features, in due time, need to be provided in all products anyway and consequently, the most cost effective approach is to perform this directly. This instead of an alternative approach where product-specific functionality evolves and generalizes over time and is incorporated into the platform when the use of the functionality has spread sufficiently broadly.

- **New or extended components:** Especially in the case of a more composition-oriented approach, incorporation of new features/requirements occurs through the creation of new components or the extension of already existing components, frequently adding new variation points. Even new components, since these adhere to the architectural principles, can be composed with older components in cases where earlier a different component was used. The new component or the extension of an existing component can, depending on the organization, be developed either by a component team or by the product team that requires the functionality the first.

- **Open-source community:** The third model, achieving increasing respect, is the proactive embedding of commercial R&D teams in the open-source software community. The important position in this case is viewing the R&D team to be part of the community, not outside it and collaborating with the community. When successfully implemented in a suitable domain, such a collaborative approach can lead to highly responsive and productive R&D work. Finally, it is important to mention that open-source software communities do not consist primarily of individuals that program for fun, but rather that several organizations may decide together that they care to share cost and effort required for the evolution of their collective products or systems.

4   Beyond the Integration-Oriented Approach

This paper is concerned with analyzing the challenges of broadening the scope of a software product family and by discussing alternative approaches to address these challenges. This section discusses two approaches. The first approach is the hierarchical product family. In this case, the reusable software artefacts are organized in a, typically, two-layered hierarchy. The first layer contains the generic code used by all products and the second layer contains the software artefacts that are used by all products in a specific category. The second approach is the composition-oriented method where the architecture is primarily principle-oriented and the components can be freely composed because these satisfy the principles. In the sections below, each approach is discussed in more detail.

4.1 Hierarchical Software Product Family

One scenario in the evolution of a software product family is that the product family naturally develops into a limited number of clusters of products. These clusters can subsequently be used as product categories and assuming that the size of each cluster and the amount of revenue generated are sufficient, business units can be made responsible for each product category. In this case, the logical approach is to organize the set of reusable components as well as the architecture of the product family in a hierarchical fashion. Thus, the functionality and features that are shared by all products in all product categories is developed as a platform by a shared R&D team. This platform is used as a basis by each business unit to build a software product family, again consisting of an extended architecture and a set of reusable components. Using the reusable product family artefact, the business unit can rapidly and easily develop new products and evolve new ones.

The hierarchical software product family occupies one point in the five dimensional space described in the previous section:

- **Business strategy - maximizing product family scope:** Although R&D cost and time-to-market are obviously relevant factors for any technology driven organization, the most important rationale for the hierarchical product family is that it facilitates the creation of a much broader set of products.

- **Architecture - micro-kernel architecture, optional elements:** The key challenge in the case of a hierarchical family is to architect the whole system such that that the business units can extend the platform architecture with the elements needed by their product category.
- **Components - internal integration-oriented components:** Although the hierarchical approach addresses several of the concerns discussed in section 2, it fundamentally takes the same integration-oriented approach as the original approach. The main difference is that the integration takes place in two stages, i.e. once for the basic platform and once for the reusable components for each product category.
- **Product creation - product-specific code based on pre-integrated platform:** As all reusable software artefacts are pre-integrated, product creation is primarily focused on the adding of product specific code on top of the reusable artefacts. This approach is excellent as long as the product specific requirements do not affect the shared components, which is often the case in embedded systems.
- **Evolution - platform organization:** Although business units add product specific code on top of the shared components, it is the responsibility of the platform organization to bring new functionality and features to the business units. This evolution typically follows the vertical path, i.e. over time product-specific code matures and becomes part of the business unit specific shared artefacts. Subsequently, the functionality becomes part of the base platform and only then it becomes available to other business units. This means that there is no effective mechanism to share specific functionality between two business units.

4.2 Composition-Oriented Method

The second approach that we discuss in this paper is the composition-oriented method. The composition-oriented approach can be viewed as being positioned close to the one end of a continuum whereas the integration-oriented approach is close to the other end. The predominant factor defining the continuum is the balance between the organization developing the reusable software and the teams creating products using, among others, the reusable software. Depending on the organizational approach chosen, component teams may not even exist, as in the case of HP Owen [Toft et al. 00], and even if these exist, they are not necessarily part of a software reuse R&D team. The key challenge in this model is the architecture of the product family. As the architecture is based on principles, rather than a structural architecture, maintaining a consistent and productive architecture-centric environment is more difficult than in other approaches.

The composition-oriented approach can be pinpointed in one location in the five dimensional space presented earlier:

- **Business strategy - maximizing product family scope:** Although R&D cost and time-to-market are obviously relevant factors for any technology driven organization, the most important rationale for the composition-oriented approach is that it facilitates the creation of a much broader set of products.
- **Architecture - architectural principles guaranteeing compositionality:** As discussed earlier, the key difference between this approach and other approaches is

that the architecture is not described in terms of components and connectors, but rather in terms of the architectural principles, design rules and design constraints.

- **Components - internal compositional components:** Obviously, the composition-oriented approach is very much driven by components. However, these components are not developed ad-hoc, but are constrained in their implementation by the architecture. As each component satisfies the architectural principles compositionality of these components is guaranteed.
- **Product creation - composing components in product specific configurations:** The explicit goal of this approach is to facilitate the derivation of a broad range of products that may need to compose components in an equally wide range of configurations. Product creation is, consequently, the selection of the most suitable components, the configuration of these components according to the product requirements, the development of glue code in places where the interaction between components needs to be adjusted for the product specific requirements and the development of product-specific code on top of the reusable components.
- **Evolution - new or extended components:** Product teams as well as component teams can be responsible for the evolution of the code base. Product teams typically extend existing components with functionality that they need for their product but is judged to be useful for future products as well. Product teams may also create new components for the same purpose. Component teams, if used, are more concerned with adding features that are required by multiple products. A typical example is the implementation of a new version of a communication protocol.

4.3 Analysis and Comparison

When analyzing the three approaches discussed in this paper, it is important to note that we have purposely excluded an open-source software based approach. The reason for this is that open-source software components can be included in each of the discussed approaches and as such are orthogonal to the approach chosen. However, the use of open-source software (OSS) has an impact on the software development organization as the evolution of OSS components can not be controlled to the same extent as internal components and most OSS licenses demand that the organization offers its addition to the OSS community. Nevertheless, despite these challenges, the use of OSS allows for very rapid creation of extensive software systems and can be a major factor in the reduction of R&D effort.

In the table below, we summarize the three approaches discussed in the paper. Each approach has a specific area of applicability. As discussed earlier in the paper for the case of the integration-oriented approach, applying an approach outside its area of applicability typically leads to several problems. As it for most organizations is difficult to replace an approach to software reuse that has been very successful in the past but has lost its applicability, this paper aims to discuss alternative approaches so organizations can, in a more explicit and objective manner, select the best approach. The table below summarizes the three approaches discussed in this paper.

Table 1. Comparison of the alternative approaches

Factors	Integration-oriented	Hierarchical	Composition-oriented
When applicable?	Well scoped family of highly related products	Broad family with a number of focused product categories	Broad family of products with significant unique requirements and features
Strategy	R&D cost minimization and/or time-to-market	maximizing product family scope	maximizing product family scope
Architecture	Fixed structural architecture	Micro-kernel architecture, optional elements	Architectural principles guaranteeing compositionality
Components	Internal integration-oriented components	Internal integration-oriented components	Internal compositional components
Product creation	Product-specific code based on pre-integrated platform	Product-specific code based on pre-integrated platform	Composing components in product specific configurations
Evolution	Platform organization	Platform organization	new or extended components
When used outside area of applicability	Unresponsiveness of platform leading to long lead times	Complicated alignment between hierachical platform organizations leading to long lead times	High R&D cost due to significant overlap of development and integration efforts

5 Related Work

This paper discusses two approaches that can be considered as alternatives to the integration-oriented approach that traditionally has been used frequently in the embedded systems industry. These alternative approaches are important because successful product families may broaden their scope considerably due to their success and this requires a conscious adjustment of the overall approach to software reuse. Although we believe that discussing these approaches in this context is a contribution of this paper, the approaches themselves have been discussed by other authors.

The paper by [Toft et al. 00] discusses the approach taken by Hewlett Packard's printer division where extensive sharing of software for the firmware of their products was achieved without the creation of a central domain engineering unit. A second author that has been promoting the notion of product populations and the consequences for the way software development is performed is Rob van Ommering [Ommering 02]. These concepts were further explored in a later publication jointly with the author of this paper [Ommering & Bosch 02].

6 Conclusions

Software product families have found broad adoption in the embedded systems industry, as well as in other domains. Due to their success, product families at several companies experience a significant broadening of the scope of the family. This may be due to the fact that the cost of product creation has decreased significantly or because of management decisions that bring previously unrelated products under the umbrella of the product family. However, a broadly scoped product family requires a different approach than the traditional integration-oriented approach proliferated in many embedded systems companies. If the organization fails to adjust its approach, one can identify several problems that may result from this, including unresponsiveness of the platform, difficulties related to dealing with cross-cutting features and architectural challenges due to the need for over-engineering.

To address these concerns, we presented the five main aspects of product families that are relevant in this context, i.e. business strategy, architecture, components, product creation and evolution. Based on these dimensions, we have presented two alternative approaches. The first is the notion of hierarchical product families where the shared software artefacts are organized in, typically, two layers. The first layer captures the functionality that is common to all products in the product family whereas the artefacts at the second layer are specific to a product category. Secondly, we presented the composition-oriented approach that is different in that it is based on a set of architectural principles, design rules and design constraints rather than an architectural structure consisting of components and connectors. The components in this approach are relatively freely composable as these satisfy the same set of architectural principles. Finally, we analysed and summarized the three approaches discussed in the paper. The contribution of this paper is that it analyses the problems of and presents alternative approaches that are better suited for broad-scoped product families.

In future work, we intend to study the identified problems in more detail as well as evaluate the proposed alternative approaches more carefully as could be achieved in this paper. In addition, potentially further alternatives can be developed, especially approaches that cross organizational boundaries and/or involve the open-source software community.

References

[Bosch 00] J. Bosch, Design and Use of Software Architectures: Adopting and Evolving a Product Line Approach, Pearson Education (Addison-Wesley & ACM Press), ISBN 0-201-67494-7, May 2000.

[Ommering & Bosch 02] R. van Ommering, J. Bosch, Widening the Scope of Software Product Lines - From Variation to Composition, Proceedings of the Second Software Product Line Conference (SPLC2), pp. 328-347, August 2002.

[Ommering 02] R. van Ommering, Building product populations with software components, Proceedings of the 24th International Conference on Software Engineering, pp. 255 – 265, 2002.

[O'Reilly 05] http://www.oreillynet.com/pub/a/oreilly/tim/news/2005/09/30/what-is-web-20.html

[Toft et al. 00] Peter Toft, Derek Coleman, and Joni Ohta, HP Product Generation Consulting, A Cooperative Model for Cross-Divisional Product Development for a Software Product Line, Proceedings of the First Software Product Lines Converence (SPLC1), Kluwer Academic Publishers, August 2000 (Ed. Patrick Donohoe), pages 111-132.

Author Index

Printing: Mercedes-Druck, Berlin
Binding: Stein+Lehmann, Berlin

Lecture Notes in Computer Science

For information about Vols. 1–4308

please contact your bookseller or Springer

Vol. 4357: L. Buttyán, V. Gligor, D. Westhoff (Eds.), Security and Privacy in Ad-Hoc and Sensor Networks. X, 193 pages. 2006.

Vol. 4355: J. Julliand, O. Kouchnarenko (Eds.), B 2007: Formal Specification and Development in B. XIII, 293 pages. 2006.

Vol. 4354: M. Hanus (Ed.), Practical Aspects of Declarative Languages. X, 335 pages. 2006.

Vol. 4353: T. Schwentick, D. Suciu (Eds.), Database Theory – ICDT 2007. XI, 419 pages. 2006.

Vol. 4352: T.-J. Cham, J. Cai, C. Dorai, D. Rajan, T.-S. Chua, L.-T. Chia (Eds.), Advances in Multimedia Modeling, Part II. XVIII, 743 pages. 2006.

Vol. 4351: T.-J. Cham, J. Cai, C. Dorai, D. Rajan, T.-S. Chua, L.-T. Chia (Eds.), Advances in Multimedia Modeling, Part I. XIX, 797 pages. 2006.

Vol. 4349: B. Cook, A. Podelski (Eds.), Verification, Model Checking, and Abstract Interpretation. XI, 395 pages. 2007.

Vol. 4348: S.T. Taft, R.A. Duff, R.L. Brukardt, E. Ploedereder, P. Leroy (Eds.), Ada 2005 Reference Manual. XXII, 765 pages. 2006.

Vol. 4347: J. Lopez (Ed.), Critical Information Infrastructures Security. X, 286 pages. 2006.

Vol. 4346: L. Brim, B. Haverkort, M. Leucker, J. van de Pol (Eds.), Formal Methods: Applications and Technology. X, 363 pages. 2007.

Vol. 4345: N. Maglaveras, I. Chouvarda, V. Koutkias, R. Brause (Eds.), Biological and Medical Data Analysis. XIII, 496 pages. 2006. (Sublibrary LNBI).

Vol. 4344: V. Gruhn, F. Oquendo (Eds.), Software Architecture. X, 245 pages. 2006.

Vol. 4342: H. de Swart, E. Orłowska, G. Schmidt, M. Roubens (Eds.), Theory and Applications of Relational Structures as Knowledge Instruments II. X, 373 pages. 2006. (Sublibrary LNAI).

Vol. 4341: P.Q. Nguyen (Ed.), Progress in Cryptology - VIETCRYPT 2006. XI, 385 pages. 2006.

Vol. 4340: R. Prodan, T. Fahringer, Grid Computing. XXIII, 317 pages. 2007.

Vol. 4339: E. Ayguadé, G. Baumgartner, J. Ramanujam, P. Sadayappan (Eds.), Languages and Compilers for Parallel Computing. XI, 476 pages. 2006.

Vol. 4338: P. Kalra, S. Peleg (Eds.), Computer Vision, Graphics and Image Processing. XV, 965 pages. 2006.

Vol. 4337: S. Arun-Kumar, N. Garg (Eds.), FSTTCS 2006: Foundations of Software Technology and Theoretical Computer Science. XIII, 430 pages. 2006.

Vol. 4335: S.A. Brueckner, S. Hassas, M. Jelasity, D. Yamins (Eds.), Engineering Self-Organising Systems. XII, 212 pages. 2007. (Sublibrary LNAI).

Vol. 4334: B. Beckert, R. Hähnle, P.H. Schmitt (Eds.), Verification of Object-Oriented Software. XXIX, 658 pages. 2007. (Sublibrary LNAI).

Vol. 4333: U. Reimer, D. Karagiannis (Eds.), Practical Aspects of Knowledge Management. XII, 338 pages. 2006. (Sublibrary LNAI).

Vol. 4332: A. Bagchi, V. Atluri (Eds.), Information Systems Security. XV, 382 pages. 2006.

Vol. 4331: G. Min, B. Di Martino, L.T. Yang, M. Guo, G. Ruenger (Eds.), Frontiers of High Performance Computing and Networking – ISPA 2006 Workshops. XXXVII, 1141 pages. 2006.

Vol. 4330: M. Guo, L.T. Yang, B. Di Martino, H.P. Zima, J. Dongarra, F. Tang (Eds.), Parallel and Distributed Processing and Applications. XVIII, 953 pages. 2006.

Vol. 4329: R. Barua, T. Lange (Eds.), Progress in Cryptology - INDOCRYPT 2006. X, 454 pages. 2006.

Vol. 4328: D. Penkler, M. Reitenspiess, F. Tam (Eds.), Service Availability. X, 289 pages. 2006.

Vol. 4327: M. Baldoni, U. Endriss (Eds.), Declarative Agent Languages and Technologies IV. VIII, 257 pages. 2006. (Sublibrary LNAI).

Vol. 4326: S. Göbel, R. Malkewitz, I. Iurgel (Eds.), Technologies for Interactive Digital Storytelling and Entertainment. X, 384 pages. 2006.

Vol. 4325: J. Cao, I. Stojmenovic, X. Jia, S.K. Das (Eds.), Mobile Ad-hoc and Sensor Networks. XIX, 887 pages. 2006.

Vol. 4323: G. Doherty, A. Blandford (Eds.), Interactive Systems. XI, 269 pages. 2007.

Vol. 4322: F. Kordon, J. Sztipanovits (Eds.), Reliable Systems on Unreliable Networked Platforms. XIV, 317 pages. 2007.

Vol. 4320: R. Gotzhein, R. Reed (Eds.), System Analysis and Modeling: Language Profiles. X, 229 pages. 2006.

Vol. 4319: L.-W. Chang, W.-N. Lie (Eds.), Advances in Image and Video Technology. XXVI, 1347 pages. 2006.

Vol. 4318: H. Lipmaa, M. Yung, D. Lin (Eds.), Information Security and Cryptology. XI, 305 pages. 2006.

Vol. 4317: S.K. Madria, K.T. Claypool, R. Kannan, P. Uppuluri, M.M. Gore (Eds.), Distributed Computing and Internet Technology. XIX, 466 pages. 2006.

Vol. 4316: M.M. Dalkilic, S. Kim, J. Yang (Eds.), Data Mining and Bioinformatics. VIII, 197 pages. 2006. (Sublibrary LNBI).

Vol. 4314: C. Freksa, M. Kohlhase, K. Schill (Eds.), KI 2006: Advances in Artificial Intelligence. XII, 458 pages. 2007. (Sublibrary LNAI).

Vol. 4313: T. Margaria, B. Steffen (Eds.), Leveraging Applications of Formal Methods. IX, 197 pages. 2006.

Vol. 4312: S. Sugimoto, J. Hunter, A. Rauber, A. Morishima (Eds.), Digital Libraries: Achievements, Challenges and Opportunities. XVIII, 571 pages. 2006.

Vol. 4311: K. Cho, P. Jacquet (Eds.), Technologies for Advanced Heterogeneous Networks II. XI, 253 pages. 2006.

Vol. 4310: T. Boyanov, S. Dimova, K. Georgiev, G. Nikolov (Eds.), Numerical Methods and Applications. XIII, 715 pages. 2007.

Vol. 4309: P. Inverardi, M. Jazayeri (Eds.), Software Engineering Education in the Modern Age. VIII, 207 pages. 2006.